Published by Charles E. Merrill Publishing Co.
A Bell & Howell Company
Columbus, Ohio 43216

This book was set in Serifa and Universal Dot
Matrix.
Production Editor: Jeffrey Putnam
Cover Design Coordination: Tony Faiola
Cover Photograph by Larry Hamill

Reviewers:
Thomas Bingham, Saint Louis Community College at
Florissant Valley
William Kleitz, Tompkins-Cortland Community
College
Paul Maini, Suffolk Community College
William Schroder, DeVry Institute of Technology-
Atlanta

Library of Congress Catalog Card Number: 83-43159
International Standard Book Number: 0–675–20161–6
Printed in the United States of America
1 2 3 4 5 6 7 8 9 10   88 87 86 85 84

BASIC Applied
to Circuit Analysis

Louis Nashelsky
Robert L. Boylestad

Queensborough Community College

CHARLES E. MERRILL PUBLISHING COMPANY
A BELL & HOWELL COMPANY
COLUMBUS • TORONTO • LONDON • SYDNEY

Preface

The advent of relatively inexpensive microcomputer units since the late 1970s has made them readily available to engineering and technology students. This manual introduces the BASIC programming language and uses it to solve problems in essentially all areas that will be studied in the introductory dc and ac circuit analysis courses. The content is ideally suited for electrical and electronics students either as their main BASIC programming text or as a supplement, and may be introduced in a lecture or laboratory setting. There are some 150 programs, extending from the simple to the more complex, that cover a wide variety of subject areas of interest to technology students.

Although this manual is not intended to be used as a primary text for learning dc and ac, it does permit an examination of how the computer can be used to perform many of the more important calculations and demonstrates the use of the computer to solve typical problems in circuit analysis. While any collateral dc/ac text may be used, optional cross-references to *Introductory Circuit Analysis, 4th edition* by Robert L. Boylestad (Charles E. Merrill, 1982) are included.

The text has twenty-four chapters—three to introduce the BASIC language, seven on dc circuit topics, eight on ac circuit topics, and six on miscellaneous topics.

The presentation includes an introduction to the problem to be solved, the method of writing and describing the programs, typical runs showing results of each program and underlined input responses, to ensure a clear understanding of the entire process of establishing useful, functional programs.

Although the BASIC language was originally developed by faculty and students for use on time-sharing terminals, it has been readily adopted by microcomputer manufacturers as a standard language. Each microcomputer company, however, provides somewhat modified versions of the language so that a program written for one computer may not run properly on another. There are a number of BASIC commands which do work on *all* the popular microcomputers, while some only require a slight modification in form. This text is limited to those programming commands used by the majority of the popular microcomputers. Most of the BASIC commands can be used identically on all the computers, while some need only a slight modification in form as described in the text. The text avoids using commands that are out of the ordinary or unique to a particular system. While this somewhat limits the programming, it does allow the student to operate on a variety of computer units and permits the presentation of the essential programming and mathematical process without undue confusion due to the equipment at hand.

Programs can be nicely divided into individual sections or modules. A module is essentially self-contained and can be carried over intact to other programs. Modular programming is used throughout the text to help subdivide each program into a number of smaller program sections—an aid in understanding the overall program.

The programs selected in each chapter represent a range of topics for that area of study. While some description of the problem to be solved is included, suitable texts on dc and ac circuit theory should first be used to learn and study the particular subject area. The programs provided on that subject area will then show the student the type of problems that can be solved using the computer and how to write the programs. It is hoped that the students can then modify these programs for their own desired results or write similar programs to help in their study of that area. Problems in each chapter range from the simple to the more complex. Most chapters also contain one or two programs which can be used by the student as a testing program for use in practicing calculations in that subject area. The computer randomly selects typical values of data, asks for the student's calculated value and compares the student's answer with the correctly calculated value.

The authors have worked diligently to develop and test all programs presented in the text. In all programming work however, there are always some possible conditions of data for which the program results are not sufficiently accurate, or for which the program calculations fail. It is hoped that any unusual results can be easily corrected by the user, if such situations do occur. While we hope no such problems result, however, we would appreciate any suggestions or comments that will ultimately improve the quality of the presentation. We both thoroughly enjoyed developing the material for the text and sincerely hope some of our enthusiasm for the subject matter will rub off on the user.

Instead of a traditional instructor's manual, the publisher will provide a disk containing a selection of the major programs of each chapter available for faculty use. This disk, which should avoid the need to keyboard every program from scratch, can be copied for student use, provided, however, that each student has purchased a book, that the disks are not offered for sale or transferred to another institution, and that written permission has been obtained from the publisher. The first disk, which the authors are producing themselves, is for use on the Apple IIe™. Contact your Merrill representative concerning other possible disks (IBM PC™, TRS-80™, etc.).

Note to the student

Many of the programs in this book refer to examples and problems in *Introductory Circuit Analysis, 4th Edition* by Robert Boylestad (Columbus: Charles E. Merrill, 1982). The references are optional; you do not need to refer to *Introductory Circuit Analysis* to understand the programs in this book.

Merrill's International Series in Electrical and Electronics Technology

Contents

1

Introduction to the BASIC Language   1

2

Programming   11

3

Writing Program Modules   23

4
Resistance   29

5
Ohm's Law, Power and Energy   49

6
Series and Parallel DC Networks   67

7
Series-Parallel DC Networks   83

8
Determinants, Ladder Networks and Delta-Wye
Conversions   99

14
Complex Numbers   211

15
Ohm's Law-AC   227

16
Series and Parallel AC Networks   243

17
Methods of AC Analysis   263

18
Network Theorems-AC   293

19
Power-AC   307

20
Resonance   321

21
Three-Phase Networks   343

22
Non-Sinusoidal Signals   357

23
Transformers   375

24
Two-Port Parameters   389

To Katrin, Kira, Larren
and Else Marie, Eric, Alison, Stacey Jo

Introduction to the BASIC Language

1

1.1 INTRODUCTION

Computers have played an important role as a scientific tool since the early 1940s and as a business machine since the mid-1950s. Computer capability has continued to increase and programming languages have been developed to enhance the user's ability to solve particular problems: COBOL for business use; FORTRAN and ALGOL for scientific use; and a host of special application languages. Eventually, greater versatility and lowered costs made the less powerful mini-computer configurations competitive with larger units. In the mid-1970s, microcomputers were developed that were particularly suited for a variety of personal, small business, and technical applications. Although the BASIC language was first developed at Dartmouth in the mid-1960s for time-shared computer applications, it was quickly adopted by microcomputer manufacturers as the appropriate language for their systems. Since BASIC is suitable for programming technical problems and is used extensively in microcomputer systems, we will use it as the language for programming the solutions to the electrical network problems in this text.

The microcomputer normally employs a self-contained program called an *interpreter* which ''interprets'' or analyzes each line of instruction entered by the user and then performs the desired operation.[1] Since each computer manufacturer has its own interpreter program, the BASIC languages, as implemented by various computers, are not identical. Apple, IBM, and Radio Shack, among others, provide one version of BASIC in Read Only Memory (ROM), while an extended version can be loaded into Random Access Memory (RAM) from floppy disk.

An additional BASIC interpreter, which extends the BASIC language of that computer, is available to floppy disk users. The Apple computer provides a ROM version of Applesoft BASIC which operates when the power is turned on. Another version of BASIC available for the Apple computer is Microsoft BASIC, which is compatible with Radio Shack's. The IBM personal computer (PC) has a BASIC interpreter which is ROM-based and operates as soon as the power is turned on and a self-check completed, as well as a few disk versions which can be selected by the user.

This text avoids any special disk and graphic operations particular to each of the above computers. We use standard commands in BASIC throughout to permit running a program on any of the above machines with only the smallest program changes. Differences explained in Chapter 3 should allow you to use the text programs, with necessary modifications, on the Radio Shack, Apple, IBM PC machines, or a time-shared terminal.

1.2 USING BASIC ARITHMETIC COMMANDS

To introduce a number of arithmetic commands that exist in BASIC, we describe various operations using the immediate or calculator mode. When operating in BASIC, the usual response of[2]

```
READY
```

indicates that the computer is waiting for input in the form of a command or line of input. For example, the command

```
A=5
```

---

[1] Most programming languages are translated into machine language by a compiler program, while most BASIC languages are implemented by an interpreter program.

[2] The IBM PC responds with OK

internally sets the variable A equal to the value (constant) 5. The command

　　B=3.5

would then set a second variable B equal to the value 3.5. The command

　　PRINT A

would then display the value of A as 5, while

　　PRINT A,B

would result in

　　5　　　3.5

　　The use of a comma separating the variables will result in the values of A and B being printed in separate print zones. In the Apple II computer, that would be 16 spaces; in other systems, it is a fixed number of spaces, typically between 10 and 20. Some examples of addition($+$) and subtraction($-$) arithmetic operations using the values of A and B above are:

　　PRINT A+B
　　8.5

instructs computer to print sum of A plus B.

　　PRINT B,B-A
　　3.5　　　-1.5

instructs computer to print value of B, then value of difference $B-A$.

　　PRINT A+A+B,B-8
　　13.5　　　-4.5

instructs computer to print sum of A plus A plus B, then value of $B-8$.

　　PRINT 3.8+B,A,B,C
　　7.3　　　5　　　3.5　　　0

instructs computer to print values of $B+3.8$, then $A,B,$ and C. Note that C is equal to 0, since a value has not been assigned.

　　The multiply operation uses *, while divide uses /. A few examples are:

　　PRINT 3*A
　　15

　　PRINT 3*A+B
　　18.5

　　PRINT 2*A,3*B
　　10　　　10.5

　　PRINT 3*B+10
　　20.5

(In the above calculations BASIC will first perform the multiplication operation, followed by the indicated addition.)

　　PRINT A/2
　　2.5

　　PRINT 10.5/B
　　3

　　PRINT 8/2,A/5,B/2
　　4　　　1　　　1.75

The BASIC language also includes an operation to raise a number to a power, or to an exponent value:

　　PRINT 5^2
　　25

　　PRINT A^2
　　25

　　PRINT B^2
　　12.25

　　PRINT 6*A^2
　　150

(In the above, BASIC first raises to a power, then multiplies.)

　　These arithmetic operations can also be grouped according to usual arithmetic rules. This includes the use of parentheses when necessary, as the following examples demonstrate.

　　PRINT 5*3+6
　　21

　　PRINT 5*(3+6)
　　45

```
PRINT A*(B+2)
27.5

PRINT A*(7+2)/(B+1.5)
9

PRINT A^2*3+4
79
```

The hierarchy rule is: When the operations `+,-,*,/,^` are used, the computer follows a hierarchy or order of operation in any one line. In an arithmetic statement the computer:

1. First raises a number to an exponent, e.g., `A^2` resulting in `5^2=25`

2. Then the computer multiplies or divides (selecting operations from left to right), e.g., `A*B/2.5` is `A*B`, the result of which is divided by 2.5 `5*3.5/2.5 = 17.5/2.5 = 7`

3. Then the computer adds or subtracts (selecting operations from left to right), e.g., `A+B-2*A = 5 + 3.5 -10 = -1.5`

Since BASIC permits the use of parentheses, they can be used whenever necessary, or whenever their use will help specify the order of operations. In any one line, the computer first performs the operations within the parentheses. Note the results of the following operation:
For $A = 5$,

```
PRINT (13+3)/(A+3)
2

PRINT 6/3*A
10

PRINT 6/(3*A)
.4
```

After entering $X = 3$ $Y = 4$,

```
PRINT X^2+Y^2
25
```

```
PRINT (X^2+Y^2)^0.5
5

PRINT ((1+2)*6)^2
324
```

1.3 USE OF FUNCTIONS IN ARITHMETIC CALCULATIONS

BASIC provides a large number of built-in function calculations, a few of which will be introduced in this section.

Trigonometric functions Among the trigonometric functions in BASIC are:

```
SIN(R), COS(R), TAN(R), ATN(X)
```

where R is an angle in radians and X is a numeric value. The computation

```
PRINT SIN(45)
```

would result in the computer printing the sine of 45 radians (not 45 degrees), which is 0.8509.[3]
To compute the sine of 45 degrees, the radians measure must first be determined from Radians = (pi/180) * Angle(in degrees):

```
PRINT SIN(3.14159*45/180)
0.7071
```

where pi/180 converts degrees into radians. To calculate the cosine of 30 degrees (cos 30),

```
PRINT COS(3.14159*30/180)
0.866
```

and to calculate the tangent of 30 degrees,

```
PRINT TAN(3.14159*30/180)
0.5773
```

---

[3] While the accuracy inherent in most computer systems includes more than 4 places (for example, the Apple II would output 0.850903524 in the above calculation), this text will provide a shortened accuracy of 4 places for simplicity and clarity.

or

```
PRINT TAN(30/57.29583)
0.5774
```

where $180/3.14159 = 57.29583$.

The arctangent of a numerical value, such as 1, is determined using

```
PRINT ATN(1)
```

The output would be 0.7854, which is 0.7854 radians. To obtain the output in degrees

```
PRINT 57.29583*ATN(1)
45
```

Trigonometric functions can be combined with other mathematical operations.

```
PRINT 8*SIN(40/57.29583)
5.1423
```

```
PRINT 8*SIN(40/57.29583)^3
2.1247
```

```
PRINT (8*SIN(40/57.29583))^3
135.9789
```

```
PRINT 2*(57.29583*ATN(0.5))
53.1301
```

ABS(X),SGN(X),INT(X) Among the operations used to handle the sign of a number is the absolute operation ABS(X), which provides the magnitude of the quantity and drops the associated sign:

```
PRINT ABS(-0.85)
.85
```

The function SGN(X) is algebraically set to $+1$ for a positive value X, to -1 for a negative value X, and to 0 if the value of X is zero.

```
PRINT SGN(-8.5)
-1
```

```
PRINT SGN(7)
1
```

A BASIC operation which extracts the integer part of a number is INT(X):

```
PRINT INT(9.8)
9
```

```
PRINT INT(8.1)
8
```

Notice that the integer operation does not round off the answer—it simply extracts the integer part, regardless of the value of the fractional component. These operations are useful as part of other mathematical calculations:

```
B=3.7845
PRINT 5*INT(B)
15
```

```
A=5
PRINT A*INT(ABS(-7.8))
35
```

LOG(X), EXP(X) Two related functions which include calculation of e^x where $e = 2.71828$, are $\log_e (x)$ and e^x:

```
PRINT EXP(3)
20.0855
```

```
PRINT 5*EXP(1.5)
22.4084
```

```
PRINT 10*(1-EXP(-3))
9.5021
```

The calculation of $ln(x) = \log_e(x)$ can be demonstrated by the following:

```
PRINT LOG(2.71828)
0.9999
```

```
PRINT LOG(EXP(1))
1
```

```
PRINT LOG(10)
2.3026
```

```
PRINT LOG(100)/2.302585
2
```

```
PRINT (2*LOG(5))^2
10.3612
```

Note in the above examples that LOG(X) calculates the natural log or log to the base e.

Obtaining calculations of the log to the base 10 only requires division by the factor 2.302585:

```
LOG(X)/2.302585
```

That is, $Log_{10}x = Log_e x/2.302585$

Random Numbers, RND A popular function used in BASIC randomly selects a number. This random number is typically either a fractional value between 0 and 1 or an integer number. In TRS-80 Level-II BASIC, for example, the function

```
RND(N)
```

selects an integer number between 1 and the integer value N. You select the value of N and the computer will calculate an integer number between 1 and N. For example:

```
PRINT RND(6)
3

PRINT RND(6)
2

PRINT RND(6)
6
```

The calculation of RND(6) thus results in an integer number between 1 and 6, the value being randomly selected by the RND function. If the number in the parentheses is 0, then a value between 0 and 1 results:

```
PRINT RND(0)
0.8155

PRINT RND(0)
0.1085
```

Some versions of BASIC use this same random function, while others define RND(X) or RND(1) as a random number between 0 and 1. In those versions of BASIC, an integer number between 0 and N can still be obtained using:

Apple II
```
INT(1+N*RND(1))
```

or

IBM PC
```
INT(1+N*RND)
```

or

```
INT(1+N*RND(X))
```

For a random number in the range 1 to 6,

```
PRINT INT(1+6*RND(1))
5
```

or

```
PRINT INT(1+6*RND)
5
```

or

```
PRINT INT(1+6*RND(X))
5
```

For a random number in the range of 1 to 52,

```
PRINT INT(1+52*RND(1))
35
```

or
```
PRINT INT(1+52*RND)
35
```

or

```
PRINT INT(1+52*RND(X))
35
```

The Apple II computer uses RND(1) to randomly select a number between 0 and 1 so that the first expression above would be used with that computer. The IBM personal computer uses RND to compute a random number between 0 and 1, so that the second expressions would be used when programming BASIC in that computer.[4]

---

[4] While the exact form of the RND function differs from one computer to another, the form RND(X) will be used in this text and you can substitute the appropriate RND expression used in your computer.

1.4 USING BASIC FOR VARIOUS CALCULATIONS AND CONVERSIONS

BASIC can be used to perform a desired calculation or conversion using either a single line or many lines of instruction. In this section we will limit our investigation to single line instructions.

The conversion from Celsius to Fahrenheit can be programmed using:

```
F=(9/5)*C+32
```

As an example:

```
C=40
PRINT (9/5)*C+32
104
```

That is, 40° C = 104° F.

Conversion from Fahrenheit to Celsius can be obtained using:

```
C=(5/9)*(F-32)

F=122
PRINT (5/9)*(F-32)
50
```

122° F converts to 50° C.

The conversion of feet to inches is made like this:

```
FT=4
PRINT 12*FT
48
```

4 feet converts to 48 inches.

Conversion of inches to centimeters (using 2.54 cm/in) could be accomplished using:

```
IN=8
CM=2.54*IN
PRINT CM
20.32
```

8 inches converts to 20.32 centimeters. Notice in this last example that CM was calculated first, then printed out. The last line could have been more complete using:

```
PRINT IN,CM
8     20.32
```

where the left value is the value of inches (IN) and the right value is centimeters (CM). Conversion of feet to centimeters takes two steps:

```
FT=4
IN=12*FT
CM=2.54*IN
PRINT FT,IN,CM
4     48     121.92
```

Printing message strings To further improve the output display, you may print message strings along with the calculated value:

```
IN=8
CM=2.54*IN
PRINT "CENTIMETERS =",CM
CENTIMETERS =     20.32
```

where the displayed value is not just the resulting number, but also a message string used to identify that value. Notice that the value of C is spaced some distance from the message string. Using a comma to separate items in a PRINT command results in a wide spacing between items. Using a semicolon between items in a PRINT command will result in close spacing between items, as the next example demonstrates.

```
IN=8
CM=2.54*IN
PRINT IN;" INCHES IS EQUAL TO
";CM;" CENTIMETERS"
8 INCHES IS EQUAL TO 20.32 CEN-
TIMETERS
```

Some versions of BASIC, TRS-80 BASIC for example, allow you to leave out the semicolon:

```
PRINT "CENTIMETERS"CM
```

being equivalent to

```
PRINT "CENTIMETERS";CM
```

A useful conversion in audio amplifier measurements uses the units of decibels (dB). For example, voltage gain (in dB) is defined by

Voltage Gain (dB) $= 20 \log_{10}(Vo/Vi)$, where $Vi =$ input voltage and $Vo =$ output voltage.

```
VI=1.5E-3
VO=6
VG=20*(LOG(VO/VI)/2.302585)
PRINT "Voltage Gain in dB = ";VG
Voltage Gain in dB = 72.0412
```

We noted earlier that since trigonometric functions do calculations directly using radian angles, conversion between degrees and radians is often required. The sine of 40 requires the following operations:

```
AN=40
RA=(3.14159/180)*AN
PRINT AN; SIN(RA)
4  0.6428
```

If the value *pi* is predefined in the BASIC language[5] you can use:

```
AN=40
RA=(PI/180)*AN
PRINT AN;COS(RA)
40  0.7660
```

or

```
RA=(&PI/180)*AN
```

As another example, the conversion of arctan results in an angle in radians. To get the result in degrees:

```
X=0.5
RA=ATN(X)
AN=(180/3.14159)*RA
PRINT For X= ";X;" ARCTAN(X)=
";AN;" DEGREES"
For X= .5 ARCTAN(X)= 26.56505
DEGREES
```

Variable Names Variable names used in BASIC can be one or two characters in most computer interpreters.[6] Proper variable names could be: *A, AB, A5, RL, X9, Y, Y1*. These variable names must start with an alphabetic character and may contain a second alphabetic character or number from 1 to 9. Some systems allow variable names of more than two characters, but only use the first two to identify the variable. In this case, the variable names *RL1* and *RL2* represent the same variable—*RL*. Proper use of variable names of more than two characters would be THETA, VOLTAGE, SUM, which can also be programmed as TH, VO, and SU respectively.

A variable name in BASIC may be used to represent a few types of data. The variable *A*, for example, represents a real value which may contain both an integer part *and* a fractional part which can be printed in either standard form, 1271.453 or in exponential format, 1.271453E3. For most computers the range of real numbers is $1E \pm 38$. Typically, these real numbers can be stored to hold 9 digits:

```
-1.23456789, 129.721118,
21.7215789E12
```

This restriction on stored numbers can limit computer accuracy and can even result in some calculation error. For example, 1/3 is stored as 0.333333333. If that number were repeatedly added, the result would be 0.999999999, not 1, as expected. This, however, is not quite the full picture. The value 1/3 is not an exact binary multiple. Since numbers are stored as binary values, many numbers are stored as close to the decimal value, but not necessarily exactly equal to the decimal value.

While a real number such as 128.729314E6 requires the full capacity to store a real number, this storage space is wasted when storing a number such as 4, which is better stored as an integer value. BASIC allows you to streamline calculation and storage of program values when you know for certain

---

[5] Some versions of BASIC provide the constant pi (= 3.141592654) as PI or &PI, etc., this value being predefined to the full accuracy available in that version of BASIC.

[6] The IBM PC allows 40 characters.

whether a particular variable will be real or integer. Some BASICs distinguish between the real variable *A* and the integer variable *A%*, since these are two different values stored by the computer.

A third type of data value can be used to store alphanumeric characters. The variable name must then be terminated with a $:

        A$, AB$, P1$, R9$, and NAME$

(the last variable being the same as NA$). The length of the stored data string varies from one computer to another, but is typically up to 255 characters.

EXERCISES

1. For *A*=5, *B*=3.5, and *C*=−2, what values are printed for the following:
a. PRINT A+B-C
b. PRINT A*B-C
c. PRINT C-B+2*A
d. PRINT A*B+C
e. PRINT A+B*C
f. PRINT A*A+B*B+C*C
g. PRINT (A+B)*(A+C)
h. PRINT A*B↑2-C

2. For *X*=3, *Y*=−4, and *Z*=7.5, what values are calculated for the following:
a. X↑2-Y↑2
b. (X-Z)^2+3*Y
c. (Y+Z/3)↑2
d. (1+Y+Z/6+X)
e. X*Y+Z-X*Y+Z

3. For *A*=30, *R*=30/57.296, what values would be printed for the following:
a. PRINT SIN(A)
b. PRINT SIN(2*A/57.296)
c. PRINT SIN(A/57.296)+COS
   (A/57.296)
d. PRINT SIN(R)↑2
e. PRINT ATN(0.5)*SIN(R↑2)
f. PRINT SIN(R)↑2+COS(R↑2)
g. PRINT 5*COS(2*R)

4. For *X*=3, *Y*=−2.5, what values are calculated for the following:
a. Z=5*ABS(-2*Y)
b. A=X+3*SGN(Y)
c. W=ABS(5*Y)+3*INT(Y)
d. Z=SGN(Y)*SIN(Y)
e. A=2.5*INT(X^2+3*ABS(Y))

5. For *X*=1, *Y*=15, what values are printed for the following:
a. PRINT LOG(Y)
b. PRINT LOG(Y)/2.3
c. PRINT 10*LOG(Y)
d. PRINT EXP(12*X)+LOG(Y^2)
e. PRINT EXP(LOG(Y/X))

6. What printed output results from the following:
a. C=3.5
   F=(9/5)*C+32
   PRINT "FOR C OF";C;"F IS";F
b. IN=3.5
   CM=2.54*IN
   PRINT "IN; "INCHES CONVERTS
   TO";CM; "CENTIMETERS"
c. VI=1000
   VO=20000
   AV=20*LOG(VO/VI)/2.3
   PRINT "FOR INPUT OF";VI;"OUTPUT
   OF";VO
   PRINT "VOLTAGE GAIN IS";AV

7. For the BASIC version introduced in this chapter, determine the proper variable names for the following:
a. RA   f. SIGMA3
b. OMEGA   g. X↑2
c. T3   h. 2X
d. X25   i. FV
e. SUM+1   j. L

8. Indicate which of the following are correct:
a. X3=Y2+Z1
b. 3*X=Y
c. A=SINE(B/57.296)
d. Y*Y-3=T
e. SUM=A+B-C

Programming

2

2.1 WRITING A PROGRAM

In this chapter, you will use the commands introduced in Chapter 1, and a number of new commands, to write a program. The program is first entered and stored by line number and then run to provide the desired output. While the BASIC language includes a considerable number of commands which permit the writing of a program, this text will confine program commands to those contained in the most popular microcomputer units, such as the APPLE II, IBM PC, and Radio Shack Models I or III. While this text will explain how various instructions are used, additional details about specific instructions can be obtained from the manufacturers' manuals.

2.2 USE OF PRINT, INPUT

This chapter will expand the list of programming commands and define the structure of a computer program. While the BASIC commands can be immediately executed as described in Chapter 1, a sequence of instructions, called a program, is first stored in the computer memory and then executed by the computer in the indicated line number order. To store these instructions in BASIC, you need only assign a line number at the beginning of each line. When BASIC "sees" a line number at the beginning of a line, it stores that line and does not execute the program instructions. A RUN command will then instruct the computer to execute the stored program, starting with the lowest line number (not necessarily the order in which it was typed in), using the input data provided. For this manual program, line numbers generally start at 10 and increase in increments of 10, to permit the addition of line

numbers before line 10 or between any other line numbers.

INPUT INPUT is a command which allows the user to enter data as requested by the program.[1] For example, consider a program designed to accept X and Y as input values and then output the sum and product of these values.

```
10 PRINT "ENTER VALUES OF X AND
   Y"
20 INPUT X
30 INPUT Y
40 S=X+Y
50 P=Y*Y
60 PRINT "SUM OF INPUTS IS";S
70 PRINT "PRODUCT OF INPUTS
   IS";P
80 END
```

To execute a program type the command RUN.

```
ENTER VALUES OF X AND Y
?10
?5
SUM OF INPUTS IS 15
PRODUCT OF INPUTS IS 50

READY
```

When the program is finished, the computer outputs READY and then awaits a new command or line of program input. The instructions on lines 20 and 30 could be entered as one line of instruction:

```
20 INPUT X,Y
```

resulting in an output of

```
ENTER VALUES OF X AND Y
?10,5
SUM OF INPUTS IS 15
PRODUCT OF INPUTS IS 50

READY
```

---

[1] To further clarify the programming process, this text will underline the data entered as requested by an INPUT command.

Note that the data requested by line 20 is then input as two values, separated by a comma.

As an alternate, the input instruction can include some initial printout:

```
20 INPUT "X VALUE=";X
30 INPUT "Y VALUE=";Y
```

which results in

```
ENTER VALUES OF X AND Y
X VALUE=?10
Y VALUE=?5
SUM OF INPUTS IS 15
PRODUCT OF INPUTS IS 50

READY
```

The PRINT instruction can be used in a number of interesting ways. For example,

```
60 PRINT "FOR X OF";X;"AND Y
   OF";Y;"THE SUM IS";S
```

which would result in output of

```
FOR X OF 10 AND Y OF 5 THE SUM IS
15
```

Using the semicolon keeps the minimum spacing, while using commas would result in wider spacing between numbers.

While the PRINT command allows a good amount of flexibility in the format of the output line, a greater degree of control is achieved using the command PRINT USING. Some examples will help describe the use of this instruction.

```
55 F$="##.# ##.## ##.##"

60 PRINT USING F$,X,Y,S
```

The resulting output when this program is run would be[2]

```
ENTER VALUES OF X AND Y

VALUE OF X=?10
```

```
VALUE OF Y=?5

10.0  5.0    15.00
```

If line 55 were

```
55 F$="FOR X OF ##.# AND Y OF
   ##.## THE SUM IS ###.##"
```

The resulting output is then

```
FOR X OF 10.0 AND Y OF 5.00 THE
SUM IS 15.00
```

Here are a few examples of how the # sign is used.

###.## results in a number with three integer places, followed by a decimal point and two fractional digits (e.g., 123.45).

##.## results in a number with two integer places, followed by a decimal point and two fractional digits (e.g., 12.34).

#.### specifies two numbers, the first a two place integer and the second a one integer place, followed by a decimal point and a three place fractional value (e.g., 42 1.234)

Note that within the defined field, F$ in this example, no further quotes or commas are needed to separate the output quantities.

Further examples of the versatility of the PRINT USING command appear in the following sections.

2.3 USE OF FOR/NEXT

Since a computer program is most useful when it provides repeated output over a range of values, the FOR/NEXT instruction is used to

---

[2] In some computers the correct format is

```
60 PRINT USING F$; X,Y,S
```

create a program loop. Consider the following program to output a table of X, X^2, and X^3 for the range of X from 1 to 25.

```
10 FOR X=1 TO 8
20 PRINT X,X^2,X^3
30 NEXT X
40 END
```

Lines 10 and 30 set up the FOR/NEXT loop; they provide the range of X and the increment of X between loops. All the lines within this loop are performed for each value of X specified in line 10. In this example, line 20 is performed for $X=1$, $X=2$, $X=3$, etc., up to and including $X=8$. When this program is run, we obtain

```
1     1      1
2     4      8
3     9      27
4     16     64
5     25     125
6     36     216
7     49     343
8     64     512
```

As another example, consider the program

```
10 FOR X=1 TO 12
20 S=SQR(X)
30 PRINT X,S
40 NEXT X
50 END

RUN

1     1
2     1.414213
3     1.732051
4     2
5     2.236068
6     2.449490
7     2.645751
8     2.828427
9     3
10    3.162278
11    3.316625
12    3.464102
```

In this program lines 20 and 30 are repeated for

$X=1$, $X=2$, up to $X=12$. Notice that the output does not align very well; it can be improved like this.

```
10 FOR X=1 to 12
20 S=SQR(X)
30 PRINT USING "## #.####",X,S
40 NEXT X
50 END

RUN

 1   1.0000
 2   1.4142
 3   1.7321
 4   2.0000
 5   2.2361
 6   2.4495
 7   2.6458
 8   2.8284
 9   3.0000
10   3.1626
11   3.3166
12   3.4641
```

Notice how the PRINT USING instruction provides a fixed place size for X and SQR(X), resulting in a uniform tabulation.

Consider next a program to provide a tabulation of angles from 0 to 90 degrees with listing of sine and cosine.

```
 5 F$="## #.##### #.#####"
10 FOR A=0 TO 90 STEP 10
20 R=(3.14159/180)*A :REM Con-
   vert angle to radians
30 S=SIN(R)
40 C=COS (R)
50 PRINT USING F$,A,S,C
60 NEXT A
70 END

RUN

 0   0.00000  1.00000
10   0.17365  0.98481
20   0.34202  0.93969
30   0.50000  0.86603
40   0.64279  0.76604
50   0.76604  0.64279
60   0.86603  0.50000
```

```
70   0.93969   0.34202
80   0.98481   0.17365
90   1.00000   0.00000
```

Notice that lines 20 through 50 are repeated for angles *A* from 0 to 90 in steps of 10 degrees. When no STEP is specified, the instruction uses a STEP of 1. Any other increment or step must be explicitly stated in the instruction. Notice also that while the body of the program uses the variable *R*, the FOR A and NEXT A lines set up the boundaries of the loop. The variable *A* assumes the incremental values defined by the range of the loop routine.

2.4   USE OF IF . . . THEN AND IF . . . THEN . . . ELSE

A BASIC instruction which allows either setting up a loop or testing a desired condition is IF . . . THEN. For example, consider the following program that will accept input of any value of *X* other than zero, and provide output of the value *Y*, where $Y = X^2 - 4X + 8$. The program will end when $X = 0$ is entered.

```
10 INPUT "VALUE OF X IS";X
20 IF X=0 THEN STOP
30 Y=X^2-4*X+8
40 PRINT "FOR X=";X;" Y=";Y
50 GOTO 10

RUN

VALUE OF X IS?10
FOR X= 10 Y= 68
VALUE OF X IS?0

READY
```

A second example tests an input value of *X*, allowing calculation of SQR(X) only if *X* is positive. The program ends when $X = 0$ is entered.

```
10 INPUT "ENTER VALUE FOR SQUARE
   ROOT, X=";X
```

```
20 IF X<0 THEN PRINT "ONLY POSI-
   TIVE VALUES OF X ALLOWED"
   :GOTO 10
30 IF X=0 THEN END
40 PRINT "SQUARE ROOT OF";X;
   " IS";SQR(X)
50 GOTO 10

RUN

ENTER VALUE FOR SQUARE ROOT,
X=?25
SQUARE ROOT OF 25 IS 5
ENTER VALUE FOR SQUARE ROOT,
X=?-8
ONLY POSITIVE VALUES OF X AL-
LOWED
ENTER VALUE FOR SQUARE ROOT,
X=?0

READY
```

Here is a program to test whether an answer is correct or not.

```
10 PRINT "HOW MUCH IS";
20 X=RND(50)      :REM or
   X=INT(1+50*RND(1))
30 Y=RND(50)      :REM or
   Y=INT(1+50*RND(1))
40 PRINT X;"+";Y;"=";
50 INPUT A
60 IF A=X+Y THEN PRINT "Correct"
   :GOTO 10
70 PRINT "No, it's";X+Y
80 GOTO 10

RUN

HOW MUCH IS 41 + 5?45
No, it's 46
HOW MUCH IS 16 + 30?46
Correct
```

In the above program, lines 60 and 70 could also have been written as

```
60 IF A<>X+Y THEN PRINT "No,
   it's";X+Y :GOTO 10
70 PRINT "Correct"
```

with the program operating identically.

The test for an exact input is not always possible. If the above example were modified to test for X/Y, for example, line 40 would become

```
40 PRINT X;"/";Y;"=";
```

Then a test such as

```
60 IF A=X/Y THEN PRINT "Correct"
   :GOTO 10
```

may be difficult to achieve.

If the computer asked

```
2/3=?
```

the response 0.66667 would not match a calculated value of, say, 0.6666667. Even though the answer is *essentially* correct, it may not be *exactly* correct. A better method for handling such test is to determine if the answer is, say, within 1% of the correct answer or if it is rounded off to the nearest tenths or hundreths. Accomplish the former in this example using

```
60 IF ABS(X/Y-A)<ABS
   (0.01*(X/Y)) THEN PRINT
   "Correct" :GOTO 10
```

If $X/Y = 0.666667$ and the input response is $A = 0.666$, the test will prove correct, since $0.666667 - 0.666 < 0.01*0.666667$. If the absolute difference between X/Y and A is within 1% of the value X/Y, the test is true and the computer is instructed to print ''Correct'' and then transfer control to step 10. If not, the program goes on to the next step in the program (step 70 in the above example).

If the test involves positive or negative numbers, the test should include the sign as well.

```
10 PRINT "HOW MUCH IS 10*SIN(";
20 A=INT(1+180*RND(1))    :REM
   or A=RND(180)
30 PRINT A;") degrees";
40 INPUT AN    :REM AN is the in-
   put answer
50 V=10*SIN((3.14159/180)*A)
60 IF SGN(V)<>SGN(AN) THEN
   GOTO 80
```

```
70 IF ABS(V-AN)<ABS(0.01*V)
   THEN PRINT "Correct" :GOTO 10
80 PRINT "No, it's";V
90 GOTO 10
```

```
RUN
```

```
HOW MUCH IS 10*SINE(86) de-
grees?9.08
No, it's 9.9756
HOW MUCH IS 10*SINE(24) de-
grees?4.07
Correct
```

Notice that if the sign of V is not the same as the sign of the input answer, line 60 directs the program to line 80. If the absolute difference between V and the input answer is not less than 1% of the correct value, the program goes on to line 80 after the program test of line 70. Many versions of BASIC support the decision sequence IF . . . THEN . . . ELSE, which can be used like this:

```
10 PRINT "How old are you";
20 INPUT AG
30 IF AG<=18 THEN PRINT "That's
   great." ELSE PRINT "You're
   only as old as you feel."
```

```
40 END
```

```
RUN
```

```
How old are you?18
That's great.
```

```
READY
```

```
RUN
```

```
How old are you?40
You're only as old as you feel.
```

```
READY
```

If, however, the version of BASIC you are using allows only IF . . . THEN, the above program could still be programmed as

```
10 INPUT "How old are you";AG
20 IF AG<=18 THEN PRINT "That's
   great."
```

```
30 IF AG>18 THEN PRINT "You're
   only as old as you feel."
40 END
```

2.5 USE OF STRING VARIABLES

Alphanumeric or string data are handled nicely using BASIC. A variable terminated with a $ indicates a string variable. For example,

```
10 PRINT "What is your name";
20 INPUT NA$
30 PRINT "Hello ";NA$
40 PRINT "Glad to meet you"
50 END
```

RUN

```
What is your name?BOB
Hello BOB
Glad to meet you
```

READY

In the following example the user is asked whether the program should be repeated.

```
50  PRINT "Do you wish to repeat
    input of name (YES or NO)";
60  INPUT AN$
70  IF AN$="YES" THEN GOTO 10
80  IF AN$="NO" THEN END
90  PRINT "Answer YES or NO
    only" :REM This comment
    printed only if YES or NO not
    entered
100 GOTO 50
```

BASIC provides a number of string functions, such as those to separate out a part of a string—either left, right, or middle. For example,

```
10 INPUT "What city is the capi-
   tal of Colorado";AN$
20 IF AN$="DENVER" THEN PRINT
   "Correct" :GOTO 60
```

```
30 IF LEFT$(AN$,1)="D" THEN
   PRINT "The first letter is
   correct - try again":GOTO 10
40 IF RIGHT$(A$,3)="VER" THEN
   PRINT "The last three letters
   are correct - try again":GOTO
   10
50 IF MID$(AN$,2,1)="E" THEN
   PRINT "The second letter is
   correct -try again":GOTO 10
60 END
```

RUN

```
What city is the capital of Col-
orado?DAVOR
The first letter is correct -
try again
What city is the capital of Col-
orado?SILVER
The last three letters are cor-
rect - try again
What city is the capital of Col-
orado?DENVER
Correct
```

READY

The command LEFT$(AN$,1) or more generally

```
LEFT$(string variable, number
of string characters), or
LEFT$(A$,N)
```

allows extracting a part of a string (A$) for a desired number of characters (*N*). The command RIGHT$(AN$,3), or more generally

```
RIGHT$(string variable, number
of string characters), or
RIGHT$(A$,N)
```

allows extracting a part of a string (A$) for a desired number of characters (*N*). The command MID$(AN$,2,1), or more generally

```
MID$(string variable, starting
position, number of characters)
MID$(A$,P,N)
```

allows extracting a part of string (*A*$) from its

middle, starting at character position P for N characters.

2.6 USE OF DIM IN ARRAYS

The ability to handle data (both numeric and string) in an orderly way is very important. The DIM (dimension) instruction defines a group or array of data. For example,

```
10 DIM V(4)
20 V(1)=12
30 V(2)=2.5
40 V(3)=-7
50 INPUT "V(4)=",V(4)
60 PRINT "VALUE V(3)=";V(3)
70 S=V(1)+V(4)
80 PRINT "SUM OF ELEMENTS 1 AND 4
   IS ";S
90 END
```

RUN

```
V(4)=?10
VALUE V(3)=-7
SUM OF ELEMENTS 1 AND 4 IS 22
```

READY

Notice that the variable name V is used to represent four different values, as specified by the subscript (value in parentheses) number.

The following example inputs four values of voltage and a fixed resistance value, and asks for the resulting current in each case, using Ohm's law. The current for each case using Ohm's law is:

```
10 DIM V(4),I(4)
20 FOR K=1 TO 4
30 INPUT "V(";K;")=";V(K)
40 NEXT K
50 INPUT "RESISTANCE, R=";R
60 PRINT
70 FOR K=1 TO 4
80 I(K)=V(K)/R
```

```
90 PRINT "FOR A VOLTAGE OF
   ";V(K);"THE CURRENT IS
   ":I(K);"AMPERE"
100 NEXT K
110 END
```

RUN

```
V( 1 )=?30
V( 2 )=?12
V( 3 )=?6
V( 4 )=?24
RESISTANCE, R=?6
FOR A VOLTAGE OF 30 THE CURRENT
IS 5 AMPERE
FOR A VOLTAGE OF 12 THE CURRENT
IS 2 AMPERE
FOR A VOLTAGE OF 6 THE CURRENT IS
1 AMPERE
FOR A VOLTAGE OF 24 THE CURRENT
IS 4 AMPERE
```

READY

Arrays of more than one dimension are also possible. For example, you can store the grades in an array GR for 10 students, each having 3 exam grades.

```
10 DIM GR(10,3)
20 PRINT "ENTER exam grades for
   the 10 students."
30 PRINT
40 FOR I=1 TO 10
50 INPUT "For student
   ";I;"grades are";
60 FOR J=1 TO 3
70 PRINT "GRADE ";J;"=";
80 INPUT GR(I,J)
90 NEXT J
100 PRINT
110 NEXT I
120 END
```

In the above program, three exam grades are entered for each of 10 students, with the program stopping at line 120. The program could continue from line 120 to compute and output the average of the grades for each student and

19

the class average for each exam—we leave this exercise for you.

Two-dimensional arrays are helpful for describing the coefficients of simultaneous equations, such as in mesh or nodal formulations. You can define the elements of an array and calculate the determinant of the array.[3] For example, for the following 2x2 array

$$\begin{array}{rr} -1 & 3 \\ 7 & -5 \end{array}$$

the general form described mathematically as

$$\begin{array}{cc} a11 & a12 \\ a21 & a22 \end{array}$$

can be represented in BASIC as

```
        DIM A(2,2)
with    A(1,1)  A(1,2)
        A(2,1)  A(2,2)
```

representing the elements of the array. To enter the array element values, you could use

```
A(1,1)=-1
A(1,2)=3
A(2,1)=7
A(2,2)=-5
```

or more generally

```
10 INPUT A(1,1),A(1,2)
20 INPUT A(2,1),A(2,2)

RUN

?-1,3
?7,-5
```

Once the element values are entered, the determinant could be calculated using

```
D=A(1,1)*A(2,2) -
A(2,1)*A(1,2)
```

The following program requests the data from a 2x2 array and then calculates its determinant value.

```
10 DIM A(2,2)
20 PRINT "Enter the first row
   elements, separated by a
   comma"
30 INPUT A(1,1),A(1,2)
40 PRINT "And the second row el-
   ements, separated by a comma"
50 INPUT A(2,1),A(2,2)
60 D=A(1,1)*A(2,2)-
   A(2,1)*A(1,2)
70 PRINT "Determinant=";D
80 END

RUN
```

Enter the first row elements, separated by a comma and the second row elements, separated by a comma

```
?-1,3
?7,-5
Determinant= -16

READY
```

2.7   USE OF READ-DATA

The BASIC language provides a command which allows defining DATA which can be READ under program control. As one example, the data of a 2x2 array could be defined and read as follows:

```
READ
A(1,1),A(1,2),A(2,1),A(2,2)
DATA -1,3,7,-5
```

The READ statement can also operate more generally, as demonstrated by this example.

```
FOR I=1 TO 2
READ A(I,1),A(I,2)
NEXT I
DATA -1,3,7,-5
```

As another example, the elements of the resistor color code could be read using the following program steps.

---

[3] More advanced versions of BASIC include MATRIX instructions which provide instructions to INPUT a whole array, calculate the determinant, OUTPUT a whole array, etc.

```
10 DIM C$(10)
20 FOR I=0 TO 9
30 READ C$(I)
40 NEXT I
50 DATA BLACK,BROWN,RED,
   ORANGE,YELLOW
60 DATA GREEN,BLUE,VIOLET,
   GRAY,WHITE
```

In this example $C\$(0) = \text{BLACK}$, $C\$(1) = \text{BROWN}$, up to $C\$(9) = \text{WHITE}$.

2.8 GRAPHING TECHNIQUES

It is possible to plot data on a CRT or printer using BASIC. CRT display commands vary considerably between computers; some have only black and white, while others offer a range of colors. There are systems with 40 characters per line, while others have as many as 80 characters per line. Some have only coarse graphic operation, while others provide a much finer degree of graphic presentation. The graphics mode we describe will operate on both the CRT and the character printer.

A limited form of plot can be achieved by outputing a character, such as *,.,, at a specific point on a line. For example, consider plotting a straight horizontal line.

```
10 FOR P=1 TO 20
20 PRINT TAB(P);"*";
30 NEXT P

RUN
```

```
********************
```

Using the PRINT TAB(P) causes the printer to tab to position P, before printing the character *. Note that printing stays on the same line because of the semicolon at the end of the print line (Line 20). A vertical line could be obtained by running the following program. In this case the computer moves to the next line, tabs to column 15, and then prints a period.

```
10 FOR I=1 TO 10
20 PRINT TAB(15);"."
30 NEXT I

RUN
```

```
              .
              .
              .
              .
              .
              .
              .
              .
              .
              .
```

```
READY
```

A plot of a sine wave could be obtained using the following program.

```
10 FOR A=0 TO 360 STEP 10
20 V=20*SIN((3.14159/180)*A)
30 P=INT(V+21)
40 PRINT TAB(P);"."
50 NEXT A
60 END

RUN
```

```
READY
```

Line 20 calculates the value, *V*, of sinusoidal signal

$$V = 20\sin(A)$$

having values of $A = 10,20,30,\ldots$up to 360 degrees. Since the values of *V* will vary between $(+/-)20$ and the plot position can only be a positive number, line 30 obtains a position *P*, where $P=1$ if $V=-20$, up to $P=41$ if $P=+20$, with *P* being 21 at $V=0$.

EXERCISES

1. What output results from the following program?

```
10 X=5.5
20 PRINT 2*X
30 PRINT X+X+X
40 PRINT "X=",X
```

2. What output results from the following program?

```
100 T=25
110 T1=55
120 PRINT"FOR T OF";T;
130 PRINT "RESULT IS";
140 PRINT 3*(T+T1)
```

3. What output results from the following program?

```
100 X=6.56
110 Y=3.2
120 F$="##.### #.## #.#"
130 PRINT USING F$,X,Y,X+Y
```

4. Show the output resulting from the following program:

```
100 FOR B=1 TO 6
110 A=3*B-2
120 PRINTB;A
130 NEXT B
```

5. Show the output resulting from the following program:

```
100 L=8
110 FOR K=1 TO L-1
120 PRINT K,2*K,K↑2
130 NEXT K
```

6. Show the output resulting from the following program:

```
10 F$="## 0.####"
20 FOR A=0 TO 45 STEP 5
30 R=A/57.296
40 PRINT USING F$,A,SIN(R)
50 NEXT A
```

7. Show the output resulting from the following program:

```
100 FOR K=1 TO 22
110 IF INT(K/2)=K/2 THEN PRINT K
120 NEXT K
```

8. Write a BASIC program to output a list of 50 randomly selected integer numbers ranging between 10 and 20, inclusive.

9. Write a BASIC program to ask for input of value *K* and then list $K\uparrow 2$ and $K\uparrow 3$ for values ranging from *K*/2 to 5 times *K*, in steps of *K*/5.

10. Write a BASIC program to ask for input of a surname and then output the messages:

```
The first letter of your surname
is _
The last letter of your surname
is _
```

11. Write a BASIC program to accept input of 20 grades, then output the sum and average of the grades (store data in an array *G*(20)).

12. Write a BASIC program to read the following data and output the average of that data:

```
400 DATA 1,2,-1,7
410 DATA 3,3,9,5,6
420 DATA 2,2
```

Writing Program Modules

3

3.1 INTRODUCTION

This chapter introduces the concept of writing program modules which we will use extensively throughout this manual. A program module is a subprogram or subroutine which performs a specific part of the total program. This approach of dividing a component into a main program and smaller subsets is very helpful in writing the overall program. In addition, if the module is essentially self-contained, you can carry it over and use it exactly as written in another program.

A section of a program can be written as a subroutine, using a GOSUB command to direct the program to the first line of the subroutine. A RETURN command then terminates the subroutine and returns the program to the next line of the main program (just after the GOSUB command). A subroutine program is entered by a GOSUB command and exited using a RETURN command.

To help organize the programs of this manual, modules that appear more than once will have line numbers higher than 2000. Modules used only within a chapter are written using line numbers below 2000. To help identify each program, we use the line numbers 10-90 as REM statements to describe the program to follow. The actual program then starts at line 100.

3.2 USE OF GOSUB

The GOSUB is a very useful instruction which permits breaking a program up into smaller parts or modules. For example, a routine to convert a number from rectangular to polar coordinates need only appear once in a program and can then be used whenever the conversion is required.

As a subroutine or module, the conversion could be written as:

```
1000 REM Module to convert from
     rectangular to polar
1010 REM Enter with X,Y - Exit
     with Z,TH (eta)
1020 Z=SQR(X^2+Y^2)
1030 IF X=0 THEN TH=90 :GOTO
     1050
1040 TH=(180/3.14159)*
     (ATN(Y/X) :REM TH in de-
     grees
1050 RETURN
```

The module from 1000 to 1050 can then be used by the main program. The following program asks for the lengths of the sides of a right triangle and calculates and prints the value of the hypotenuse and the triangle acute angles as follows:

```
100 PRINT "Enter the values of
    the sides of a right tri-
    angle:"
110 INPUT "X=";X
120 INPUT "Y=";Y
130 GOSUB 1000     :REM Go to con-
    version module
140 PRINT "Hypotenuse is ";Z
150 PRINT "and acute angles
    are";TH;"and";
    90-TH;"degrees"
160 END
```

When both main program and subroutine are entered, the resulting RUN is:

```
RUN

Enter the values of the sides of
a right triangle: X=?10
Y=?20
Hypotenuse is 22.3607
and the acute angles are 63.4349
and 26.5651 degrees

READY
```

As a second example, consider a module designed to ask for positive inputs only:

```
100 PRINT "Enter first value"
110 GOSUB 500
120 A=X  :REM Save first input
```

```
130 PRINT "Enter second value"
140 GOSUB 500
150 B=X  :REM Save second input
160 Program continues here......
      .
      .
280 END (Program stops before
      subroutine)
500 REM Module to accept input
510 INPUT "Enter data value of
      X, X=";X
520 IF X<=0 THEN PRINT "only
      positive values " :GOTO 510
530 RETURN
```

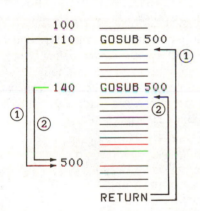

FIGURE 3.1 Use of GOSUB/RETURN

3.3 USE OF ON . . . GOTO, AND ON . . . GOSUB

A technique for choosing one of a few selected program sections or subroutines is the use of the ON . . . GOTO or ON . . . GOSUB commands. For example, the ON . . . GOTO statement is used in the following program to perform the desired conversion. Since there is a choice of 3 conversions, 3 distinct subroutines have to be available.

```
10 PRINT "Do you want to con-
      vert:"
```

```
20 PRINT TAB(5);"(1) INCHES TO
      FEET"
30 PRINT TAB(5);"(2) METERS TO
      YARDS"
40 PRINT TAB(5);"(3) CENTIME-
      TERS TO INCHES"
50 INPUT "CHOICE=";C
60 ON C GOTO 100,500,750
70 Program continues here...
      .
      .
      .
100 REM Module to convert inches
      to feet
110 INPUT "inches=";IN
120 FT=IN/12
130 PRINT IN;"inches =
      ";FT;"feet"
140 GOTO 10

500 REM Module to convert meters
      to yards
510 INPUT "meters = ";MT
520 YD=1.093613*MT
530 PRINT MT;"meters
      =";YD;"yards"
540 GOTO 10

750 REM Module to convert cen-
      timeters to inches
760 INPUT "Centimeters =";CM
770 IN=0.393701*CM
780 PRINT IN;"inches
      =";CM;"centimeters"
790 GOTO 10
```

Note that on line 60 the case of $C=1$ sends the program to line 100, on the case of $C=2$ the program continues at line 500, and on the case of $C=3$ the program continues on line 750. In this example the subroutines all return to line 10.

The following program demonstrates the ON...GOSUB command by choosing which subroutine will be used to determine the unknown parameter of a right triangle.

```
10 PRINT "Do you wish to deter-
      mine"
```

```
20  PRINT TAB(5);(1) the hypot-
    enuse and one acute angle,
    or"
30  PRINT TAB(5);(2) sides of the
    right triangle"
40  INPUT "Choice =",C
50  ON C GOSUB 800,900
60  Program continues here...
    .
    .
800 REM Module to accept input
    of triangle sides
810 INPUT "X=",X
820 INPUT "Y=",Y
830 RETURN
900 REM Module accepts input of
    hypotenuse and one acute an-
    gle
910 INPUT "Z=",Z
920 INPUT "and angle, theta
    =",TH
930 RETURN
```

Further use of the ON . . . GOSUB command will appear often in the chapters to follow.

3.4 WRITING PROGRAM MODULES

The GOSUB or ON . . . GOSUB commands allow the user to divide a program into separate parts or modules. The result is an easier path toward a properly written, accurate program. In addition, modules can be copied intact from one program to another. For example, the following modules can accept data in either the rectangular or polar form and convert to the other form. The three subroutines at 500, 700, and 900 will appear in a number of programs involving the analysis of ac systems.

```
500 PRINT "Enter input data:"
510 PRINT TAB(5);"(1) Rectangu-
    lar form"
520 PRINT TAB(5);"(2) Polar
    form"
530 INPUT "Your choice";C
```

```
540 ON C GOSUB 700,900
    .
    .
700 REM Module to accept rec-
    tangular input data
710 INPUT "X=";X
720 INPUT "Y=";Y
730 Z=SQR(X^2+Y^2)
740 IF X=0 THEN TH=90 :GOTO 760
750 TH=(180/3.14159)*ATN(Y/X)
760 RETURN
900 REM Module to accept polar
    input data
910 INPUT "Magnitude, Z=";Z
920 INPUT "at angle, theta=";TH
930 X=Z*COS((3.14159/180)*TH)
940 Y=Z*SIN((3.14159/180)*TH)
950 RETURN
```

The same set of modules can also be used to add two vectors, as shown below. Note that only the beginning of the subroutine at 500 appears in the program. The subroutines at 700 and 900 are referenced within the subroutine between lines 500 and 540.

```
100 PRINT "To add two vector
    numbers"
110 PRINT "enter first vector:"
120 GOSUB 500
130 X1=X :Y1=Y
140 PRINT "enter second vec-
    tor:"
150 X2=X :Y2=Y
160 XS=X1+X2
170 YS=Y1+Y2
180 PRINT "The sum of these vec-
    tors is ";
190 PRINT XS;"+j";YS
200 STOP
    .
    .
500 .....
    .
700 .....
    .
900 .....
    .
```

Notice how straightforward the main program is. This approach permits you to write the main frame of the program directly and simply, using the subroutines when required. The result is a structured format that is quite popular. A program to compute a current in an ac circuit using Ohm's law (as a simple example) could be structured as follows:

```
100 PRINT "This program accepts
    inputs of voltage and re-
    sistance"
110 PRINT "and calculates the
    resulting current using
    Ohm's law."
120 PRINT
130 GOSUB 1000 :REM Get values
    of V and R
140 GOSUB 1200 :REM Calculate I
150 GOSUB 1300 :REM Print answer
160 GOSUB 1500 :REM Ask if more
    input
170 Print "Enter new data"
180 GOTO 120
    .
    .
1000 REM Input module
1010 INPUT "Enter voltage,
     V=";V
1020 INPUT "Enter resistance,
     R=";R
1030 RETURN

1200 REM Module to calculate I
1210 I=V/R
1220 RETURN

1300 REM Module to print answer
1310 PRINT
1320 PRINT "For V of";V;"volts"
1330 PRINT "and R of";R;"ohms"
1340 PRINT "current calculated
     using Ohm's law, I=V/R is";
1350 PRINT I;"amperes"
1360 RETURN

1500 REM Ask if more input de-
     sired
1510 PRINT "Do you wish to do any
     more calculations(YES or
     NO)";
```

```
1520 INPUT A$
1530 IF A$="YES" THEN RETURN
1540 PRINT "Done for now"
1550 END
```

After setting up the main program like this, the programmer can concentrate separately on the details of the individual modules, writing them one at a time and entering them with the main program.

3.5 GRAPHICS

A more structured approach to graphic technique is presented next, to supplement the material in Chapter 2. To structure data graphing, you might first use a module to calculate all the data to be plotted, storing this data in an array. Then, write a second module to plot this data on a printer, after initially determining the needed scaling of the plot to fit on a printed line of, for example, 64 characters or 80 characters. We first consider how to write the data calculation module.

```
100 DIM D(20)
110 N=20
120 FOR K=1 TO N
130 D(K)=2*K :REM Calculate a
    data point
140 NEXT K
    .
    .
```

Having stored 20 points in the array D(I), a plot routine can now be called to print the graph

```
 10 REM ***** PROGRAM *****
 20 REM Program to plot data in
    array D( )
 30 REM
100 DIM D(50)
110 N=20
120 FOR K=1 TO N
130 D(K)=2*K :REM Calculate a
    data point
```

```
140 NEXT K
150 FM=2*N
160 PRINT "Data now stored for
    ";N;"points" :PRINT
170 GOSUB 1600 :REM Do Plot
180 END
1600 REM Module to plot data in
     array D( )
1610 FS=40 :REM Allow for line
     width of 40
1620 FOR K=1 TO N
1630 P=INT(D(K)*FS/FM)
1640 PRINT TAB(P);"*"
1650 NEXT K
1660 PRINT:PRINT
1670 RETURN
```

READY

RUN

Data now stored for 20 points

```
*
  *
    *
      *
        *
          *
            *
              *
                *
                  *
                    *
                      *
                        *
                          *
                            *
                              *
                                *
                                  *
                                    *
                                      *
```

READY

EXERCISES

1.   Write a program module (starting at 1200) to calculate the sine and cosine of an input angle(in degrees).

2.   Write a program module (starting at 2200) to calculate N factorial (N!) for a given value of N.

3.   Write a program module to calculate the parallel resistance (Rp) for given values of $R1$ and $R2$.

4.   Write a program module to select a random number between 6 and 13, inclusive.

5.   Write a BASIC program module to obtain the sum of the N values in an array T( ).

6.   Write a program module to test the values of X, Y, Z and store the largest in B.

7.   Write a module to test an array D( ) for the smallest and largest entries.

8.   Write a program module to change each of the N entries in array D( ) into twice their value.

Resistance

4

4.1 INTRODUCTION

The introduction to BASIC in chapters 1, 2, and 3 was brief and limited primarily to those commands of particular importance to the current area of study. We emphasize that this manual is not intended to cover the language in the detail normally found in a manual devoted solely to the language or the instructional material normally provided with the machine. We assume that you can find this information if you need it; this manual concentrates on the application of the commands to the areas of dc and ac circuit analysis.

The manual is designed to build on the information provided in the earlier chapters. We are assuming that you read and understood the earlier chapters; we will not repeat most of the comments in those chapters. It is very important to proceed through the manual in a sequential manner because specific subroutines and modules are introduced and then used with limited support or comment in a later chapter. Of course, if you have prior experience with the language the detail will be sufficient to examine any particular section of the manual without proceeding through all the earlier chapters. You may possibly have built up enough confidence in the first few chapters to be able to look ahead and understand some of the later programs.

The programs progress from the simple to the more complex within each chapter and throughout the manual, as do the problems at the end of the chapter. Be aware that there is usually more than one way to obtain a desired result. In fact, the programs may be longer than necessary to be more instructional. However, your first goal is developing your skills so that you can write a program properly and obtain the correct results. With practice, you will become more skillful in using various commands and the available memory. You will learn the vocabulary of computer programs, which may not be the best English but is efficient and clearly defines the operation to be performed.

This language is universally understood and unifies the effort to develop the most efficient software packages.

For many students, writing a computer program is their first exposure to an open-ended problem—a problem which may have more than one path toward the solution (we will always be optimistic and assume a solution exists for any given problem). It is an excellent opportunity for you to develop your creative skills and experience the satisfaction of developing a routine that is unique and particularly efficient. It is especially important for you to realize that the computer is limited in its response to any improper command or input. If you do not fully understand the applicable concept, law, or theorem, you cannot expect to properly program it. The truth of this statement will become evident when you begin to write programs in areas where your knowledge or experience is limited; you will need a number of trial runs to obtain the results you desire. However, this hit-and-miss experience can have an enormous positive impact on the development of your computer skills.

Most lengthy problems can be solved by bringing together several subroutines or modules developed individually for narrower applications. For longer programs, the *flow chart*, a pictorial description of the program using particular shapes to represent the operations and subroutines, can be particularly useful. It maintains some control over the flow of data and the sequence of operations.

Chapters with more programs should not be reviewed as more important than those with fewer. Rather, we chose the programs to insure that the computer control statements were properly introduced and applied to sequentially more difficult areas of application.

It is virtually impossible to investigate all the areas of dc and ac circuit analysis in one manual. However, we selected areas for reasons indicated in the previous paragraphs; we also wanted to insure that those areas not covered could be easily investigated. For instance, the steps to determine the Thevenin and Norton equivalent circuits are very similar. We cover the former in detail while leaving the other as an end-of-chapter exercise.

Although the format of the commands your computer system uses may not perfectly match those that appear here, be assured that the similarities are so broad that applying these programs to your system should be relatively easy. In those areas where additional comment is necessary, we will provide it to insure a direct transfer to your system. We are anxious to hear of any difficulty you may have applying these routines, and will do our best to respond to inquiries or suggestions.

The traditional way of presenting the concepts of dc and ac circuit analysis lends itself well to developing the computer skills necessary to analyze the most difficult of systems. There is a natural progression from a few relatively simple mathematical relationships to the more complex theorems and methods that require additional control and care in formulating the best sequence of commands to properly analyze the system.

4.2 $R = \rho\dfrac{\ell}{A}$

The first program determines the resistance of a conductor as determined by the material, area, and length. Note in Program 4-1 that we have added brackets to the left-hand margin to define the components of the program. Isolating particular regions of a program can help in analyzing a lengthy complex program for the first time. Entries to the right of the program will clarify the step or equation being performed.

Throughout, the manual locations 1 through 99 will be reserved for REM statements to define the program. In addition, we will use an increment of 10 between commands unless program length or modifications require intermediate locations. Any input provided by the user, such as the required RUN statement or input data, will be underlined.

In program 4-1 statement 60 simply insures space between the REM statements and the beginning of the program. Statements 130 (PRINT) and 170 (PRINT) insure an additional space following the RUN command and the displaying of the data. In both cases, there is

```
          ┌10 REM *****   PROGRAM 4-1   *****
          │20 REM   ********************************
Program   │30 REM   Calculation of wire resistance
Description│40 REM   using R=pl/A
          │50 REM   ****************************
          └60 REM
          ┌100 P=10.37
Data      │110 D=0.020
          └120 L=100
          ┌130 PRINT
          │140 PRINT "For p=";P;"CM-ohms/ft"
Display   │150 PRINT "    d=";D;"inches"
Data      │160 PRINT "and l=";L;"feet"
          └170 PRINT
          ┌180 DM=D*1000
Calc.     │190 PRINT "d=";DM;"mils"
&         │200 A=DM^2
Output    │210 PRINT "A=";A;"CM"
Control   │220 R=P*L/A
          └230 PRINT "R=";R;"ohms"
           240 END
```

$d_{mils} = d_{inches} \times 1000$

$A_{CM} = (d_{mils})^2$

$R = \rho\ell/A$

READY

```
RUN

*

For p= 10.37 CM-ohms/ft
    d= .02 inches
and l= 100 feet

d= 20 mils
A= 400 CM
R= 2.5925 ohms


READY
```

*Example 3.1, ICA. Many of the programs in this book refer to examples and problems in *Introductory Circuit Analysis, 4th Edition* by Robert Boylestad (Columbus: Charles E. Merrill, 1982). These references are optional; you do not need to refer to *Introductory Circuit Analysis* to understand the programs in this book.

nothing to print and the computer skips to the next line. This kind of detailed description will not appear throughout the manual. From this point on, we expect that you will recognize an isolated PRINT command as there to provide spacing between the lines of the computer response.

Note in the program that the data is set equal to a variable defined by a capital letter such as P, D, or L and is actually part of the program. If the input data should change, it would be necessary to recall that particular line and change the value before running the program.

Statements 140 through 160 will display the data to be operated on when the program is RUN. Note in statements 140, 150, and 160 that anything in these quotes will appear exactly in that form in the response below. If an incorrect spelling appears or an equal sign is left out, it would appear that way in the output; the computer prints out whatever is between the quotes whether it is correct or not. In each case, of course, the values of P, D, and L will appear in the output response.

The operations performed by statements 180, 200, and 220 appear to the right of the program. Statements 190, 210, and 230 will print out the results as shown below the program. Although the results were set for fourth

place accuracy, the answer of $R = 2.5925$ ohms is exact. If set for fifth or sixth place accuracy, the result would appear the same since zeros are not included to the right of an exact result. Note that RUN is the only underlined quantity since it is the only input from the user once the program is set in motion.

Although the program is short, if you are a new student of the art of writing programs, you will probably find the program difficult enough at this stage in your development. Try the first problem at the end of the chapter. It is of similar difficulty and will help you better appreciate the effort required to write such programs.

As indicated above, a change in input data requires that statements 100, 110, or 120 be listed and changed as required. Program 4-2 is designed to request the input data and avoid the necessity of listing the entire program for each change in input data. The program is simply rerun and the new data inserted as requested by the program. It is never necessary to relist the entire program unless you want a change in format or results. Note in this case that the input data is underlined, since it is now input provided after the RUN statement is initiated. The results obtained are the same as obtained for Program 4-1, since the input data is the same. However, the second RUN employed a different set of data to demonstrate that it is

```
Program
Description   ┌10 REM *****   PROGRAM 4-2  *****
             │20 REM   *******************************
             │30 REM   Calculation of wire resistance
             │40 REM   using R=pl/A
             │50 REM   *******************************
             └60 REM
Request      ┌100 PRINT:PRINT "Enter the following data:"
Input        │110 INPUT "p(CM-ohms/ft)=";P
Data         │120 INPUT "d(inches)=";D
             └130 INPUT "l(feet)=";L
Display      ┌140 PRINT
Input        │150 PRINT "For p=";P;"CM-ohms/ft"
Data         │160 PRINT "    d=";D;"inches"
             │170 PRINT "and l=";L;"feet"
             └180 PRINT
Calc.        ┌190 DM=D*1000
&            │200 PRINT "d=";DM;"mils"
Output       │210 A=DM^2
Control      │220 PRINT "A=";A;"CM"
             │230 R=P*L/A
             └240 PRINT "R=";R;"ohms"
              250 END
```

READY

RUN

```
*
Enter the following data:
p(CM-ohms/ft)=? 10.37
d(inches)=? 0.020
l(feet)=? 100

For p= 10.37 CM-ohms/ft
    d= .02 inches
and l= 100 feet

d= 20 mils
A= 400 CM
R= 2.5925 ohms
```

READY

RUN

```
Enter the following data:
p(CM-ohms/ft)=? 10.37
d(inches)=? 0.005
l(feet)=? 50

For p= 10.37 CM-ohms/ft
    d= 5E-03 inches
and l= 50 feet

d= 5 mils
A= 25 CM
R= 20.74 ohms
```

READY

*Example 3.1, ICA

unnecessary to relist the data statements before rerunning the program. It is a tremendous time saving element for future calculations of the same RUN to have the request for data in the program. Once the program is properly entered there is no need to ever recheck the intermediate calculations. The result will be accurate to the degree requested, independent of the magnitude of the input data. Do you remember occasions when you knew the equation well, but, when it came time to substituting the numerical values, you made a foolish error and ended up with a ridiculous result? Once the program is

properly entered and the input data properly provided, the result will be accurate and you no longer have to worry about the numerical calculations. What a relief! For both runs, note the absence of the zeros to the right of the last decimal digit. Also note that an exact answer occurred in the hundredths place for the second RUN, while an exact answer for the first RUN required fourth place accuracy.

Program 4-3 is a slight variation of Program 4-2, in that the input data is the resistance, material, and length, and the output is the area and the diameter in mils and inches.

```
10 REM *****  PROGRAM 4-3  *****
20 REM *******************************
30 REM Calculation of the wire area
40 REM and diameter using R=pl/A.
50 REM *******************************
60 REM
100 INPUT "P=";P
110 INPUT "R=";R
120 INPUT "L=";L
130 PRINT
140 PRINT "For p=";P;"CM-ohms/ft"
150 PRINT "    R=";R;"ohms"
160 PRINT "and l=";L;"feet"
170 PRINT
180 A=P*L/R                          A = pℓ/R
190 PRINT "A=";A;"CM"
200 DM=SQR(A)                        d_mils = √A_CM
210 PRINT "d=";DM;"mils"
220 D=DM/1000                        d_inches = d_mils/1000
230 PRINT " =";D;"inches"
240 END
```

$$A = p\ell/R$$
$$d_{mils} = \sqrt{A_{CM}}$$
$$d_{inches} = d_{mils}/1000$$

```
READY

RUN
*
P=? 17
R=? 2.5
L=? 78

For p= 17 CM-ohms/ft
     R= 2.5 ohms
and l= 78 feet

A= 530.4 CM
d= 23.0304 mils
 = .023 inches


READY
```

*Problem 3.6, ICA

4.3 TEMPERATURE EFFECTS

Program 4-4 determines the resistance of the conductor at a particular temperature if its resistance is known at any other temperature. It is similar in many respects to the program introduced in the previous section. Line 100 requests that the output skip a line before requesting the input data to insure some isola-

tion between RUNS. It is essentially a combination of statements 130 and 140 in Program 4-3. The colon within line 100 provides the required separation between command statements. Note in line 160 the use of the ABS(T) command, which insures that the value of T is positive in the equation of line 160. Note again those quantities that are underlined when the program is RUN. In this case, an exact answer was not obtained with four-place accuracy.

```
        ┌10 REM *****  PROGRAM 4-4  *****
Program │20 REM ***********************************************
Description│30 REM Determining R2 using (T+t1)/R1 = (T+t2)/R2
        │40 REM ***********************************************
        └50 REM
        ┌100 PRINT:PRINT"Enter the following data:"
        │110 INPUT "T=";T
 Input  │120 INPUT "R1=";R1
 Data   │130 INPUT "t1=";T1
        │140 INPUT "t2=";T2
 Calc.  └150 PRINT
 &      ┌160 R2=(ABS(T)+T2)*R1/(ABS(T)+T1)
 Output └170 PRINT "R2=";R2;"ohms"
 Control 180 END
```

$$R_2 = (T + t_2)R_1/(T + t_1)$$

```
READY

RUN


*
Enter the following data:
T=? 234.5
R1=? 50
t1=? 20
t2=? 100

R2= 65.7171 ohms


READY
```

*Example 3.4, ICA

Program 4-5 provides the temperature at which a particular resistance is attained. Students frequently have more difficulty with this operation than that performed in Program 4-4 because of the associated mathematical manipulations. This program, once properly written, will avoid the many errors that seem to result from the loss of a sign or the improper calculation. Note in the second RUN that the minus

sign will automatically be associated with the solution when required.

4.4 TABULATION OF *R* VERSUS *L*

A list of resulting values of *R* for a range of values for *L* (with ρ and *A* fixed) will provide an

```
10 REM *****  PROGRAM 4-5  *****
20 REM *********************************************
30 REM Determining t2 using (T+t1)/R1 = (T+t2)/R2
40 REM *********************************************
50 REM
100 PRINT:PRINT"Enter the following data:"
110 INPUT "T=";T
120 INPUT "R1=";R1
130 INPUT "R2=";R2
140 INPUT "t1=";T1
150 PRINT
160 T2=R2*(ABS(T)+T1)/R1 - ABS(T)
170 PRINT "t2=";T2;"Celsius"
180 END
```

$$t_2 = R_2(T + t_1)/R_1 - T$$

```
READY

RUN


*
Enter the following data:
T=? 234.5
R1=? 0.92
R2=? 1.06
t1=? 4

t2= 40.2935 Celsius


READY

RUN


*
Enter the following data:
T=? 234.5
R1=? 0.92
R2=? 0.15
t1=? 4

t2=-195.6141 Celsius


READY


*Problem 3.19, ICA
```

opportunity to investigate the procedure of tabulating data. For fixed values of ρ and d, as established by lines 100 and 140 of Program 4-6, the value of R will be determined for 10 values of L extending from 100 to 1000 feet, in increments of 100 feet. Note the brevity of statement 170 for setting such a routine in motion. Every LOOP routine requires a concluding statement, such as on line 200, to tell the routine to repeat the sequence for the next value of L. The TAB

statement of lines 120 and 130 establishes the space (10) between the left-hand margins and the column headings. Even though the column headings of length and (feet) have the same TAB command as the output variable L in statement 190, the semicolon appearing in statement 190 before the variable L will add an additional space before printing out the result. For the BASIC language used with this computer, the presence of the semicolon adds an

```
┌10 REM *****   PROGRAM 4-6  *****
│20 REM ********************************
Program      │30 REM Table to list R vs l for
Description  │40 REM length (l) from 100 to 1000 feet
│50 REM ********************************
└60 REM
┌100 PRINT:PRINT "For p=10.37 CM-ohms/ft, d=20 mils"
Table        │110 PRINT
Heading      │120 PRINT TAB(10);"length";TAB(20);"resistance"
└130 PRINT TAB(10);"(feet)";TAB(22);"(ohms)"
Data  140 P=10.37 :D=0.020
┌150 DM=D*1000
Calc. └160 A=DM^2
┌170 FOR L=100 TO 1000 STEP 100
ℓ     │180 R=P*L/A
loop  │190 PRINT TAB(10);L;TAB(20);R
└200 NEXT L
210 END
```

READY

RUN


```
For p=10.37 CM-ohms/ft, d=20 mils

        length      resistance
        (feet)        (ohms)
         100         2.5925
         200         5.185
         300         7.7775
         400         10.37
         500         12.9625
         600         15.555
         700         18.1475
         800         20.74
         900         23.3325
        1000         25.925
```

READY

additional space before a printout results. In this program the additional space was desirable because the resulting data was indented, as shown in the printout. The same applies to the "resistance" heading and the resistance R. For each value of L, the computer will automatically shift to the next line before printing out the results. When $L = 1000$ feet, all values of L have been applied and the program will automatically move on to line 210 and END the operation. Note in the resistance column those resistance values that are exact because there are fewer than four digits to the right of the decimal sign. In the above program, the LOOP routine only includes statements 170 through 200. For each value of L, the resistance is calculated on line 180 and the value of L and R printed out by line 190. The next L command will send the program back to line 170 to calculate the next value of R for $L = 100$ feet longer. The value of P (which is rho, ρ) and the area A are fixed by the data provided in statement 140. If the material or the diameter were changed, statement 140 would have to be relisted and the values of ρ and the diameter changed appropriately. The LOOP would remain from 100 to 1000 feet unless statement 170 was recalled and some changes made in the range of the LOOP or the increment. If you run the program, you will be amazed how quickly the computer prints out the different values of resistance for the indicated length. Try to appreciate the amount of time and energy expended to perform the same calculations by hand and achieve the same accuracy.

4.5 $R = \rho\frac{\ell}{A}$ REVISITED

The following program (4-7) is an expanded version of an earlier effort. It will permit a choice of conductors from a provided list and provide the value of ρ for the calculation to follow. In addition, a number of commands will be used for the

first time. The DIM P(7) of line 100 alerts the memory that sufficient space must be set aside for the seven values that the variable P (actually ρ) will assume. All subscripted variables such as $P(1)$, $P(2)$ . . . $P(4)$ should have a dimension statement or the computer may display an ER-ROR signal.[1] When a quantity has a range of values defined by the subscripted notation indicated above, a dimension statement must be included. If, in a particular program, there is more than one quantity with a list of subscripted values, each must have a separate dimension statement. In this program, statement 110 establishes three lines between each RUN of the program. Statements 140 through 200 provide a list of materials to choose from for the calculations to be performed. Statement 220 re-

```
10 REM *****  PROGRAM 4-7  *****
20 REM *******************************************************
30 REM Expanded program to demonstrate calculation
40 REM of resistance using R=pl/A.
50 REM *******************************************************
60 REM
100 DIM P(7)
110 PRINT:PRINT:PRINT
120 PRINT "The resistance of a wire will now be determined."
130 PRINT "Select from the following wire materials:"
140 PRINT "(1) Silver"
150 PRINT "(2) Copper"
160 PRINT "(3) Aluminum"
170 PRINT "(4) Tungsten"
180 PRINT "(5) Nickel"
190 PRINT "(6) Nichrome"
200 PRINT "(7) Carbon"
210 PRINT TAB(15);
220 INPUT "Wire type=";W :REM W is the number of the selected wire
230 PRINT
240 INPUT "Wire length (in feet) = ";L
250 INPUT "and wire diameter (in mils) = ";D
260 GOSUB 500 :REM Obtain resistivity of selected wire type
270 REM Now calculate wire area
280 A=D^2
290 REM Next, calculate wire resistance
300 R=P*L/A
310 PRINT
320 PRINT "Wire resistance is ";R;"ohms."
330 PRINT:PRINT
340 INPUT "More (YES or NO)";A$
350 IF A$="YES" THEN GOTO 110
360 PRINT "Thanks and Goodbye"
370 END
```

Choose material: 130–230
ℓ 240
d 250
Calc.: 270–300
Output: 310–330
Continue?: 340–360

---

[1] Some versions of BASIC only require the programmer to specify dimensions greater than a specific size—10 for the Radio Shack TRS–80, for example.

Determine
ρ
```
500 REM  Subroutine to obtain wire resistivity, p
510 RESTORE
520 FOR I=1 TO 7
530 READ P(I)
540 DATA 9.9,10.37,17,33,47,600,21000
550 NEXT I
560 P=P(W)
570 RETURN
```

READY

RUN

```
*
The resistance of a wire will now be determined.
Select from the following wire materials:
(1) Silver
(2) Copper
(3) Aluminum
(4) Tungsten
(5) Nickel
(6) Nichrome
(7) Carbon
            Wire type=? 2

Wire length (in feet) = ? 100
and wire diameter (in mils) = ? 20

Wire resistance is  2.5925 ohms.


More (YES or NO)? YES



The resistance of a wire will now be determined.
Select from the following wire materials:
(1) Silver
(2) Copper
(3) Aluminum
(4) Tungsten
(5) Nickel
(6) Nichrome
(7) Carbon
            Wire type=? 3

Wire length (in feet) = ? 500
and wire diameter (in mils) = ? 35

Wire resistance is  6.9388 ohms.


More (YES or NO)? NO
Thanks and Goodbye
```

READY

*Example 3.1, ICA

quests the number associated with the material in the list above. Note that a fixed numerical value has been assigned to the variable W, but the computer is still totally unaware of the impact of the choice. The listing provided was solely for the user's information, to permit choosing the appropriate value of W. Note the use of the REM statement in 220 to define (within the program) the significance of the variable W. Lines 240 and 250 request the value of L and D, while 260 requires that the program jump down to location 500 to obtain the value of ρ as defined by W. Locations 500 through 570 insure that the values of ρ are available and provide the correct value for the chosen material. The RESTORE statement insures that the data pointer returns to the first data point (9.9 of line 540) before using the input value of W to select the appropriate value of $P(I)$. It will repeat this RESTORE operation during each run of the program so that the proper value of $P(I)$ is chosen. Without it, the data pointer will not return to the beginning of the data list each time the read loop is run. Whenever a string of data is to be reread, it is necessary to use the RESTORE statement so that the data pointer is on the first entry. The 500 subroutine will establish the values of $P(1)$, $P(2)$, and $P(3)$ and so on as 9.9, 10.37 and 17, etc. respectively. Statement 560 will identify the value of P as determined by the value of W chosen above. For instance, if $W = 2$, $P = P(W) = P(2) = 10.37$. Statement 570 directs the routine back to 260 once the value of P is established. It will then proceed to determine the value of R and print the result as defined by statement 320. Line 340 asks whether you would like to RUN the program again for a different set of input data. Recall from Chapter 2 that the combination of a capital letter and a dollar sign, such as $A\$$ in statement 340, is referred to as a *string variable*. It is for those occasions when a character string, such as YES or HELLO, is to be defined as equal to $A\$$. For example, if a request was made to print out $A\$$, the result would be YES or HELLO. If the user should inadvertently type YS instead of YES, the computer would not recognize it as equivalent to YES on line 350 and proceed to the next line, which types out "Thanks and Goodbye" and ends the program on line 370. Note the underlined quantities in the two RUNS of the program representing the input data, and the clean, concise format of the output data as established by the TAB statements and the PRINT commands within the program.

4.6   COLOR CODE

This chapter will conclude with the introduction of three programs that deal with the color code as applied to fixed carbon resistors. Program 4-8 generates the color code for resistor values greater than 10 ohms. Statement 130 repeats the request for the resistor value if the entered value is less than 10 ohms.

Note that two subroutines or modules appear in this program, one at 1500 reading color band information and the other at 1600, to determine the color bands for the input resistor value.

Subroutine 1500 extends through 1560 before the RETURN command sends it back to line 110. Because each data value is a string of alpha characters, the dimension statement must use $C\$$ as its designation. The 10 in the dimension statement reveals the number of colors employed in the labeling technique. Actually, for a dimension of 10 there are 10 values for subscripts 1 through 10 and an additional value at subscript 0. Since the routine in module 1500 uses subscripts from 0 to 9, the dimension in line 1510 could have been DIMC(9). For each value of J starting with zero $J(0)$, $J(1)$, $J(2)$, etc., the color associated with each subscripted variable will be determined. For instance, during the first pass through the loop $C\$(0) = $ BLACK, during the second pass $C\$(1) = $ BROWN, and

```
       10 REM *****  PROGRAM 4-8  *****
       20 REM ********************************
       30 REM Program determines color code
       40 REM for standard resistor values.
       50 REM ********************************
       60 REM
      100 GOSUB 1500 :REM Read color band information
Input 110 INPUT "Value of resistor";R
  R   120 IF R=0 THEN END
      130 IF R<10 THEN PRINT "Value too low" :GOTO 110
      140 GOSUB 1600 :REM Obtain band colors
Output 150 PRINT "Color bands are ";C1$;" ";C2$;" ";C3$
      160 PRINT
Repeat 170 GOTO 110
     1500 REM Module to read color band data
     1510 DIM C$(10)
Read 1520 FOR J=0 TO 9
Band 1530 READ C$(J)
Data 1540 NEXT J
     1550 DATA BLACK,BROWN,RED,ORANGE,YELLOW,GREEN,BLUE,VIOLET,GRAY,WHITE
     1560 RETURN
     1600 REM Module to obtain color bands given resistance R
     1610 IF R<100 THEN I=0:R1=R:GOTO 1660
     1620 FOR I=1 TO 7
     1630 IF R/10^I+0.5<100 THEN GOTO 1650
     1640 NEXT I
     1650 R1=INT(R/10^I+0.5) :REM R1 has 2 digit value of R
     1660 V1=INT(R1/10+0.05)
     1670 V2=R1-10*V1
     1680 V3=I
     1690 C1$=C$(V1)
     1700 C2$=C$(V2)
     1710 C3$=C$(V3)
     1720 RETURN
```

(left margin labels: Input R; Output; Repeat; Read Band Data; Determine Color Bands)

```
READY

RUN

Value of resistor? 220
Color bands are RED RED BROWN

Value of resistor? 3.3E3
Color bands are ORANGE ORANGE RED

Value of resistor? 5.6E6
Color bands are GREEN BLUE GREEN

Value of resistor? 110
Color bands are BROWN BROWN BROWN

Value of resistor? 11
Color bands are BROWN BROWN BLACK

Value of resistor? 4.7E4
Color bands are YELLOW VIOLET ORANGE

Value of resistor? 5.6
Value too low
Value of resistor? 3.6E4
Color bands are ORANGE BLUE ORANGE

Value of resistor? 0

READY
```

so on. Later in the program, if $C\$(1)$ is called for, it will spell out the color BROWN.

As indicated by the REM statement, the module extending from 1600 through 1720 will determine the color band for the input resistor value. For values of R less than 100, the third color band will always be black, permitting the statement of line 1610, which automatically sets $I = 0$. The result is that $V3$ in line 1680 will always equal zero and $C3\$$ of line 1710 will equal $C\$(0)$. The subscript 3 in $V3$ and $C3\$$ identifies which band is being determined. For values less than 100, $R1$ is set equal to R (the input value) and the program jumps to line 1660 to determine the remaining color bands. You will recall from Chapter 2 that INT(X) will result in the integer part of X, while totally ignoring the fractional part, even if it is greater than 0.5.

The impact of line 1660 can best be demonstrated through an example using $R = R1 = 30$ ohms. Statement 1660 will result in $V1 = INT(30/10 + 0.05) = INT(3.0 + 0.05) = INT(3.05) = 3$. You may question the value of 0.05, if the result is still 3. In fact, the program appears to provide all the correct answers without having the 0.05 factor at all. However, keep in mind that all mathematical operations of the computer are performed in binary, and it is possible that the result of the division by 10 would result in 2.999----- which would provide an integer value of only two. By adding the 0.05 to the above (2.999---- + 0.05 = 3.04999----) result, we can insure that the number resulting from the binary arithmetic will provide the proper integer value. The same reasoning can be applied to 0.5, appearing in lines 1630 and 1650.

In the above example, the three (of 30) has now been isolated as the value of $V1$ which will result in $C1\$ = C\$(V1) = C\$(3) = $ ORANGE. The next step results in $V2 = 30 - 10(3) = 0$ and $C2\$ = C\$(V2) = C\$(0) = $ BLACK, as it should be for the second band.

In cases where R is greater than 100 (such as perhaps 8200), we move on to line 1630,

where $I = 1$ (for the first pass) and $R/10^1 + 0.5 = 8200/10 + 0.5 = 8200.5$, which is greater than 100. A return to 1620 is required and the operation of line 1630 repeated to result in $8200/10^2 + 0.5 = 82.5$ which is now less than 100, and the program will move to line 1650, where $R1 = INT(8200/100 + 0.5) = INT(82.5) = 82$. The value of $V1$ is $V1 = INT(82/10 + 0.05) = INT(8.25) = 8$ and $C1\$ = C\$(8) = $ GRAY. The value of $V2$ equals $82 - 10(8) = 2$ and $C2\$ = C\$(2) = $ RED. Finally, $V3 = I = 2$ and $C3\$ = C\$(2) = $ RED also.

The RETURN command of line 1720 will send the program back to line 150, where the colors associated with the subscripted variables will be printed out after the descriptive statement. The program will then return to 110, where an input value of 0 will END the program.

Program 4-9 will randomly select values of resistance among the 20% tolerance values. The 24 data points appearing on lines 1270 through 1290 require the dimension statement of line 100, while line 110 reveals that the program will generate 10 random values of resistance. Based on previous discussion, the need for the RESTORE command and the function performed by steps 1240 through 1290 should be clear. However, the routine appearing on lines 1300 through 1350 is new and will be described in some detail. Recall that the command RND(X) will randomly generate a number in the range $0 \le X < 1$. On line 1320 the maximum value within the parenthesis is $(1 + 8 \times 0.99999. . .)$, which will always be less than 9, resulting in a maximum value of 8 for INT(1 + 7.99999. . .). For the future, remember that the maximum value of INT(1 + $N \times$ RND(X)) is always N. The minimum value is 1 when RND(X) = 0. For statement 1330, the maximum value is 24, while on line 1340, the multiplying factor $R(J)$ is determined by the random value of J generated by line 1330. The multiplying factor is then multiplied by a power of 10 that is in a range of 10^0 through 10^8, as determined by the value of E randomly gen-

```
          10 REM *****   PROGRAM 4-9  *****
          20 REM ********************************************************
          30 REM Program to randomly select the values of resistance
          40 REM from among the 20% tolerence components.
          50 REM ********************************************************
          60 REM
10        100 DIM R(24)
values   ┌110 FOR T=1 TO 10
of       │120 GOSUB 1200
R        │130 PRINT T;" ";R;"ohms"
( T      └140 NEXT T
  Loop)   150 END
          1200 REM Module to randomly select a value of resistance
          1210 REM from the standard 20% tolerence values
         ┌1220 REM Read standard component values
         │1230 RESTORE
         │1240 FOR J=1 TO 24
Define   │1250 READ R(J)
R(J)     │1260 NEXT J
         │1270 DATA 0.10,0.11,0.12,0.13,0.15,0.16,0.18
         │1280 DATA 0.20,0.22,0.24,0.27,0.30,0.33,0.36,0.39
         └1290 DATA 0.43,0.47,0.51,0.56,0.62,0.68,0.75,0.82,0.91
       E ┌1300 REM Randomly select exponent from 1 to 8
         └1310 E=INT(1+8*RND(X))
       J ┌1320 REM Randomly select one of 24 resistor values
         └1330 J=INT(1+24*RND(X))
       R  1340 R=R(J)*10^E
          1350 IF R<10 THEN 1310
          1360 RETURN
```

READY

<u>RUN</u>

```
     1    2.4E+05 ohms
     2    3.3E+05 ohms
     3    240 ohms
     4    9.1E+04 ohms
     5    1.5E+05 ohms
     6    1.5E+06 ohms
     7    4700 ohms
     8    300 ohms
     9    3.9E+04 ohms
    10    3.9E+07 ohms
```

READY

erated by line 1310. Line 1360 will then direct the program back to line 130 to print out the randomly generated value. The loop will then be repeated for the next value of T until all 10 values have been generated; at this point the program will END. Note the wide range or values generated in the printout. It is certainly a random generation of numbers, as pointed out by the gap between 240 ohms and 39 megohms.

The last program (4-10) of the chapter is designed to test your knowledge of the color code by generating standard resistor values and asking for the color code. It will then check your answer and indicate if whether you are correct. As indicated by the dimension statement, there

```
 10 REM *****   PROGRAM 4-10  *****
 20 REM ******************************************
 30 REM Test knowledge of the resistor color code
 40 REM ******************************************
 50 REM
100 DIM R(24),C$(10)
110 PRINT "Your knowledge of the resistor color code "
120 PRINT "will be tested by this program."
130 PRINT
140 PRINT "For each resistor value listed below enter"
150 PRINT "the band colors (enter QUIT when done)"
160 PRINT
170 C=0 :REM Initially set number of correct answers=0
180 FOR N=1 TO 5 :REM Repeat test for up to 5 tries
190 GOSUB 1200 :REM Select resistor value
200 PRINT N;".For a resistance of";R;"ohms"
210 INPUT "Band 1=";A1$
220 IF A1$="QUIT" THEN 380
230 INPUT "Band 2=";A2$
240 INPUT "Band 3=";A3$
250 PRINT
260 GOSUB 1500 :REM Test answer
270 IF C1$<>A1$ THEN GOTO 330
280 IF C2$<>A2$ THEN GOTO 330
290 IF C3$<>A3$ THEN GOTO 330
300 PRINT "Correct" :PRINT
310 C=C+1 :REM Add one to number of correct answers
320 GOTO 360
330 PRINT "No, correct answer is:   ";
340 PRINT C1$;"  ";C2$;"  ";C3$
350 PRINT
360 PRINT "You've correctly answered";C;"thus far" :PRINT
370 NEXT N
380 PRINT:PRINT "You had";C;"correct answers."
390 END
1200 REM Module to randomly select a value of resistance
1210 REM from the standard 20% tolerence values
1220 REM Read standard component values
1230 RESTORE
1240 FOR J=1 TO 24
1250 READ R(J)
1260 NEXT J
1270 DATA 0.10,0.11,0.12,0.13,0.15,0.16,0.18
1280 DATA 0.20,0.22,0.24,0.27,0.30,0.33,0.36,0.39
1290 DATA 0.43,0.47,0.51,0.56,0.62,0.68,0.75,0.82,0.91
1300 REM Randomly select exponent from 1 to 8
1310 E=INT(1+8*RND(X))
1320 REM Randomly select one of 24 resistor values
1330 J=INT(1+24*RND(X))
1340 R=R(J)*10^E
1350 IF R<10 THEN 1310
1360 RETURN
1500 REM Moldule to read color band data
1510 FOR J=0 TO 9
1520 READ C$(J)
1530 NEXT J
1540 DATA BLACK,BROWN,RED,ORANGE,YELLOW,GREEN,BLUE,VIOLET,GRAY,WHITE
1600 REM Module to obtain color bands given resistance R
1610 IF R<100 THEN I=0:R1=R:GOTO 1660
1620 FOR I=1 TO 7
1630 IF R/10^I+0.5<100 THEN GOTO 1650
1640 NEXT I
```

Select R

Input Sol.

Test & Print Answer

Subroutine

```
1650 R1=INT(R/10^I+0.5)  :REM R1 has 2 digit value of R
1660 V1=INT(R1/10+0.05)
1670 V2=R1-10*V1
1680 V3=I
1690 C1$=C$(V1)
1700 C2$=C$(V2)
1710 C3$=C$(V3)
1720 RETURN
```

READY

RUN

Your knowledge of the resistor color code
will be tested by this program.

For each resistor value listed below enter
the band colors (enter QUIT when done)

 1 .For a resistance of 750 ohms
Band 1=? VIOLET
Band 2=? GREEN
Band 3=? BROWN

Correct

You've correctly answered 1 thus far

 2 .For a resistance of 200 ohms
Band 1=? RED
Band 2=? BLACK
Band 3=? BLACK

No, correct answer is: RED BLACK BROWN

You've correctly answered 1 thus far

 3 .For a resistance of 300 ohms
Band 1=? ORANGE
Band 2=? BLACK
Band 3=? BROWN

Correct

You've correctly answered 2 thus far

 4 .For a resistance of 13 ohms
Band 1=? BROWN
Band 2=? ORENGE
Band 3=? BLACK

No, correct answer is: BROWN ORANGE BLACK

You've correctly answered 2 thus far

 5 .For a resistance of 62 ohms
Band 1=? BLUE
Band 2=? BROWN
Band 3=? BLACK

No, correct answer is: BLUE RED BLACK
```
```

```
You've correctly answered 2 thus far

You had 2 correct answers.

READY
```

are 24 data points for the subscripted value of R and 10 for subscripts of $C\$$. Line 170 will initially set the number of correct answers (held by the variable C) to zero. Statement 180 reveals that the program will test your results for five different resistor values.

Note that the subroutines appearing between 1200 and 1360 and between 1500 and 1720 were introduced in earlier programs, removing any further need to describe them in detail. In fact, once they are stored in memory they can be called for, rather than having to re-enter every step and, possibly, introduce errors. Lines 190 and 200 will select and printout the randomly chosen resistor value, while line 210 will begin the process of requesting the color for each band. Remember that if the color is correct, but is mispelled when entered (note the provided run), the computer will treat it as an incorrect answer and proceed accordingly.

Once the chosen colors are entered as $A1\$$, $A2\$$, and $A3\$$, the program (through the subroutine at 1500) will generate the correct answers as $C1\$$, $C2\$$, and $C3\$$ and then compare them on lines 270 through 290. The symbol $<>$ appearing on each line between the subscripted variables means "not equal to" and requires that the program go to 330 for an indication of an incorrect answer if the wrong color is typed in. If correct, it will simply move on to line 300 for a "Correct" indication, followed by an increase by one in the number of correct answers stored as C on line 310. Line 360 will then print out the number of correct answers thus far and move on to the next test on line 180. If the answer is incorrect, lines 330 and 340 will print out the correct answers. Once the five

tests are completed ($N = 5$), line 380 will reveal how many correct answers were obtained for the five tests. Review the resulting run of the program and the range of values appearing in the test. For each section of this chapter there is a selection of problems to follow that progress from the simple to the more complex. A solution is not provided at the end of the text, but a correct result and a clean output will suggest that the program is properly written. Good luck!

EXERCISES

Write a program to perform the following operations.

1. Request R, ρ and d (in inches) and calculate the length l of a conductor.

2. Calculate the resistance of a copper busbar after requesting the length (l), width (w) and height (h). In addition, print out the area in square inches and square mils.

3. Request t_1, R_1 and T and calculate R_2 for a particular t_2 (greater than t_1) using $(T + t_1)/R_1 = (T + t_2)/R_2$. Then proceed to determine the slope of the resistance versus temperature curve from $m = $ slope $= \Delta_y/\Delta_x = (R_2 - R_1)/(t_2 - t_1)$. The slope will reveal the change in resistance per degree Celsius. Be sure the units are applied to the slope in the printout.

4. For the general equation of problem 4-3, provide a choice of whether R_2 or t_2 is to be determined and request the necessary data before calculating the result.

5. Given $\rho = 10.37$ and $l = 200$ feet, tabulate R, d, and A for d from 0.02 inches to 0.2 inches in increments of 0.02 inches.

6. Given R_1 and t_1, use the general equation of problem 4-3 to tabulate R_2 and t_2 for a range of t_2 from $2t_1$ to $20t_1$ in increments of $2t_1$. Choose $t_1 > 0$ and $t_2 > t_1$.

7. Given l and d, tabulate R and the material for four materials chosen (in the program) from the list of seven as appearing in Program 4-7.

8. Given the color code, determine the resistance value of the carbon resistor.

9. Develop a testing routine for the $R = \rho l / A$ equation. That is, assume a particular ρ and randomly generate l and d.

10. Develop a testing routine for the general temperature equation of problem 4-3.

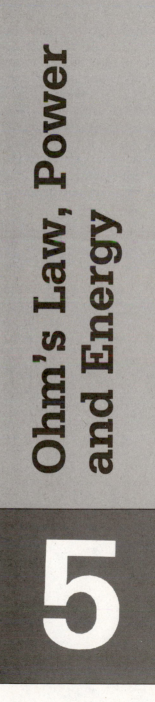

Ohm's Law, Power and Energy

5

5.1   INTRODUCTION

This chapter will examine Ohm's law in its various forms, perform a number of power and energy calculations, and determine the cost of using a number of appliances over a specified period of time. We will concentrate on tabulating techniques and introduce a few plots.

The plot routines introduced in this manual can be applied to virtually all of the computer systems noted earlier. Some systems may have special software plotting routines, but we leave it to you to become familiar with their special requirements. This chapter will be an important first step in developing your plotting skills since it will require that you fully understand how the spacing, labeling, plotting, etc. is performed with the fundamental list of BASIC commands. When particular quantities are used or requested in a software package you will better understand the significance of the quantities and how they impact on the resulting plot.

5.2   OHM'S LAW

The first program (5-1) will solve for the voltage across a resistor, given the current and resistance, and then calculate the power and en-

```
               10 REM *****   PROGRAM 5-1   *****
               20 REM ********************************************
               30 REM Calculate voltage, power, and energy
               40 REM ********************************************
               50 REM
       ┌ 100 INPUT "I=";I
Input  │ 110 INPUT "R=";R
Data   └ 120 INPUT "T=";T
Display  ┌ 130 PRINT"For I=";I;"amperes"
Input    │ 140 PRINT"    R=";R;"ohms"
Data     │ 150 PRINT "and t=";T;"hours"
         └ 160 PRINT
       ┌ 170 V=I*R
Calc.  │ 180 P=I^2*R
       │ 190 W=P*T
       └ 200 KW=W/1000
Output ┌ 210 PRINT "Voltage is ";V;"volts"
Control│ 220 PRINT "Power is ";P;"watts"
       │ 230 PRINT "Energy =";W;"watt-hours"
       └ 240 PRINT "         =";KW;"kW-hrs"
              250 END


READY

RUN

I=? 1.2
R=? 92
T=? 4.5
For I= 1.2 amperes
    R= 92 ohms
and t= 4.5 hours

Voltage is  110.4 volts
Power is  132.48 watts
Energy = 596.16 watt-hours
       = .5962 kW-hrs


READY
```

ergy dissipated. It is not a particularly difficult program to write, but it does perform a number of important basic calculations. The previous chapter should make it relatively easy to follow and understand.

Program 5-2 lets you select which form of Ohm's law to apply; the data you receive depends on the choice of equations. Line 190 is the first application of the ON---GOSUB--- command. Once the choice of C is made on line 170, the program will move to 400 if $C = 1$, 600 if $C = 2$, and 800 if $C = 3$. Note the brevity of

such a command and the power it holds. In each case, the input data is requested, the operations performed, and the results printed out. Once the program is RUN, note the clear format of the data input requests and the final output.

5.3   POWER AND ENERGY

We will now examine a number of programs to calculate the power level and energy dissipated

```
10 REM  ***** PROGRAM 5-2 *****
20 REM ***********************************************
30 REM Program demonstrates selecting various forms
40 REM of equations
50 REM ***********************************************
60 REM
```

Equation Selection
```
100 PRINT:PRINT "Select which form of Ohm's law equation "
110 PRINT "you wish to use."
120 PRINT
130 PRINT TAB(10);"(1) V=I*R"
140 PRINT TAB(10);"(2) I=V/R"
150 PRINT TAB(10);"(3) R=V/I"
160 PRINT TAB(20);
170 INPUT "choice=";C
180 IF C<1 OR C>3 THEN GOTO 100
190 ON C GOSUB 400,600,800
200 PRINT:PRINT
```

Continue?
```
210 INPUT "More (YES or NO)";A$
220 IF A$="YES" THEN 100
230 PRINT "Have a good day"
240 END
```

$V = IR$
```
400 REM Accept input of I,R and output V
410 PRINT:PRINT "Enter the following data:"
420 INPUT "I=";I
430 INPUT "R=";R
440 V=I*R
450 PRINT "Voltage is ";V;"volts"
460 RETURN
```

$I = \dfrac{V}{R}$
```
600 REM Accept input of V,R and output I
610 PRINT "Enter the following data:"
620 INPUT "V=";V
630 INPUT "R=";R
640 I=V/R
650 PRINT "Current is ";I;"amperes"
660 RETURN
```

$R = \dfrac{V}{I}$
```
800 REM Accept input of V,I and output R
810 PRINT "Enter the following data:"
820 INPUT "V=";V
830 INPUT "I=";I
840 R=V/I
850 PRINT "Resistance is ";R;"ohms"
860 RETURN
```

READY

```
RUN

Select which form of Ohm's law equation
you wish to use.

              (1) V=I*R
              (2) I=V/R
              (3) R=V/I
                        choice=? 2
Enter the following data:
V=? 12
R=? 4E3
Current is   3E-03 amperes


More (YES or NO)? YES

Select which form of Ohm's law equation
you wish to use.

              (1) V=I*R
              (2) I=V/R
              (3) R=V/I
                        choice=? 1

Enter the following data:
I=? 2E-3
R=? 5.6E3
Voltage is   11.2 volts


More (YES or NO)? YES

Select which form of Ohm's law equation
you wish to use.

              (1) V=I*R
              (2) I=V/R
              (3) R=V/I
                        choice=? 3
Enter the following data:
V=? 48
I=? 0.025
Resistance is   1920 ohms


More (YES or NO)? NO
Have a good day


READY
```

for one or several appliances over a specified period of time.

For one to N appliances, Program 5-3 will determine the total energy dissipated and the cost of using that energy. Rather than set up a separate series of statements for each appliance, a LOOP routine was established between lines 150 and 210 that asks for the power dissipation of the appliance and the hours it would be used, and then calculates the energy (W) dissipated. During the first LOOP ($K = 1$), the initial value of W is zero and the total energy

```
 10 REM *****   PROGRAM 5-3  *****
 20 REM ************************************************
 30 REM Calculate cost of energy usage
 40 REM ************************************************
 50 REM
100 PRINT:PRINT "Enter the cost of electricity in cents/kWh: ";
110 INPUT C
120 PRINT:PRINT "How many appliances are to be considered";
130 INPUT N
140 PRINT
150 FOR K=1 TO N
160 PRINT
170 PRINT "Power (in watts) used by ";K;"=";
180 INPUT P1
190 INPUT "Time (in hours)=";T
200 W=W+P1*T
210 NEXT K
220 PRINT
230 PRINT "Total kWh=";W/1000
240 TC=W*C/1000
250 PRINT "Total cost (in cents) is ";TC
260 END
```

Input Data — lines 100–140
K Loop — lines 150–210
Output Control — lines 220–250

READY

RUN

Enter the cost of electricity in cents/kWh: ? 4.6

How many appliances are to be considered? 3

Power (in watts) used by 1 =? 1200
Time (in hours)=? 1.5

Power (in watts) used by 2 =? 600
Time (in hours)=? 40

Power (in watts) used by 3 =? 4500
Time (in hours)=? 2.5

Total kWh= 37.05
Total cost (in cents) is 170.43

READY

dissipated is $P1$ x T. During the second loop, the energy dissipated by the first appliance is added to the $P1$ x T product to establish the new total energy demand. Each appliance will be assigned a number determined by the value of K on line 170. Once $K = N$, the looping process is complete and the program will move on to line 230 for a printout of the total energy dissipated and the total cost of that energy

based on the cost per kilowatt-hour provided on line 110.

For a fixed value of supply voltage, Program 5-4 will generate a table of current and power levels for a range of resistance values. The TAB values of lines 110 and 120 were carefully chosen to provide the balanced printout appearing below the program. In most cases it is not obvious which TAB values will provide

```
10 REM *****   PROGRAM 5-4  *****
20 REM ********************************************
30 REM List table of resistance, current, and power
40 REM over a range of resistance values
50 REM ********************************************
60 REM
100 INPUT "Supply voltage,E=";E
110 PRINT:PRINT TAB(7);"Resistance";TAB(20);"Current";TAB(32);"Power"
120 PRINT TAB(8);"(kilohms)";TAB(20);" (mA)";TAB(30);" (watts)"
130 REM Print output table
140 FOR R=1 TO 10
150 I=E/R
160 P=I^2*R/1000
170 PRINT TAB(10);R;TAB(20);I;TAB(32);P
180 NEXT R
190 END
```

Table Heading brackets lines 110–120.
R Loop brackets lines 140–180.

READY

<u>RUN</u>

Supply voltage,E=? <u>50</u>

Resistance (kilohms)	Current (mA)	Power (watts)
1	50	2.5
2	25	1.25
3	16.6667	.8333
4	12.5	.625
5	10	.5
6	8.3333	.4167
7	7.1429	.3571
8	6.25	.3125
9	5.5556	.2778
10	5	.25

READY

the best format. You will usually have to run a program a few times and modify it to obtain the desired appearance of the output data. The routine from 140 to 180 defines the value of R, calculates the current and power, and then prints out the three quantities as defined by statement 170. Note that the spacing between the numbers of the TAB statements is not fixed, but chosen to provide the best output. Although R only has a magnitude of 1 through 10 on line 140, the fact that R is in kilohms in the table is covered by dividing the equation of 160 by 1000. Note in the printout that the resistance column is even on the left edge, but the decimal points of the current and power columns are not aligned, and that 1, 2, and 3 place accuracy appear in the last two columns. Program 5-5 was designed to improve on the format of the output and permit a selection of the range of resistance values to be examined. The initial, final, and increment levels of R are first provided by statements 110 through 130. Line 170 will then establish the range of resistance values to be employed in the calculations of lines 180 and 190. The format statement of line 200 defines the manner in which the data is to be printed out. The first column will now be aligned on the right edge, and the decimal points of the last two columns will be lined up as shown in the output. The format statement defines the number of places that must be filled to the right of the decimal point even if zeros

```
    10 REM *****  PROGRAM 5-5  *****
    20 REM *******************************************
    30 REM List table of resistance, current, and power
    40 REM over a range of resistance values.
    50 REM *******************************************
    60 REM
   100 INPUT "Supply voltage,E=";E
   110 INPUT "R from initial value of";RI
   120 INPUT "to final value of";RF
   130 INPUT "for increments of";RX
   140 PRINT
   150 PRINT:PRINT TAB(9);"Resistance";TAB(22);"Current";TAB(34);"Power"
   160 PRINT TAB(9);"(kilohms)";TAB(22);"(mA)";TAB(34);"(mW)"
   170 FOR R=RI TO RF STEP RX
   180 I=E/R
   190 P=I^2*R
   200 F$="        #####     ##.###     ###.##"
   210 PRINT USING F$,R,I,P
   220 NEXT R
   230 END
```

Labels on left margin:
- Input Data (lines 100–140)
- Table Heading (lines 150–160)
- R Loop (lines 170–220)

READY

RUN

```
Supply voltage,E=? 40
R from initial value of? 100
to final value of? 1000
for increments of? 100
```

Resistance (kilohms)	Current (mA)	Power (mW)
100	0.400	16.00
200	0.200	8.00
300	0.133	5.33
400	0.100	4.00
500	0.080	3.20
600	0.067	2.67
700	0.057	2.29
800	0.050	2.00
900	0.044	1.78
1000	0.040	1.60

READY

are used. Even though the current or power levels may not be exact in the accuracy level requested, the result cannot have an accuracy level beyond that defined by statement 200. Realize also that the spacing between columns on line 200 will also define the spacing between output values. Line 210 will then request that the format defined by $F\$$ be used and the data printed out in the order of R, I, and the P.

The following program (5-6) will tabulate the energy dissipated by each device over a 24-hour period and then print out the total cost for the same period of time. Statements 10 through 230 are self-explanatory. Lines 240 through 260 print out UNIT1, UNIT2, etc. with a distance between starting points equal to the product of the unit number and 10. The TAB will be 10 for UNIT1, 20 for UNIT2, and so on. Line 270 will then add another column 10 spaces after the last unit setting for the total cost. Line 280 will simply print-out ''Time'' three spaces from the left hand column, as designated within the

```
 10 REM *****   PROGRAM 5-6   *****
 20 REM *********************************************************
 30 REM Program to provide a table of cost for 1 to 10
 40 REM devices over a 24 hour period
 50 REM *********************************************************
 60 REM
100 DIM P(10)
110 PRINT:PRINT
120 INPUT "How many devices are to be considered";N
130 PRINT
140 IF N>5 THEN PRINT "No more than 5 please" :GOTO 110
150 FOR K=1 TO N
160 PRINT "Enter power dissipated by device ";K;
170 INPUT P(K)
180 NEXT K
190 PRINT:INPUT "What is the cost of electricity in cents/kWh";R
200 PRINT
210 PRINT:PRINT "The following table shows the cost of electricity"
220 PRINT "over 24 hours (in cents) for the ";N;" devices"
230 PRINT
240 FOR K=1 TO N
250 PRINT TAB(10*K);"Unit";K;
260 NEXT K
270 PRINT TAB(10*(K+1));" Total"
280 PRINT "    Time";
290 FOR K=1 TO N
300 PRINT TAB(10*K);"kWh";
310 NEXT K
320 PRINT TAB(10*(K+1));"(cents)"
330 FOR T=1 TO 24
340 PRINT TAB(4);T;
350 FOR K=1 TO N
360 WH=P(K)*T/1000
370 C=R*WH
380 TC=TC+C
390 PRINT TAB(10*K);WH;
400 NEXT K
410 PRINT TAB(10*(K+1));TC
420 TC=0
430 NEXT T
440 END
```

The left margin brackets are labelled: **Input Data** (lines 120–200), **Table Heading** (lines 210–320), **T Loop (Hours)** (lines 330–430).

```
READY

RUN



How many devices are to be considered? 4

Enter power dissipated by device  1 ? 1200
Enter power dissipated by device  2 ? 3600
Enter power dissipated by device  3 ? 300
Enter power dissipated by device  4 ? 1000

What is the cost of electricity in cents/kWh? 4.6


The following table shows the cost of electricity
over 24 hours (in cents) for the  4  devices
```

Time	Unit 1 kWh	Unit 2 kWh	Unit 3 kWh	Unit 4 kWh	Total (cents)
1	1.2	3.6	.3	1	28.06
2	2.4	7.2	.6	2	56.12
3	3.6	10.8	.9	3	84.18
4	4.8	14.4	1.2	4	112.24
5	6	18	1.5	5	140.3
6	7.2	21.6	1.8	6	168.36
7	8.4	25.2	2.1	7	196.42
8	9.6	28.8	2.4	8	224.48
9	10.8	32.4	2.7	9	252.54
10	12	36	3	10	280.6
11	13.2	39.6	3.3	11	308.66
12	14.4	43.2	3.6	12	336.72
13	15.6	46.8	3.9	13	364.78
14	16.8	50.4	4.2	14	392.84
15	18	54	4.5	15	420.9
16	19.2	57.6	4.8	16	448.96
17	20.4	61.2	5.1	17	477.02
18	21.6	64.8	5.4	18	505.08
19	22.8	68.4	5.7	19	533.14
20	24	72	6	20	561.2
21	25.2	75.6	6.3	21	589.26
22	26.4	79.2	6.6	22	617.32
23	27.6	82.8	6.9	23	645.38
24	28.8	86.4	7.2	24	673.44

READY

quotes. Line 290 through 310 will add kWh under each unit column, while 320 will add (cents) under the "Total" column. Lines 330 and 340 will list the hour four spaces from the left hand margin, while 350 and 360 will calculate the power dissipated by each device for the period of time, T. Line 370 will calculate the cost for each unit and the total cost will be stored by line 380. For each value of T (and note that the LOOP defined by T extends from 330 to 430), the kilowatts dissipated and the total cost will

be printed out by lines 390 and 410. Line 420 sets the total cost to zero before incrementing the time by one hour. Even though the increment between the hourly costs of the last column is fixed at 28.06 cents, note how quickly the cost has risen to over $6.00. It clearly demonstrates the need to conserve electricity by turning off appliances when they are not in use.

Program 5-7 determines the energy dissipated by a single device for a specified period of time and provides the choice of tabulating the

```
     10 REM *****  PROGRAM 5-7  *****
     20 REM ********************************************************
     30 REM Program calculates cost of energy using
     40 REM     E(kWh)=P(watts)*T(hrs)/1000
     50 REM ********************************************************
     60 REM
    100 PRINT "To provide calculation of the energy used by a device"
    110 PRINT "Input power in watts and time of usage in hours"
    120 INPUT "P(watts)=";P
    130 INPUT "T(hours)=";T
    140 INPUT "What is cost of electricity in cents/kWh";R
    150 E=P*T/1000
    160 C=E*R
    170 PRINT:PRINT "Energy=";E;"kilowatt-hours"
    180 PRINT "at a cost of";C;"cents"
    190 PRINT
```

Input Data — lines 100–140
Calc. — lines 150–160
Output Control — lines 170–190

(Continued)

```
┌200 PRINT "Do you want an additional comparison using the same"
│210 PRINT "device over a range of 24 hours (YES or NO)";
│220 INPUT A$
│230 IF A$<>"YES" THEN END
└240 PRINT:PRINT
 250 PRINT "T(hours)              kWh              Cost(cents)"
 260 FOR T1=1 TO 24
┌270 E=P*T1/1000
└280 C=E*R
 290 PRINT T1,E,C
 300 NEXT T1
 310 PRINT
 320 PRINT "How's that!"
 330 END
```

(labels at left: Continue?, Heading, Calc.)
(label at right of lines 260-280: T1 Loop (Hours))

READY

RUN

```
To provide calculation of the energy used by a device
Input power in watts and time of usage in hours
P(watts)=? 1200
T(hours)=? 6
What is cost of electricity in cents/kWh? 4.6

Energy= 7.2 kilowatt-hours
at a cost of 33.12 cents

Do you want an additional comparison using the same
device over a range of 24 hours (YES or NO)? YES

T(hours)        kWh          Cost(cents)
1               1.2          5.52
2               2.4          11.04
3               3.6          16.56
4               4.8          22.08
5               6            27.6
6               7.2          33.12
7               8.4          38.64
8               9.6          44.16
9               10.8         49.68
10              12           55.2
11              13.2         60.72
12              14.4         66.24
13              15.6         71.76
14              16.8         77.28
15              18           82.8
16              19.2         88.32
17              20.4         93.84
18              21.6         99.36
19              22.8         104.88
20              24           110.4
21              25.2         115.92
22              26.4         121.44
23              27.6         126.96
24              28.8         132.48

How's that!


READY
```

cost for a 24-hour period. Statements 200 through 230 of the program determine whether the tabulation will be performed. The time interval is not fixed, but a function of the value of T provided in response to statement 130. If the input provided on line 230 is anything other than YES, the program will end.

5.4 INTRODUCTORY PLOTTING ROUTINES

As indicated earlier, some computers have a graphics plotting capability that goes beyond the plots to be examined here. However, the following programs, will, at the very least, introduce some appropriate considerations and terminology associated with graphics plotting. The first factor of universal importance in any plotting routine is the *scale factor*, normally abbreviated *SF*. It determines how many divisions of a scale will be assigned to each unit of measure of the quantity to be plotted. For instance, in Program 5-8 the horizontal scale (across the printout) will be current in milliampere (mA) and the vertical scale (down the printout) will be volts (V). In a later program the scales will be defined on the printout. The computer printout for the system has a maximum width of 64 divisions or characters. The scale-factor was chosen in this example with $R = 1$ kilohm (note line 150) to have maximum displacement when the current is its maximum value. In equation form, this is accomplished by setting the 64 divisions equal to the product of the maximum

current and the scale factor: $(I_{max})(SF) = 64$. Since $I_{max} = V_{max}/R = V_{max}/1000$, $SF = 64/V_{max}$ x 1000 as appearing on line 150. And since $I_{max} = V_{max}/1000 = 40/1000 = 40$mA, it is defining $SF = 64$ divisions/$I_{max} = 64$ divisions/40mA $= 1.6$ divisions/mA, or 1.6 divisions will be displaced for each mA of current resulting from the calculations of the program. The placement of a data point on a plot is determined by line 180.

For $I = V/R = 18/1000 = 1.8$mA, the product I x SF will result in (1.8 mA) (1.6 divisions/mA) $= 2.88$ divisions on the scale or 2.88/64 x 100% = 5% of the total width of the horizontal line. However, the placement of a data point must correspond with a particular division of the horizontal scale; it must be exactly two or three divisions. To insure that a result such as 2.88 will result in a three division shift rather than two, 0.5 is added within the parenthesis of statement 180 to increase 2.88 to 3.38, resulting in INT(3.38) = 3. Numbers with fractional parts less than 0.5, such as 2.25 will result in $PO = 2$. Statement 200 will then TAB to position PO and print a data point. Following the plotting of the data point for $V = 40$ volts, the program will automatically move on to line 220 and conclude the program.

In the first run, the chosen value of resistance equals the value included in the scale factor ($R = 1$ kilohm). The result is a plot that has exactly 21 data points (0 through 40) within the 64 horizontal divisions. If the chosen resistor value is less than the value used in the scale factor, the number of data points will be

```
        10 REM ***** PROGRAM 5-8 *****
        20 REM ************************************************
        30 REM Plot of I vs V demonstrating Ohm's law
        40 REM ************************************************
        50 REM
Input  ┌100 INPUT "Resistance=";R
Data   └110 VM=40 :REM Vmax=40
       ┌120 PRINT "The following is a plot of I vs V for"
Title  │130 PRINT "V from 0 to";VM;"volts"
       └140 PRINT
```

(Continued)

Plot
Routine
```
150 SF=(64/VM)*1E3 :REM Set scale factor for R=1E3
160 FOR V=0 TO VM STEP 2
170 I=V/R : REM Calculate I
180 PO=INT(I*SF+0.5)   :REM Determine position to plot
190 IF PO>64 THEN 210
200 PRINT TAB(PO);"*"
210 NEXT V
220 END
```

READY

RUN

```
Resistance=? 1E3
The following is a plot of I vs V for
V from 0 to 40 volts
```

READY

RUN

```
Resistance=? 500
The following is a plot of I vs V for
V from 0 to 40 volts
```

READY

```
RUN

Resistance=? 2E3
The following is a plot of I vs V for
V from 0 to 40 volts
```

```
READY
```

less because the current will exceed the maximum value employed in determining the scale factor. For instance, in the second plot, the value of R is 0.5 kilohms, resulting in a maximum value of 40/500 = 80 mA, rather than the 40 mA that resulted with R = 1 kilohm. The result is a plot with only 11 data points, since the full width of the plot is reached at V = 20 volts, since I = 20/500 = 40 mA. Note that the curve approaches the horizontal or current axis as the value of R is decreased. Choosing a value of R greater than 1 kilohm would generate 21 data points, but the plot would not extend across the full width of the available divisions.

In Program 5-9 the current axis will be indicated and a printout of I at the 63rd divi-sion. The loop appearing from 150 to 170 would print a dash at each position determined by the value of K. The semicolon at the end of line 160 will insure that the dashes remain on the same line. In fact, not until after line 180 is executed will the program move to the next line. Line 220 provides an initial indent of four to insure suffi-cient room for the voltage levels shown in the resulting printout. The indent, however, re-duces the full width of the plot to only 60 divi-sions. The scale factor of line 190 is therefore changed accordingly, so that the maximum value will occur at the 60th division. For all values of PO less than 64, the program will move on to line 240, where the value of V will be printed using the format indicated. The dash

```
        10 REM *****  PROGRAM 5-9  *****
        20 REM *********************************************************
        30 REM Plot of I vs V demonstrating Ohm's law
        40 REM *********************************************************
        50 REM
Input  ┌100 INPUT "Resistance=";R
Data   └110 VM=40 :REM Set maximum voltage=40
```

(Continued)

```
Title      ┌ 120 PRINT "The following is a plot of I vs V for"
           │ 130 PRINT "V from 0 to";VM;"volts"
           └ 140 PRINT
           ┌ 150 FOR K=1 TO 64
           │ 160 PRINT "-";
           │ 170 NEXT K
           │ 180 PRINT "I"
           │ 190 SF=(60/VM)*1E3 :REM Maximum position is 60 + indent of 4
Plot       │ 200 FOR V=0 TO VM STEP 2
Routine    │ 210 I=V/R : REM Calculate I
           │ 220 PO=INT(4+I*SF+0.5) :REM PO is plot POsition with indent of 4
           │ 230 IF PO>64 THEN PRINT USING "##",V; :PRINT " -" :GOTO 270
           │ 240 PRINT USING "##",V;
           │ 250 PRINT " -";
           │ 260 PRINT TAB(PO);"*"
           │ 270 NEXT V
           └ 280 PRINT "   V(volts)"
           ┌ 290 PRINT:PRINT
           │ 300 PRINT "In the above plot, I=";VM/R;"amperes at";VM;"volts"
Label      │ 310 PRINT "and the change in current between plot points is";
           └ 320 PRINT 2/R;"amperes"
             330 END
```

READY

<u>RUN</u>

```
Resistance=? 500
The following is a plot of I vs V for
V from 0 to 40 volts


----------------------------------------------------------------I
   0 -*
   2 -        *
   4 -            *
   6 -                *
   8 -                    *
  10 -                        *
  12 -                            *
  14 -                                *
  16 -                                    *
  18 -                                        *
  20 -                                            *
  22 -
  24 -
  26 -
  28 -
  30 -
  32 -
  34 -
  36 -
  38 -
  40 -
     V(volts)


In the above plot, I= .08 amperes at 40 volts
and the change in current between plot points is 4E-03 amperes
```

READY

```
RUN

Resistance=? 1E3
The following is a plot of I vs V for
V from 0 to 40 volts


-------------------------------------------------------------------I
   0 -*
   2 -    *
   4 -      *
   6 -        *
   8 -          *
  10 -            *
  12 -             *
  14 -               *
  16 -                *
  18 -                  *
  20 -                    *
  22 -                     *
  24 -                       *
  26 -                        *
  28 -                         *
  30 -                          *
  32 -                           *
  34 -                            *
  36 -                             *
  38 -                              *
  40 -                                *
     V(volts)


In the above plot, I= .04 amperes at 40 volts
and the change in current between plot points is 2E-03 amperes


READY
```

will then be printed, followed by the data point at the TAB value of *PO*. We need line 230 to make sure that even though the printout may be beyond the number of divisions available on a horizontal axis, the voltage scale will continue through 40 volts. Statement 280 will then label the axis three spaces indented from the left hand margin. The concluding statement on the printout insures that the horizontal axis had some numerical scaling. Note in the second run of the program that the use of the resistance of 0.5 kilohm results in a curve that approaches the current axis and has a maximum value of 80 milliampere at 40 volts.

The concluding plot (Program 5-10) is one of the power-versus-voltage for a single resistive element. In this case, the value of the *R* is an input quantity that makes sure that the scale factor of line 140 results in the full 64 divisions when maximum conditions are applied. There are only 61 divisions for the plot, due to the indent of three on line 190. Note the nonlinear curve of the printout and the fact that the last data point corresponds exactly with $V = 40$ volts. A second run of the program using a resistance of 2.2 kilohms reveals that, again, the final data point corresponds exactly with 40 volts and extends across the full width of the plotting region. In fact, the two curves are identical except that maximum value of the power in the second curve is 0.727 watts rather than the 3.2 watts obtained for the resistance of 500 ohms. The resulting curve is therefore actually a *universal* curve, requiring that only the

```
 10 REM *****  PROGRAM 5-10  *****
 20 REM ****************************
 30 REM Plot of P vs V
 40 REM ****************************
 50 REM
100 INPUT "Resistance=";R
110 PRINT "The following is a plot of P vs V for"
120 PRINT "V from 0 to 40 volts"
130 REM Set scale factor for maximum power at position 64 with indent of 3
140 SF=61*R/(40)^2
150 PRINT "with scale factor=";SF
160 PRINT
170 FOR V=0 TO 40 STEP 2
180 P=V^2/R
190 PO=INT(3+P*SF+0.5) :REM PO is plot POsition with indent of 3
200 PRINT USING "##",V;
210 PRINT TAB(PO);"*"
220 NEXT V
230 END
```

READY

<u>**RUN**</u>

```
Resistance=? 0.5E3
The following is a plot of P vs V for
V from 0 to 40 volts
with scale factor= 19.0625

 0 *
 2 *
 4  *
 6  *
 8   *
10    *
12     *
14      *
16       *
18        *
20         *
22          *
24           *
26            *
28             *
30              *
32               *
34                *
36                 *
38                  *
40                   *
```

READY

<u>**RUN**</u>

```
Resistance=? 2.2E3
The following is a plot of P vs V for
V from 0 to 40 volts
with scale factor= 83.875
```

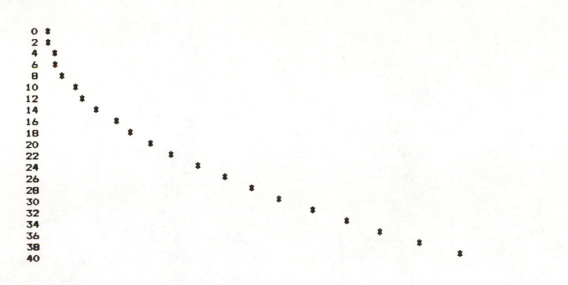

```
 0  *
 2  *
 4   *
 6    *
 8     *
10       *
12         *
14           *
16            *
18             *
20               *
22                *
24                  *
26                   *
28                     *
30                       *
32                        *
34                          *
36                           *
38                             *
40
```

READY

maximum value be changed for each plot and the scale factor determined so that the individual data points can be computed.

A number of additional plots will appear in the following chapters.

EXERCISES

Write a program to perform the following operations:

1. Request V and P and calculate R and I for a single resistive element.

2. Request I, R, and t and choose V, P, or W as the quantity to be determined. Print out the results with the appropriate units.

3. Determine the total efficiency of a cascaded system of any length, given the input and output power of each stage.

4. Determine the efficiency of a stage of a cascaded system necessary to provide a specified total efficiency η_T. That is, given η_1 through η_4 for a five-stage system, determine η_5 to provide the desired level of efficiency. If the result is greater than 100%, indicate in the printout that specified total efficiency is impossible to attain.

5. Determine the total cost (at 4.4¢/kWh) for any number of elements used for any length of time.

6. Request R (in kilohms) and tabulate I, V, and P for a range of I extending from 1 to 10mA in increments of 1mA.

7. Tabulate the time (in hours) and the cost (at 4.4¢/kWh) for the use of a particular system for $T = 1$ to N hours in increments of 1 hour. N is an input quantity in the range 5 to 15.

8. Plot cost versus hours for a single dissipative load. Use 4.4¢/kWh, $P = 500$ watts and a range of T from 1 to 20 hours. There is no need to label the cost or hours axis.

9. Plot W versus hours for the same system of problem 5-8, but label both axes in the manner indicated in the programs of this chapter.

10. Develop a testing routine for Ohm's law in any one of its three forms, with a random generation of the two required quantities.

Series and Parallel DC Networks

6

6.1   INTRODUCTION

This chapter will examine the basic series and parallel dc networks. General solutions, which permit altering the number of elements and the quantities to be determined, are provided for each configuration. We will emphasize tabulating the output data for each configuration. Because of the general nature of the programs, place them in memory so you can recall them when needed.

6.2   SERIES RESISTOR NETWORK

Program 6-1 will provide a general solution for the series dc circuit of Figure 6.1 with three resistors. In the absence of a resistor, a zero is substituted in response to the request for input data on 140. The program will then determine

the total resistance R_T, the circuit current I, and the voltage across and the power to each resistor. Kirchhoff's voltage law is verified by line 300, and the power supplied by the source is determined by line 310. The program is fairly straightforward, with its major effort directed toward providing a clear list of output results. Once the program is properly entered and the RUN command executed, consider all the information generated by providing the circuit components. In the second run, an element was left out to demonstrate that the program is still applicable for fewer than three components.

FIGURE 6.1

```
        10 REM *****  PROGRAM 6-1  *****
        20 REM **********************************************
        30 REM Analysis of a series resistor network
        40 REM **********************************************
        50 REM
       100 PRINT:PRINT "Enter resistor values for up to 3 resistors"
       110 PRINT "in series (enter 0 if no resistor):"
       120 INPUT "R1=";R1
       130 INPUT "R2=";R2
       140 INPUT "R3=";R3
 R_T   150 RT=R1+R2+R3
       160 PRINT:PRINT "The total resistance is RT=";RT;"ohms"
       170 PRINT:INPUT "Enter value of supply voltage, E=";E
 I     180 I=E/RT
       190 PRINT
       200 PRINT "Supply current is, I=";I;"amperes"
       210 PRINT
 V_x   220 PRINT "The voltage drop across each resistor is:"
       230 V1=I*R1:V2=I*R2 :V3=I*R3
       240 PRINT "V1=";V1;"volts    V2=";V2;"volts    V3=";V3;"volts"
 P_x   250 P1=I^2*R1 :P2=I^2*R2 :P3=I^2*R3
       260 PRINT
       270 PRINT "The power dissipated by each resistor is:"
       280 PRINT "P1=";P1;"watts","P2=";P2;"watts","P3=";P3;"watts"
       290 PRINT
       300 PRINT "Total voltage around loop is, V1+V2+V3=";V1+V2+V3;"volts"
       310 PRINT "and total power dissipated, P1+P2+P3=";P1+P2+P3;"watts"
       320 END
```

READY

```
      RUN
      *

      Enter resistor values for up to 3 resistors
      in series (enter 0 if no resistor):
      R1=? 6
      R2=? 7
      R3=? 5

      The total resistance is RT= 18 ohms

      Enter value of supply voltage, E=? 54

      Supply current is, I= 3 amperes

      The voltage drop across each resistor is:
      V1= 18 volts    V2= 21 volts    V3= 15 volts

      The power dissipated by each resistor is:
      P1= 54 watts    P2= 63 watts    P3= 45 watts

      Total voltage around loop is, V1+V2+V3= 54 volts
      and total power dissipated, P1+P2+P3= 162 watts


      READY

      RUN


      Enter resistor values for up to 3 resistors
      in series (enter 0 if no resistor):
      R1=? 1E3
      R2=? 4E3
      R3=? 0

      The total resistance is RT= 5000 ohms

      Enter value of supply voltage, E=? 50

      Supply current is, I= .01 amperes

      The voltage drop across each resistor is:
      V1= 10 volts    V2= 40 volts    V3= 0 volts

      The power dissipated by each resistor is:
      P1= .1 watts    P2= .4 watts    P3= 0 watts

      Total voltage around loop is, V1+V2+V3= 50 volts
      and total power dissipated, P1+P2+P3= .5 watts


      READY
```

*Example 5.6, ICA

6.3 PARALLEL RESISTOR NETWORK

Program 6-2 is essentially the dual of program 6-1. It provides the general solution for the parallel dc network of Figure 6.2 with up to three elements. Note in this case that the absence of

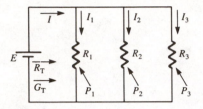

FIGURE 6.2

an element requires that an open-circuit equivalent be established on an approximate basis by a resistor of 10^{30} ohms (a very valid approximation). Again, consider the amount of effort required to do all the calculations by hand. It is certainly advantageous to have the program properly entered and only have to provide the components of the system. It is, of course, very important that the user *first* understand the analysis completely before relying on the computer for the mathematical solutions. Note in the second example that $R3 = 1E30$, with the result that $G3 = 10^{-30}$; this certainly approaches a short-circuit equivalent. The resulting $I3 = 4 \times 10^{-29}$ is essentially zero ampere as required.

```
       10 REM *****  PROGRAM 6-2  *****
       20 REM *****************************************
       30 REM Analysis of a parallel resistor network
       40 REM *****************************************
       50 REM
      100 PRINT:PRINT "Enter resistor values for up to three"
      110 PRINT "resistors in parallel (enter 1E30 if none)"
      120 PRINT
      130 INPUT "R1=";R1
      140 INPUT "R2=";R2
      150 INPUT "R3=";R3
      160 G1=1/R1 :G2=1/R2 :G3=1/R3
      170 PRINT "The component conductances are:"
      180 PRINT "G1=";G1;"Siemens    G2=";G2;"Siemens    G3=";G3;"Siemens"
G_T   190 GT=G1+G2+G3
      200 PRINT:PRINT "The total conductance is GT=";GT;"Siemens"
R_T   210 RT=1/GT
      220 PRINT "and the total resistance is RT=1/GT=";RT;"ohms"
      230 PRINT:INPUT "Enter value of supply voltage, E=";E
I     240 IR=E/RT :IG=E*GT
      250 PRINT :PRINT "The supply current calculated by I=E/RT=";IR;"amperes"
      260 PRINT "while the current calculated using I=E*GT=";IG;"amperes"
      270 PRINT :PRINT "The branch currents are:"
I_x   280 I1=E*G1 :I2=E*G2 :I3=E*G3
      290 PRINT "I1=";I1;"amperes","I2=";I2;"amperes","I3=";I3;"amperes"
      300 PRINT
P_x   310 P1=E^2*G1 :P2=E^2*G2 :P3=E^2*G3
      320 PD=E*IG
      330 PRINT "The power dissipated by each resistor is:"
      340 PRINT "P1=";P1;"watts","P2=";P2;"watts","P3=";P3;"watts"
      350 PRINT "while the total power delivered by the supply is";PD;"watts"
      360 END
```

READY

```
     RUN

     *

     Enter resistor values for up to three
     resistors in parallel (enter 1E30 if none)

     R1=? 4
     R2=? 8
     R3=? 10
     The component conductances are:
     G1= .25 Siemens    G2= .125 Siemens    G3= .1 Siemens

     The total conductance is GT= .475 Siemens
     and the total resistance is RT=1/GT= 2.1053 ohms

     Enter value of supply voltage, E=? 12

     The supply current calculated by I=E/RT= 5.7 amperes
     while the current calculated using I=E*GT= 5.7 amperes

     The branch currents are:
     I1= 3 amperes    I2= 1.5 amperes I3= 1.2 amperes

     The power dissipated by each resistor is:
     P1= 36 watts    P2= 18 watts    P3= 14.4 watts
     while the total power delivered by the supply is 68.4 watts


     READY

     RUN


     Enter resistor values for up to three
     resistors in parallel (enter 1E30 if none)

     R1=? 1E3
     R2=? 2E3
     R3=? 1E30
     The component conductances are:
     G1= 1E-03 Siemens    G2= 5E-04 Siemens    G3= 1E-30 Siemens

     The total conductance is GT= 1.5E-03 Siemens
     and the total resistance is RT=1/GT= 666.6667 ohms

     Enter value of supply voltage, E=? 40

     The supply current calculated by I=E/RT= .06 amperes
     while the current calculated using I=E*GT= .06 amperes

     The branch currents are:
     I1= .04 amperes I2= .02 amperes I3= 4E-29 amperes

     The power dissipated by each resistor is:
     P1= 1.6 watts    P2= .8 watts    P3= 1.6E-27 watts
     while the total power delivered by the supply is 2.4 watts


     READY
```

*Example 5.20, ICA

6.4 VOLTAGE DIVIDER RULE

The voltage divider rule is employed in Program 6-3 to determine the voltage across any one of the five series resistors of Figure 6.3. The dimension statement of line 100 is required to store the five subscripted values of R. Although the computer will initially set all the variables to zero when the power is turned on, line 140 will insure that the previous RUNS of the program have not left a residue value for R_T. During the first pass ($K = 1$) through the sub-routine starting at line 150, the initial value of RT to the right of the equal sign is zero as set by line 140. In addition, rather than have a separate statement requesting each value of resistance, the same module was designed to ask for each value and calculate the total resistance. The program requires that all non-zero resistors be listed last. For example, if Figure 6.3 had only three resistors, R_4 and R_5 would be zero ohms. Line 190 will then pick up on the first absent resistor and move on to line 210 without continuing to request additional values of resistance. The variable NR of line 210 will hold the number of series elements.

Line 230 will request the number of the resistor across which the voltage is to be determined and use the provided value of K in the

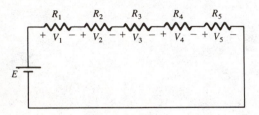

FIGURE 6.3

```
          10 REM ***** PROGRAM 6-3  *****
          20 REM *****************************
          30 REM The voltage divider rule
          40 REM *****************************
          50 REM
         100 DIM R(5)
         110 PRINT "For a network with up to 5 resistors in series"
         120 PRINT "enter resistor values (enter 0 if no value)"
         130 PRINT
         140 RT=0
    K    150 FOR K=1 TO 5
   Loop   160 PRINT "R(";K;")=";
          170 INPUT R(K)
   Input  180 RT=RT+R(K)
   Data   190 IF R(K)=0 THEN GOTO 210
   (R)    200 NEXT K
         210 NR=K-1 :REM NR is the number of resistor values entered
    E    220 PRINT:INPUT "Supply voltage is, E=";E
         230 PRINT:INPUT "Voltage will be calculated across which resistor";K
         240 IF K<1 OR K>NR THEN PRINT TAB(20);"No." :GOTO 230
         250 PRINT "Using the voltage divider rule:"
         260 PRINT "V(";K;")=R(";K;")*E/RT=";
   VDR   270 PRINT R(K);"*";E;"/";RT;"=";
         280 V=R(K)*E/RT
         290 PRINT V;"volts"
         300 PRINT :INPUT "More(YES or NO)";A$
         310 IF A$="YES" THEN 230
         320 PRINT "ok"
         330 END
```

READY

```
RUN

For a network with up to 5 resistors in series
enter resistor values (enter 0 if no value)

R( 1 )=? 100
R( 2 )=? 200
R( 3 )=? 300
R( 4 )=? 400
R( 5 )=? 500

Supply voltage is, E=? 15

Voltage will be calculated across which resistor? 4
Using the voltage divider rule:
V( 4 )=R( 4 )*E/RT= 400 * 15 / 1500 = 4 volts

More(YES or NO)? NO
ok


READY

RUN
*
For a network with up to 5 resistors in series
enter resistor values (enter 0 if no value)

R( 1 )=? 4E3
R( 2 )=? 4E3
R( 3 )=? 2E3
R( 4 )=? 0

Supply voltage is, E=? 40

Voltage will be calculated across which resistor? 2
Using the voltage divider rule:
V( 2 )=R( 2 )*E/RT= 4000 * 40 / 1E+04 = 16 volts

More(YES or NO)? YES

Voltage will be calculated across which resistor? 4
              No.

Voltage will be calculated across which resistor? 1
Using the voltage divider rule:
V( 1 )=R( 1 )*E/RT= 4000 * 40 / 1E+04 = 16 volts

More(YES or NO)? NO
ok


READY

*Example 5.7, ICA
```

printout of lines 260 and 270. Note in the printout that lines 260, 270 and 290 are part of the same line because of the semicolon at the end of lines 260 and 270. If a second RUN is requested, the program will return to line 230 because the statements above 230 remain unchanged. Note, in the second run, the absence of the resistor R_4 and the repeat of the question of line 230 when the voltage across R_4 was requested.

6.5   SERIES CIRCUIT DESIGN PROGRAM

The synthesis to be performed by Program 6-4 can best be understood by first examining the network of Figure 6.4, in which R_1, R_2 and E

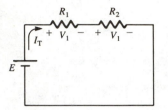

FIGURE 6.4

are specified and R_T, I, V_1 and V_2 to be determined. The question is what resistor R_x should be added, as shown in Figure 6.5, to create a specified voltage across the resistor R_x. The total resistance R_T and the circuit current I_T can then be calculated, along with the voltage drop across each resistor.

The program through line 260 is quite similar to those examined earlier in this chapter and requires no further description. The subroutine at 500 will determine the voltage drops using the voltage divider rule (530) and list the component values and the resulting voltages. Note the choice of TAB values to insure a "clean" format for the output. The RETURN

```
          10 REM ***** PROGRAM 6-4   *****
          20 REM ****************************************************
          30 REM Program to list voltages across series resistor
          40 REM circuit, and determination of additional series
          50 REM resistor to obtain a desired voltage drop.
          60 REM ****************************************************
          70 REM
         100 DIM R(10)
         110 PRINT "Enter the number of resistors in the series connection"
         120 PRINT "and the supply voltage applied"
         130 INPUT "How many resistors are is series (up to 10)";N
         140 IF N>10 THEN GOTO 130
Input  ┌ 150 FOR K=1 TO N
R(k)   │ 160 PRINT "R(";K;")=";
&      │ 170 INPUT R(K)
Calc.  │ 180 RT=RT+R(K)
R_T    └ 190 NEXT K
         200 PRINT
         210 PRINT "At present, the total input resistance, RT=";RT;"ohms"
         220 PRINT
E        230 INPUT "Supply voltage, E=";E
I_T      240 IT=E/RT
         250 PRINT "The resulting supply current IT=";IT;"amps"
         260 PRINT:PRINT "At present, the voltage across each resistor is:"
         270 GOSUB 500
         280 PRINT:PRINT "Do you wish to add an additional resistor in series";
         290 INPUT "(YES or NO)";A$
         300 IF A$="YES" THEN 310 ELSE END
         310 INPUT "What voltage do you wish to have across the added resistor";VX
         320 IF VX>E THEN PRINT "Can't be done":GOTO 310
R_x      330 RX=(RT*VX)/(E-VX)
R_T      340 RT=RT+RX
         350 PRINT "The added resistance is";RX;"ohms"
         360 PRINT "resulting in a total resistance of";RT;"ohms"
I_T      370 IT=E/RT
         380 PRINT "and an IT=";IT;"amps"
         390 PRINT:PRINT "The voltage drop across each resistor is now:"
         400 N=N+1
         410 R(N)=RX
         420 GOSUB 500
         430 END
```

```
List   ┌500 REM Subroutine to list table of component values and voltages
R(K)   │510 PRINT TAB(5);"Resistance";TAB(17);"Voltage drop"
&      │520 FOR K=1 TO N
V      │530 V=R(K)*E/RT
       │540 PRINT TAB(7);R(K);TAB(20);V
       │550 NEXT K
       └560 RETURN
```

READY

RUN

```
Enter the number of resistors in the series connection
and the supply voltage applied
How many resistors are is series (up to 10)? 3
R( 1 )=? 250
R( 2 )=? 150
R( 3 )=? 400

At present, the total input resistance, RT= 800 ohms

Supply voltage, E=? 20
The resulting supply current IT= .025 amps

At present, the voltage across each resistor is:
     Resistance  Voltage drop
        250         6.25
        150         3.75
        400        10

Do you wish to add an additional resistor in series(YES or NO)? YES
What voltage do you wish to have across the added resistor? 12
The added resistance is 1200 ohms
resulting in a total resistance of  2000 ohms
and an IT= .01 amps

The voltage drop across each resistor is now:
     Resistance  Voltage drop
        250         2.5
        150         1.5
        400         4
       1200        12
```

READY

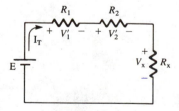

FIGURE 6.5

statement of line 560 will bring the program back to line 280 to determine if this run is to be expanded into the design mode. Obviously, the choice of voltage V_x cannot be greater than the input voltage as controlled by line 320. The equation of line 330 is the result of the following mathematical manipulations using R_T as the currently stored value of total resistance:

$$V_x = \frac{R_x E}{R_1 + R_2 + R_x} = \frac{R_x E}{R_T + R_x}$$

and

$$V_x(R_T + R_x) = R_x E$$

or

$$R_x(E - V_x) = V_x R_T$$

with

$$R_x = \frac{R_T V_x}{E - V_x}$$

The new value of stored total resistance is increased by R_x on line 340 and the value of N is increased by 1 on line 400. For N originally equal to 3, line 410 results in $R(4) = R_x$. The program will move on to subroutine 500 to calculate the voltage drops and print out the results. The program will END at line 430. Note in the provided run that the voltage across the resistors R_1 and R_2 will decrease due to the added resistor which will claim its share of the input voltage.

6.6 RESISTORS OF THE SAME VALUE IN SERIES AND PARALLEL

The next two programs will analyze networks with resistors of the same value in series and parallel. Program 6-5 tabulates the total resistance of 1 to 10 resistors of the same value in series or parallel. Note in the provided run the increasing value of the series combination and the rapidly decreasing result for the parallel

```
10 REM *****  PROGRAM 6-5  *****
20 REM **********************************************
30 REM Table of the total resistance of resistors
40 REM of the same value in series and parallel.
50 REM **********************************************
60 REM
100 INPUT "What is value of resistor to combine";R
110 PRINT
120 PRINT :PRINT TAB(5);"No. R's     R's in Series   R's in Parallel"
130 FOR NR=1 TO 10
140 PRINT TAB(7);NR;
150 RS=NR*R
160 RP=R/NR
170 PRINT TAB(20);RS;TAB(35);RP
180 NEXT NR
190 END
```

NR Loop brackets lines 130 through 180.

```
READY

RUN

What is value of resistor to combine? 400
```

No. R's	R's in Series	R's in Parallel
1	400	400
2	800	200
3	1200	133.3333
4	1600	100
5	2000	80
6	2400	66.6667
7	2800	57.1429
8	3200	50
9	3600	44.4444
10	4000	40

```
READY
```

configuration. Ten elements in series results in a total resistance ten times the original, while ten elements in parallel results in one-tenth the initial value. For each value of NR (the number

in series or parallel), loop 130 through 180 will print NR, calculate the total series and parallel resistance, and print out the results in the appropriate column. An exercise at the end of the chapter will request a program that will perform the same function, but it will ask for the number of elements to be considered in series or parallel and specify a particular format for the tabulated results.

Program 6-6 will not only calculate the total resistance of 1 through 10 resistors in series, but it will also calculate the current through the circuit, the voltage across each element, the power to each component, and the source power. Note that R_T is calculated first to permit its use in the voltage divider rule of line 190 and to determine the source power on line 210. The voltage determined by line 190 is used to calculate the power of line 200. We carefully chose the order in which the calculations were performed so that the number of command statements was as few as possible. Of course,

```
10 REM *****  PROGRAM 6-6  *****
20 REM *******************************************
30 REM List table of voltage, power and eqv't R
40 REM of 10 series resistors of equal value.
50 REM *******************************************
60 REM
100 INPUT "What is the value of the resistor to be combined in series";R
110 INPUT "What is value of the supply voltage";E
120 PRINT
130 PRINT TAB(6);"No. R's";TAB(17);"I";TAB(24);"V(each R)";
140 PRINT TAB(35);"P(each R)";
150 PRINT TAB(47);"Eqv't R";TAB(56);"P(supply)"
160 FOR NR=1 TO 10
170 PRINT TAB(8);NR;
180 RT=RT+R
190 V=R*E/RT
200 P=V^2/R        Calc.
210 PS=E^2/RT
220 I=E/RT
230 PRINT TAB(15); :PRINT USING "#.###",I;
240 PRINT TAB(25); :PRINT USING "##.##",V;
250 PRINT TAB(35); :PRINT USING "##.##",P;       Output Control
260 PRINT TAB(48); :PRINT USING "####.#",RT;
270 PRINT TAB(58); :PRINT USING "#.###",PS
280 NEXT NR
290 END
```

NR Loop (brace covering lines 160–280)
Calc. (brace covering lines 180–220)

READY

RUN

What is the value of the resistor to be combined in series? 400
What is value of the supply voltage? 50

No. R's	I	V(each R)	P(each R)	Eqv't R	P(supply)
1	0.125	50.00	6.25	400.0	6.250
2	0.063	25.00	1.56	800.0	3.125
3	0.042	16.67	0.69	1200.0	2.083
4	0.031	12.50	0.39	1600.0	1.563
5	0.025	10.00	0.25	2000.0	1.250
6	0.021	8.33	0.17	2400.0	1.042
7	0.018	7.14	0.13	2800.0	0.893
8	0.016	6.25	0.10	3200.0	0.781
9	0.014	5.56	0.08	3600.0	0.694
10	0.013	5.00	0.06	4000.0	0.625

READY

the voltage V could also be calculated from $V = IR$ if I had been calculated first, but we wanted to demonstrate the use of a number of different rules and laws to provide the desired results. It is interesting to note in the results how the power to each element and the provided source power drop with increase in load. The choice of 10 elements is an excellent one because each quantity is either one-tenth or ten times the initial value.

6.7 TEST PROGRAMS

The remaining programs of the chapter will test your skills analyzing the series and parallel dc network. They will randomly generate the number of resistors to be examined in series or parallel and then determine the value of each resistor and the magnitude of the supply voltage. They will then request specific quantities and check your input against the computer-calculated answers.

Progressing through Program 6-7, the first quantity to be randomly generated is the num-

ber of resistors in series using the subroutine starting at line 700. The resulting value of NR will then be printed out by line 150. As specified by line 710, the number to be randomly generated is an integer value in the range 2 through 5. A supply voltage in the range 1 through 100 volts is then randomly selected by the subroutine beginning at 800. Subroutine 900 will randomly generate a value of R between 1 ohm and 1 megohm. Line 910 generates the integer value of R between 1 and 1000, while line 920 randomly selects the exponent of the power of ten to be employed in 930. In equation form, $R = R \times 10^{EX}$, where R is randomly generated by line 910 and EX by line 920.

The routine from 190 to 240 will print out the value of R randomly generated and also calculate the total resistance of the network on line 230. Line 250 will randomly select a value of K in the range 1 through NR which will be used to ask for a particular voltage of the network on line 260. Line 270 asks for your response, while the next line calculates the value to be compared by lines 290 and 300. On line 290 the sign

```
           10 REM ***** PROGRAM 6-7 *****
           20 REM *******************************************************
           30 REM Program to test your skills analyzing a series circuit.
           40 REM *******************************************************
           50 REM
          100 DIM R(10)
          110 PRINT "This program will test your skills analyzing"
          120 PRINT "a series dc circuit."
          130 PRINT
          140 GOSUB 700
          150 PRINT "For the following";NR;"resistors in series"
          160 GOSUB 800 :REM select E less than 100 volts
          170 PRINT "and a supply voltage of E=";E;"volts"
          180 PRINT
          190 FOR K=1 TO NR
          200 GOSUB 900 :REM Select resistor value at random
          210 R(K)=R
          220 PRINT "R(";K;")=";R(K)
          230 RT=RT+R(K)
          240 NEXT K
          250 K=INT(1+NR*RND(X))
          260 PRINT"What is the voltage drop across R(";K;")=";R(K);"ohms"
Input V   270 INPUT "V=";VA
Calc. V   280 V=R(K)*E/RT
Testing ⎡ 290 IF SGN(I)<>SGN(IA) THEN GOTO 310
V       ⎣ 300 IF ABS(V-VA)<ABS(0.01*V) THEN PRINT "Correct" :GOTO 320
          310 PRINT "No, it's";V;"volts"
```

```
Calc. /      320  I=E/RT
Input /      330  INPUT "What is the value of the supply current";IA
Testing ┌─  340  IF SGN(I)<>SGN(IA) THEN GOTO 360
     /   └─  350  IF ABS(I-IA)<ABS(0.01*I) THEN PRINT "Correct" :GOTO 370
             360  PRINT "No, it's";I;"amps"
             370  END
         ┌─  700  REM Subroutine to select number of resistors (from 2 to 5)
         │   710  NR=INT(2+4*RND(X))   :REM Randomly select from 2 to 5 resistors
         │   720  RETURN
         │   800  REM Select supply voltage E (maximum 100 volts)
Random   │   810  E=INT(1+100*RND(X))
Gen.     │   820  RETURN
         │   900  REM Subroutine to select resistor value(maximum of 1 Megohm)
         │   910  R=INT(1+1000*RND(X))
         │   920  EX=INT(3*RND(X))
         │   930  R=R*10^EX
         └─  940  RETURN
```

```
READY

RUN

This program will test your skills analyzing
a series dc circuit.

For the following 5 resistors in series
and a supply voltage of E= 94 volts

R( 1 )= 947
R( 2 )= 7580
R( 3 )= 6120
R( 4 )= 998
R( 5 )= 34
What is the voltage drop across R( 3 )= 6120 ohms
V=? 36.7
Correct
What is the value of the supply current? 6E-3
Correct


READY

RUN

This program will test your skills analyzing
a series dc circuit.

For the following 2 resistors in series
and a supply voltage of E= 34 volts

R( 1 )= 8.55E+04
R( 2 )= 1010
What is the voltage drop across R( 2 )= 1010 ohms
V=? 0.4
Correct
What is the value of the supply current? 3.9E-3
No, it's 3.9302E-04 amps


READY
```

of the answer is compared and sends the program to line 310 if incorrect. On line 300 the difference between the computer-calculated value and the value provided by the user is determined. On line 300 the ABS (X) command converts a negative answer resulting from a solution from the user that is greater than the computer-calculated results to one that is positive. If that difference is less than one percent (or 0.01) of the computer-calculated value, the computer will indicate a correct answer. If not, it will respond with a negative indication and print out the correct answer. The program will

then proceed to test the solution for the current in the same manner before ending at line 370. Be careful in the second run to provide the proper power of ten with a solution.

Program 6-8 is essentially the dual of Program 6-7 and needs little additional description. In this case the various currents of the network are requested and compared against the calculated values. Note that the subroutines starting at lines 700, 800, and 900 are the same as those in Program 6-7. Note also how the computer picked up the error in the second run and the accuracy it required for the supply current.

```
10 REM ***** PROGRAM 6-8 *****
20 REM *************************************
30 REM Program to test your skills analyzing
40 REM a parallel dc circuit.
50 REM *************************************
60 REM
100 DIM R(10)
110 PRINT "This program will test your skills analyzing"
120 PRINT "a parallel dc resistor network."
130 PRINT
140 GOSUB 700
150 PRINT "For the following";NR;"resistors in parallel"
160 GOSUB 800 :REM Select E less than 100 volts
170 PRINT "and a supply voltage of E=";E;"volts"
180 PRINT
190 FOR K=1 TO NR
200 GOSUB 900 :REM Select resistor value at random
210 R(K)=R
220 PRINT "R(";K;")=";R(K)
230 GT=GT+1/R(K)
240 RT=1/GT
250 NEXT K
260 K=INT(1+NR*RND(X))
270 PRINT "What is the current through R(";K;")=";R(K);"ohms"
280 INPUT "I=";IA
290 I=E/R(K)
300 IF SGN(I)<>SGN(IA) THEN GOTO 320
310 IF ABS(I-IA)<ABS(0.01*I) THEN PRINT "Correct" :GOTO 330
320 PRINT "No, it's";I;"amps"
330 I=E/RT
340 INPUT "What is the value of the supply current";IA
350 IF SGN(I)<>SGN(IA) THEN GOTO 370
360 IF ABS(I-IA)<ABS(0.01*I) THEN PRINT "Correct" :GOTO 380
370 PRINT "No, it's";I;"amps"
380 END
700 REM Subroutine to select number of resistors (from 2 to 5)
710 NR=INT(2+4*RND(X))
720 RETURN
800 REM Select supply voltage E (maximum 100 volts)
810 E=INT(1+100*RND(X))
820 RETURN
900 REM Subroutine to select resistor value(maximum of 1 Megohm)
```

```
910 R=INT(1+1000*RND(X))
920 EX=INT(3*RND(X))
930 R=R*10^EX
940 RETURN
```

READY

RUN

```
This program will test your skills analyzing
a parallel dc resistor network.

For the following 5 resistors in parallel
and a supply voltage of E= 66 volts

R( 1 )= 969
R( 2 )= 571
R( 3 )= 2380
R( 4 )= 6990
R( 5 )= 2.83E+04
What is the current through R( 1 )= 969 ohms
I=? 68.11E-3
Correct
What is the value of the supply current? 0.223
Correct
```

READY

RUN

```
This program will test your skills analyzing
a parallel dc resistor network.

For the following 4 resistors in parallel
and a supply voltage of E= 32 volts

R( 1 )= 450
R( 2 )= 6280
R( 3 )= 2.9E+04
R( 4 )= 218
What is the current through R( 1 )= 450 ohms
I=? .07
No, it's .071 amps
What is the value of the supply current? .224
Correct
```

READY

EXERCISES

Write a program to perform each of the following operations.

1. For three series resistors, request R_T, I, R_1,
and R_2 and calculate R_3, E, and the power to
each resistor.

2. For three parallel resistors, request R_T, E,
R_1, and R_2 and calculate R_3, G_T, I_T, I_1, I_2, and I_3
and the power to each element.

3. Develop a general solution for the total resistance of any number of series *or* parallel resistors. The configuration and the values of
each resistor is to be requested.

4. For any number of series elements, determine the voltage across any *one* or *series connection* of resistors of the configuration using the voltage divider rule.

5. Determine the current through each of two parallel resistors using the current divider rule. Request the values R_1, R_2, and I_T.

6. For three parallel resistors, determine the current through each resistor, given only the total input current. The values of R_1, R_2, and R_3 must also be requested.

7. Tabulate N and R_T for a series and parallel connection of resistors of values R_1, $2R_1$, $3R_1$ . . . NR_1. Specify N and R_1.

8. Plot a curve of R_T versus R for 1 to 20 resistors of the same value in series. Label the axis indicating the number of resistors of magnitude R.

9. Repeat problem 6-8 for a parallel system.

10. Develop a testing routine for the general solution of a series *or* parallel dc system with two elements.

Series-Parallel
DC Networks

7

7.1 INTRODUCTION

Chapter 7 will examine a few of the methods of analysis typically applied to series-parallel dc networks. Although there are an infinite number of possibilities, the networks chosen are fairly standard in nature and are simplified versions of more complex systems. If a few fundamental series or parallel combinations are made to reduce the complex network to one that has been computer-analyzed, the results can then be used to work back to the solutions of the more complex system.

Since most series-parallel networks require a long series of sequential calculations, the computer is particularly useful because it can save time and energy once the program is properly entered. Recall all the effort required to analyze a series-parallel system with values requiring two or three place accuracy.

7.2 SERIES-PARALLEL DC NETWORKS

The network of Figure 7.1 is fundamental because it has only one parallel combination and

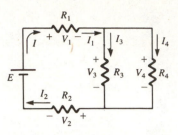

FIGURE 7.1

two series resistors, R_1 and R_2. Of course, the parallel combination of R_3 and R_4 is also in series with R_1 and R_2. In Program 7-1 the network parameters are requested by lines 130 through 170. Note the associated REM statements that permit a limited modification of the network. The absence of R_1 or R_2 was limited to inserting zero ohms (a short-circuit equivalent) and not $1E30$ ohms (10^{30} ohms), so that a complete circuit would still exist. R_2 and R_4 were limited to $1E30$ (an open-circuit equivalent), so that inserting zero ohms would not short-circuit the other parallel element. The network calculations are all performed by lines 180 through 280 and printed out by lines 290 through 460.

```
        10 REM *****  PROGRAM 7-1  *****
        20 REM *********************************************
        30 REM Calculate all currents, voltages and power
        40 REM in a series-parallel circuit.
        50 REM *********************************************
        60 REM
       100 PRINT "Enter the following data for the series-"
       110 PRINT "parallel circuit of Figure 7-1."
       120 PRINT
      ┌130 INPUT "R1=";R1   :REM Insert 0 if no resistor
       140 INPUT "R2=";R2   :REM Insert 0 if no resistor
Input │150 INPUT "R3=";R3   :REM Insert 1E30 if no resistor
Data  │160 INPUT "R4=";R4   :REM Insert 1E30 if no resistor
      └170 INPUT "and the supply voltage, E=";E
      ┌180 REM Circuit calculations
       190 RT=R1+R2+(R3*R4)/(R3+R4)
       200 I=E/RT
       210 I1=I  :I2=I
       220 I3=(R4*I)/(R3+R4)
Calc. │230 I4=I-I3
       240 V1=I1*R1  :P1=V1*I1
       250 V2=I2*R2  :P2=V2*I2
       260 V3=I3*R3  :P3=V3*I3
       270 V4=I4*R4  :P4=V4*I4
      └280 PD=E*I
```

```
290 PRINT :PRINT "The currents, voltages and power to"
300 PRINT "each part of the circuit are the following:"
310 PRINT "Currents:"
320 PRINT "I1=";I1;"amps"
330 PRINT "I2=";I2;"amps"
340 PRINT "I3=";I3;"amps"
350 PRINT "I4=";I4;"amps"
360 PRINT:PRINT "Voltages:"
370 PRINT "V1=";V1;"volts"
380 PRINT "V2=";V2;"volts"
390 PRINT "V3=";V3;"volts"
400 PRINT "V4=";V4;"volts"
410 PRINT:PRINT "Power dissipated:"
420 PRINT "P1=";P1;"watts"
430 PRINT "P2=";P2;"watts"
440 PRINT "P3=";P3;"watts"
450 PRINT "P4=";P4;"watts"
460 PRINT:PRINT "and the total power from the supply is, Pdel=";PD;"watts"
470 END
```

Output
Control

```
READY

RUN

Enter the following data for the series-
parallel circuit of Figure 7-1.

R1=? 100
R2=? 200
R3=? 300
R4=? 400
and the supply voltage, E=? 100

The currents, voltages and power to
each part of the circuit are the following:
Currents:
I1= .2121 amps
I2= .2121 amps
I3= .1212 amps
I4= .091 amps

Voltages:
V1= 21.2121 volts
V2= 42.4242 volts
V3= 36.3636 volts
V4= 36.3636 volts

Power dissipated:
P1= 4.4995 watts
P2= 8.9991 watts
P3= 4.4077 watts
P4= 3.3058 watts

and the total power from the supply is, Pdel= 21.2121 watts

READY

RUN

Enter the following data for the series-
parallel circuit of Figure 7-1.
```

(Continued)

```
R1=? 2.2E3
R2=? 0
R3=? 1E30
R4=? 5.6E3
and the supply voltage, E=? 22.5

The currents, voltages and power to
each part of the circuit are the following:
Currents:
I1= 2.8846E-03 amps
I2= 2.8846E-03 amps
I3= 1.6154E-29 amps
I4= 2.8846E-03 amps

Voltages:
V1= 6.3462 volts
V2= 0 volts
V3= 16.1538 volts
V4= 16.1538 volts

Power dissipated:
P1= .018 watts
P2= 0 watts
P3= 2.6095E-28 watts
P4= .047 watts

and the total power from the supply is, Pdel= .065 watts

READY

RUN

Enter the following data for the series-
parallel circuit of Figure 7-1.

R1=? 2.5714E3
R2=? 3E3
R3=? 10E3
R4=? 2E3
and the supply voltage, E=? 60

The currents, voltages and power to
each part of the circuit are the following:
Currents:
I1= 8.2895E-03 amps
I2= 8.2895E-03 amps
I3= 1.3816E-03 amps
I4= 6.9079E-03 amps

Voltages:
V1= 21.3156 volts
V2= 24.8685 volts
V3= 13.8158 volts
V4= 13.8158 volts

Power dissipated:
P1= .1767 watts
P2= .2061 watts
P3= .019 watts
P4= .095 watts

and the total power from the supply is, Pdel= .4974 watts

READY
```

Note in the first run the enormous amount of information provided after simply providing the network parameters. Performing the calculations by hand would be time-consuming and possibly incorrect. In the second run R_2 and R_3 are non-existent, replaced by their short-circuit and open-circuit equivalents respectively. The simple series network that results appears in Figure 7.2. The analysis will provide values of I_3 and P_3 because of the finite (although very large) value of R_3. Both can be approximated to be zero in magnitude, compared to other quantities of the network.

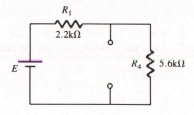

FIGURE 7.2

In the third run, the network of Figure 7.3 is first examined to determine the appropriate values of R_1 through R_4. The values are then inserted into the program to determine the quantities defined by Figure 7.1. The results are used to determine the currents and voltages for the original work.

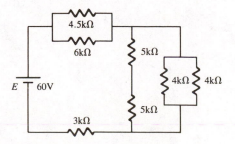

FIGURE 7.3

For Figure 7.3:

$$I_{6k\Omega} = 4.5k\Omega \ (I_1)/(4.5k\Omega + 6.0k\Omega)$$
$$= 0.4286(8.2895 \times 10^{-3}) = \underline{3.5529mA}$$

$$I_{4.5k\Omega} = I_1 - I_{6k\Omega}$$
$$= 8.2895mA - 3.5529mA = \underline{4.7366mA}$$

$$I(\text{for each } 5k\Omega) = I_3 = \underline{1.3816mA}$$

$$I(\text{for each } 4k\Omega) = I_4/2 = 6.9079mA/2$$
$$= \underline{3.4540mA}$$

The voltages and powers can then be calculated directly.

Although you may not have the background to fully understand the behavior of the transistor network of Figure 7.4, you can perform the dc analysis of the system if a few basic relationships are provided. Given the values of V_B and V_{BE}, knowing that $I_C = I_E$ permits a complete analysis of the dc behavior of the system. Lines 230, 240, 280, and 290 are applications of Kirchhoff's voltage law around the indicated loop. The resulting voltages and currents are printed out by lines 300 through 410. Since the network of Figure 7.4 is a fairly standard transistor biasing arrangement, the program can be a particularly significant time-saver in the analysis and design modes.

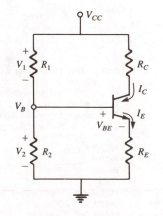

FIGURE 7.4

```
        10 REM *****   PROGRAM 7-2  *****
        20 REM ***********************************************
        30 REM General solution for the voltage-divider bias
        40 REM transistor network.
        50 REM ***********************************************
        60 REM
       100 PRINT "Referring to the network of Figure 7-4"
       110 PRINT "input the following network data:"
       120 PRINT
```
Input
Data
```
      ⎡130 INPUT "What value of VB is desired, VB=";VB
      ⎢140 INPUT "and what is the value of VBE ";BE
      ⎢150 INPUT "The supply voltage is, Vcc=";CC
      ⎢160 PRINT "and the network resistor values are:"
      ⎢170 INPUT "R1=";R1
      ⎢180 INPUT "R2=";R2
      ⎢190 INPUT "RC=";RC
      ⎣200 INPUT "RE=";RE
```
Calc.
```
      ⎡210 REM Network calculations
      ⎢220 V2=VB
      ⎢230 V1=CC-V2
      ⎢240 VE=VB-BE
      ⎢250 IE=VE/RE
      ⎢260 IC=IE :REM Basic assumption about transistor operation
      ⎢270 VC=IC*RC
      ⎢280 CE=CC-VC-VE
      ⎣290 BC=CC-VC-VB
```
Output
Control
```
      ⎡300 REM Output results of above calculations
      ⎢310 PRINT
      ⎢320 PRINT "Network voltages are:"
      ⎢330 PRINT "V(R1)=";V1;"volts"
      ⎢340 PRINT "V(R2)=";V2;"volts"
      ⎢350 PRINT "V(Re)=";VE;"volts"
      ⎢360 PRINT "V(Rc)=";VC;"volts"
      ⎢370 PRINT "Collector-emitter voltage Vce=";CE;"volts"
      ⎢380 PRINT "and base-collector voltage Vbc=";BC;"volts"
      ⎢390 PRINT
      ⎢400 PRINT "Network currents are:"
      ⎣410 PRINT "Emitter current=Collector current=";IE;"amps"
       420 END


READY


RUN
*
Referring to the network of Figure 7-4
input the following network data:

   What value of VB is desired, VB=? 2
   and what is the value of VBE ? 0.7
   The supply voltage is, Vcc=? 22
   and the network resistor values are:
   R1=? 40E3
   R2=? 4E3
   RC=? 10E3
   RE=? 1E3

   Network voltages are:
   V(R1)= 20 volts
   V(R2)= 2 volts
```

*Example 6.8, ICA

```
V(Re)= 1.3 volts
V(Rc)= 13 volts
Collector-emitter voltage Vce= 7.7 volts
and base-collector voltage Vbc= 7 volts

Network currents are:
Emitter current=Collector current= 1.3E-03 amps
```

READY

Program 7-3 will analyze the more complex network of Figure 7.5. Although Program 7.1 could be employed as a partial analysis of the system this program will provide all the quantities desired. Note how the program first analyzed each branch of the network before it considered the system in total. The reduced

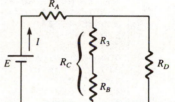

FIGURE 7.6

network, using the notation appearing in the program, is provided in Figure 7.6. Note the similarities with the network of Figure 7.1. The analysis, as defined by lines 240 through 480 is quite straightforward and should not require a detailed description. The results are printed out by lines 490 through 660.

Note in the first run for the network of Figure 7.3 that $R_7 = 0$ ohms, and in the second run $R_2 = 10^{30}$ ohms and $R_3 = 0$ ohms, demon-

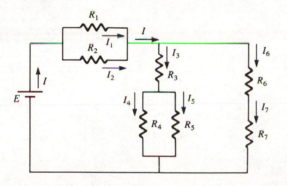

FIGURE 7.5

```
10 REM *****  PROGRAM 7-3  *****
20 REM ************************************************
30 REM This program calculates the currents and
40 REM voltages for the general network of Figure 7-5.
50 REM ************************************************
60 REM
100 PRINT "This program calculates the currents and voltages"
110 PRINT "for the general network of Figure 7-5."
120 PRINT "(Enter 0 for no series resistor"
130 PRINT " and 1E30 for no parallel resistor)"
140 PRINT
150 PRINT "Enter resistance data:"
160 INPUT "R1=";R1
170 INPUT "R2=";R2
180 INPUT "R3=";R3
190 INPUT "R4=";R4
200 INPUT "R5=";R5
210 INPUT "R6=";R6
220 INPUT "R7=";R7
230 INPUT "and the supply voltage E=";E
```

Input Data

(Continued)

```
        ┌ 240 REM Resistance calculations:
        │ 250 RA=R1*R2/(R1+R2)
        │ 260 RB=R4*R5/(R4+R5)
   R    │ 270 RC=R3+RB
  Calc. │ 280 RD=R6+R7
        │ 290 RE=RC*RD/(RC+RD)
        └ 300 RT=RA+RE
        ┌ 310 REM Current calculations:
        │ 320 I=E/RT
        │ 330 I1=R2*I/(R1+R2)
   I    │ 340 I2=I-I1
  Calc. │ 350 I3=RD*I/(RD+RC)
        │ 360 I4=R5*I3/(R4+R5)
        │ 370 I5=I3-I4
        │ 380 I6=I-I3
        └ 390 I7=I6
        ┌ 400 REM Voltage calculations:
        │ 410 V1=RA*E/RT
        │ 420 V2=V1
   V    │ 430 VX=RE*E/RT
  Calc. │ 440 V3=I3*R3
        │ 450 V4=RB*VX/RC
        │ 460 V5=V4
        │ 470 V6=I6*R6
        └ 480 V7=I7*R7
        ┌ 490 PRINT:PRINT "The source current I=";I;"amps"
        │ 500 PRINT "The network currents are:"
        │ 510 PRINT "I1=";I1;"amps"
        │ 520 PRINT "I2=";I2;"amps"
   I    │ 530 PRINT "I3=";I3;"amps"
 Output │ 540 PRINT "I4=";I4;"amps"
        │ 550 PRINT "I5=";I5;"amps"
        │ 560 PRINT "I6=";I6;"amps"
        │ 570 PRINT "I7=";I7;"amps"
        └ 580 PRINT
        ┌ 590 PRINT "and the network voltages are:"
        │ 600 PRINT "V1=";V1;"volts"
        │ 610 PRINT "V2=";V2;"volts"
   V    │ 620 PRINT "V3=";V3;"volts"
 Output │ 630 PRINT "V4=";V4;"volts"
        │ 640 PRINT "V5=";V5;"volts"
        │ 650 PRINT "V6=";V6;"volts"
        └ 660 PRINT "V7=";V7;"volts"
          670 END


        READY


        RUN
        *
        This program calculates the currents and voltages
        for the general network of Figure 7-5.
        (Enter 0 for no series resistor
         and 1E30 for no parallel resistor)

        Enter resistance data:
        R1=? 9
        R2=? 6
        R3=? 4
        R4=? 6
```

*Example 6.3, ICA

```
R5=? 3
R6=? 3
R7=? 0
and the supply voltage E=? 16.8

The source current I= 3 amps
The network currents are:
I1= 1.2 amps
I2= 1.8 amps
I3= 1 amps
I4= .3333 amps
I5= .6667 amps
I6= 2 amps
I7= 2 amps

and the network voltages are:
V1= 10.8 volts
V2= 10.8 volts
V3= 4 volts
V4= 2 volts
V5= 2 volts
V6= 6 volts
V7= 0 volts


READY

RUN

*
This program calculates the currents and voltages
for the general network of Figure 7-5.
(Enter 0 for no series resistor
 and 1E30 for no parallel resistor)

Enter resistance data:
R1=? 4
R2=? 1E30
R3=? 0
R4=? 4
R5=? 4
R6=? 0.5
R7=? 1.5
and the supply voltage E=? 10

The source current I= 2 amps
The network currents are:
I1= 2 amps
I2= 0 amps
I3= 1 amps
I4= .5 amps
I5= .5 amps
I6= 1 amps
I7= 1 amps

and the network voltages are:
V1= 8 volts
V2= 8 volts
V3= 0 volts
V4= 2 volts
V5= 2 volts
V6= .5 volts
V7= 1.5 volts


READY
```

*Example 6.2, ICA

strating that the program, if properly modified, can correctly analyze the network. The second run will actually analyze the network of Figure 7.7, which is very similar to the network of Figure 7.3.

Program 7-4 is an innovative procedure for analyzing series-parallel networks of any number of elements. It requires that each branch first be analyzed independently and the results stored before examining the system in total. If each branch has a single element (or the "most distant" branch from the source is the only branch with series-parallel combinations), the need for the intermediate storage is removed and the program can be applied directly. The program starts with those elements "most distant" from the source and works toward the source, until the total resistance is determined.

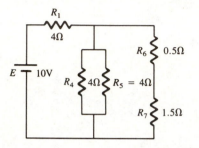

FIGURE 7.7

```
10 REM *****  PROGRAM 7-4  *****
20 REM *********************************************************
30 REM Program to calculate the equivalent resistance of
40 REM a number of resistors in a series-parallel network.
50 REM *********************************************************
60 REM
100 DIM R(10) :REM Allow for up to 10 resistors
110 R=0:G=0 :REM Set initial value of resistance or conductance to 0.
120 GOSUB 300 :REM Identify type of connection
130 INPUT "How many resistors ";N
140 GOSUB 500 :REM Input resistor values
150 ON C GOSUB 700,900 :REM IF CHOICE=1 GOSUB 700; IF CHOICE=2 GOSUB 900
160 PRINT "Total resistance thus far is";R;"ohms"
170 G=1/R :REM determine G if next combination is parallel
180 INPUT "More(YES or NO)";A$
190 IF A$="NO" THEN PRINT "Total network resistance is";R;"ohms." :END
200 GOSUB 300 :REM Ask for choice of next component branch connection
210 INPUT "Number of additional resistor components to combine";N
220 GOTO 140
300 REM Module to choose connection
310 PRINT "Enter (1) series connection"
320 PRINT "      (2) parallel connection"
330 PRINT
340 INPUT "CHOICE=";C :REM CHOOSE (1) OR (2)
350 IF C<1 OR C>2 THEN 340
360 RETURN
500 REM Module to input resistor values
510 FOR J=1 TO N
520 PRINT "R(";J;")=";
530 INPUT R(J)
540 NEXT J
550 RETURN
700 REM Module to sum resistors in series
710 FOR I=1 TO N
720 R=R+R(I)
730 NEXT I
740 RETURN
```

Choose Series or Parallel *(annotation for lines 300–360)*

Input R(J) *(annotation for lines 500–550)*

Series R *(annotation for lines 700–740)*

```
Parallel
R
   ┌900 REM Module to combine resistors in parallel
   │910 FOR I=1 TO N
   │920 G=G+1/R(I)
   │930 NEXT I
   │940 R=1/G
   └950 RETURN
```

READY

<u>RUN</u>

*
Enter (1) series connection
      (2) parallel connection

CHOICE=? <u>1</u>
How many resistors ? <u>2</u>
R(1)=? <u>0.5</u>
R(2)=? <u>1.5</u>
Total resistance thus far is 2 ohms
More(YES or NO)? <u>YES</u>
Enter (1) series connection
      (2) parallel connection

CHOICE=? <u>2</u>
Number of additional resistor components to combine? <u>2</u>
R(1)=? <u>4</u>
R(2)=? <u>4</u>
Total resistance thus far is 1 ohms
More(YES or NO)? <u>YES</u>
Enter (1) series connection
      (2) parallel connection

CHOICE=? <u>1</u>
Number of additional resistor components to combine? <u>1</u>
R(1)=? <u>4</u>
Total resistance thus far is 5 ohms
More(YES or NO)? <u>NO</u>
Total network resistance is 5 ohms.

READY

<u>RUN</u>

*
Enter (1) series connection
      (2) parallel connection

CHOICE=? <u>2</u>
How many resistors ? <u>2</u>
R(1)=? <u>6</u>
R(2)=? <u>9</u>
Total resistance thus far is 3.6 ohms
More(YES or NO)? <u>NO</u>
Total network resistance is 3.6 ohms.

READY

*Example 6.2, ICA
*Example 6.3, ICA

(Continued)

```
RUN

Enter (1) series connection
      (2) parallel connection

CHOICE=? 2
How many resistors ? 2
R( 1 )=? 3
R( 2 )=? 6
Total resistance thus far is 2 ohms
More(YES or NO)? YES
Enter (1) series connection
      (2) parallel connection

CHOICE=? 1
Number of additional resistor components to combine? 1
R( 1 )=? 4
Total resistance thus far is 6 ohms
More(YES or NO)? YES
Enter (1) series connection
      (2) parallel connection

CHOICE=? 2
Number of additional resistor components to combine? 1
R( 1 )=? 3
Total resistance thus far is 2 ohms
More(YES or NO)? YES
Enter (1) series connection
      (2) parallel connection

CHOICE=? 1
Number of additional resistor components to combine? 1
R( 1 )=? 3.6
Total resistance thus far is 5.6 ohms
More(YES or NO)? NO
Total network resistance is 5.6 ohms.


READY
```

In the network of Figure 7.8, which represents the first run of the program, the resistors of 1.5 and 0.5 ohms will be combined first. The result will then be combined with the two 4-ohm parallel resistors before it is added to the 4-ohm source resistance. Routine 700 through 740 will combine the elements in series, and routine 900 through 950 the elements in parallel. As noted by line 130, there is no limit to the number of elements in parallel. Module 500 through 550 labels the input resistor values as $R(1)$, $R(2)$ etc. for each series or parallel combination. Line 110 is required to be sure that the total R and total G is reset to zero before considering the next series or parallel combination. After each calculation, line 160 will print out the total resistance determined by the combinations considered.

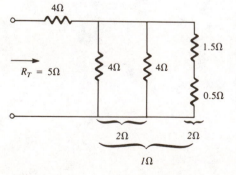

FIGURE 7.8

The program can best be examined by following the first few calculations applied to the network of Figure 7.8 in the first run of the program. Line 120 will first identify the configuration as a series or parallel combination. For the two series elements at the "end" of the network, line 130 will result in $N = 2$. Subroutine 500 will then label the two values $R(1)$ and $R(2)$, and line 150 will send the program to the addition mode, starting at line 700 as determined by $C = 1$ from line 340. The total resistance will then be printed out, followed by the question of whether the analysis is complete. Figure 7.8 will then result in three parallel elements, and N will be three on line 130 with $C = 2$ on line 340. The program will then provide the new total resistance before considering the 4-ohm source resistance. As demonstrated in Figure 7.8 and the first run, the total resistance is 5 ohms.

The second run of Program 7-4 will analyze the network of Figure 7.9. In this case the parallel combination of the 9-ohm and 6-ohm resistors is not in the "most distant" branch from the source and must be calculated first and stored for future use. As indicated by the run, the program will then return to the branch "most distant" from the source and calculate the parallel combination of the 3-ohm and 6-ohm resistors before adding the series 4-ohm resistor. The total resistance of the parallel 9-ohm and 6-ohm resistor will then be added as if it were in series with the total resistance determined by the network to the right. The usefulness of the program will be demonstrated when you perform some of the exercises at the end of the chapter.

The concluding program (7-5) will test your ability to properly analyze the network of Figure 7.1. Random values of E, R_1, R_2, R_3 and R_4 will be generated, along with a random request for a particular current and voltage. As an added highlight, the program will limit its selection to the 20% tolerance standard values. Since E on line 1320 has a range of 1 through 3 and $J1$ of line 1340 is limited to the range 1 through 24, line 1350 will result in resistance levels that can extend from $0.1 \times 10^1 = 1$ ohm to $0.91 \times 10^3 = 910$ ohms. Line 180 will randomly generate a value of E from 1 to 50 volts, while line 200 will provide a value of K that will define the current and voltage to be calculated. The testing sequence for each quantity starts at line 250 and 310. Note that module 500 will calculate all the quantities of the network before the comparison with the input (the user's solution) value is made. Of course, the value calculated could be limited to just the desired value after K was determined, but the only change in module 500 would be to change line 560 to FOR $JJ = K$. Note the random generation of resistor values in the range of 1 through 910 ohms in both runs of the program.

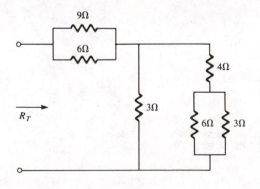

FIGURE 7.9

```
10 REM *****  PROGRAM 7-5  *****
20 REM *********************************************************
30 REM Self-test program designed to ask for the
40 REM calculation of a current or voltage in the
50 REM network of Figure 7-1.
60 REM *********************************************************
```

(Continued)

```
  70 REM
 100 DIM R(4),I(4),V(4) :REM Four values of each to be determined
 110 DIM RS(24)  _:REM 24 data points for the 20% tolerence resistor values
 120 REM Randomly select 4 resistor values
 130 FOR J=1 TO 4
 140 GOSUB 1200
 150 R(J)=R
 160 PRINT "R(";J;")=";R(J);"ohms"
 170 NEXT J
 180 E=INT(1+50*RND(X)) :REM Pick E from 1 to 50 volts
 190 PRINT "and E=";E;"volts"
 200 K=INT(1+4*RND(X)) :REM Choose the component to ask about
 210 GOSUB 500 :REM Do network calculations
 220 PRINT:PRINT "What is the value of:"
 230 PRINT "I(";K;")=";
 240 INPUT IA
 250 IF SGN(IA)<>SGN(I(K)) THEN GOTO 270
 260 IF ABS(IA-I(K))<ABS(0.01*I(K)) THEN PRINT "Correct" :GOTO 280
 270 PRINT "No, its";I(K);"amps"
 280 PRINT:PRINT "and what is the value of:"
 290 PRINT "V(";K;")=";
 300 INPUT VA
 310 IF SGN(VA)<>SGN(V(K)) THEN 330
 320 IF ABS(VA-V(K))<ABS(0.01*V(K)) THEN PRINT "Correct" :GOTO 340
 330 PRINT "Sorry, the answer is";V(K);"volts"
 340 PRINT:PRINT
 350 INPUT "More(YES or NO)";A$
 360 IF A$="YES" THEN PRINT:GOTO 120
 370 END
 500 REM Module to perform network calculations
 510 RT=R(1)+R(2)+(R(3)*R(4))/(R(3)+R(4))
 520 I=E/RT
 530 I(1)=I:I(2)=I
 540 I(3)=(R(4)*I)/(R(3)+R(4))
 550 I(4)=I-I(3)
 560 FOR JJ=1 TO 4
 570 V(JJ)=I(JJ)*R(JJ)
 580 NEXT JJ
 590 RETURN
1200 REM Module to randomly select a value of resistance
1210 REM among suitable 20% tolerence components
1220 REM Read standard component values
1230 RESTORE
1240 FOR L=1 TO 24
1250 READ RS(L)
1260 NEXT L
1270 DATA 0.10,0.11,0.12,0.13,0.15,0.16,0.18
1280 DATA 0.20,0.22,0.24,0.27,0.30,0.33,0.36,0.39
1290 DATA 0.43,0.47,0.51,0.56,0.62,0.68,0.75,0.82,0.91
1300 REM Select exponent from 1 TO 3
1310 REM E=Exponent of resistance value
1320 E=INT(1+3*RND(X))
1330 REM Select one of 24 resistor values
1340 J1=INT(1+24*RND(X))
1350 R=RS(J1)*10^E
1360 IF R<10 THEN 1320
1370 RETURN
```

READY

```
RUN

R( 1 )= 18 ohms
R( 2 )= 430 ohms
R( 3 )= 16 ohms
R( 4 )= 75 ohms
and E= 17 volts

What is the value of:
I( 4 )=? 6.5E-3
Correct

and what is the value of:
V( 4 )=? 0.5
Sorry, the answer is .4861 volts


More(YES or NO)? YES

R( 1 )= 820 ohms
R( 2 )= 390 ohms
R( 3 )= 43 ohms
R( 4 )= 200 ohms
and E= 12 volts

What is the value of:
I( 2 )=? 9.64E-3
Correct

and what is the value of:
V( 2 )=? 3.76
Correct


More(YES or NO)? NO


READY
```

EXERCISES

Write a program to perform each of the following operations.

1. For the network of Figure 7.10, develop a general solution that will print out R_T and I_T and the current, voltage, and power to each element.

2. For the network of Figure 7.10, tabulate E and the power to R_3 for E from 4→40 volts in increments of 4 volts. Use $R_1 = 4\Omega$, $R_2 = 6\Omega$, and $R_3 = 2\Omega$.

3. Given V_{BE}, V_C, V_E, V_{CC}, R_E, and R_C for the network of Figure 7–4 print out the results for V_{CE}, V_{RC}, V_B, V_{BC}, I_E and I_C. For one run, use

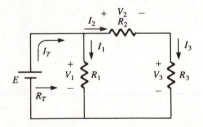

FIGURE 7.10

$V_{BE} = 0.7$ volts, $V_C = 8$ volts, $V_E = 2$ volts, $V_{CC} = 12$ volts, $R_E = 4k\Omega$, and $R_C = 8k\Omega$.

4. Develop a total solution for the network of Figure 7.11. That is, print out the results for R_T and I_T and the current, voltage, and power to each element.

5. Use the method introduced by Program 7-5 to determine the total resistance for the network of Figure 7.11.

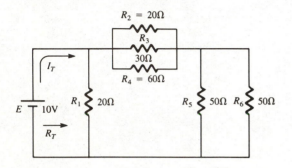

FIGURE 7.11

6. Plot I_T versus R_1 for the network of Figure 7.11 for values of R_1 that extend from 20 to 400 ohms in increments of 20 ohms.

7. Develop a test routine for the network of Figure 7.10. That is, randomly generate the values of E, R_1, R_2, and R_3 and ask for the total current, I, total resistance, R_T, and the current and power to the resistor R_3. All answers should be within 1% to be considered correct.

Determinants, Ladder
Networks and Delta-
Wye Conversions

8

8.1 INTRODUCTION

We look at the solution of second- and third-order determinants in this chapter, in preparation for the methods of analysis we will investigate in the chapters to follow. We will review two methods of solving ladder networks before proceeding to the delta and wye configurations.

8.2 THE SECOND- AND THIRD-ORDER DETERMINANT

The most direct method of evaluating the third-order determinant is to expand the solution to the format indicated by line 220 of Program 8-1. The values of the parameters are provided in the order indicated by lines 180, 190 and 200.

Although you must be very careful in writing line 220 to provide the correct input data, performing a calculation longhand will convince you of the value of such a program—how often an error due to a single sign loss can result in an impossible solution. The statement of line 240 identifies the particular solution of $D = 0$.

Program 8-2 introduces the matrix notation which defines the first number within the parenthesis as the row of the subscripted variable, while the second refers to the column. For clarity, the format of the general 3×3 matrix is provided below. The 2×2 configuration is indicated within the general structure. In Program 8-2 you must decide whether a 2×2 or 3×3 determinant is to be solved, because the substitution of zero values for the subscriptive variables outside the 2×2 configurations will result in a zero for each value of D on lines 380 through 400. Line 150 through 210 includes a

```
         10 REM *****   PROGRAM 8-1   *****
         20 REM ***********************************
         30 REM Solution of the 3rd order determinant.
         40 REM ***********************************
         50 REM
        100 PRINT "This program solves the 3rd order determinant"
        110 PRINT "of the form:"
        120 PRINT
        130 PRINT TAB(5);"A1";TAB(10);"A2";TAB(15);"A3"
        140 PRINT TAB(5);"B1";TAB(10);"B2";TAB(15);"B3"
        150 PRINT TAB(5);"C1";TAB(10);"C2";TAB(15);"C3"
        160 PRINT
 Input ┌ 170 PRINT "Enter the following data:"
 Data  │ 180 INPUT "A1,A2,A3",A1,A2,A3
       │ 190 INPUT "B1,B2,B3";B1,B2,B3
       └ 200 INPUT "C1,C2,C3";C1,C2,C3
 Calc. ┌ 210 REM Now calculate the value of the determinant
       └ 220 D=A1*B2*C3+B1*C2*A3+C1*A2*B3-A3*B2*C1-B3*C2*A1-C3*A2*B1
Output ┌ 230 PRINT "DETERMINANT=D=";D
       └ 240 IF D=0 THEN PRINT "and is singular"
        250 END


READY


RUN

This program solves the 3rd order determinant
of the form:

     A1    A2    A3
     B1    B2    B3
     C1    C2    C3
```

```
Enter the following data:
A1,A2,A3? 1,0,1
B1,B2,B3? 0,3,2
C1,C2,C3? -2,1,3
DETERMINANT=D= 13


READY

RUN

This program solves the 3rd order determinant
of the form:

     A1    A2    A3
     B1    B2    B3
     C1    C2    C3

Enter the following data:
A1,A2,A3? 1,2,3
B1,B2,B3? 4,5,6
C1,C2,C3? 7,8,9
DETERMINANT=D= 0
and is singular


READY
```

```
10 REM *****  PROGRAM 8-2  *****
20 REM *********************************************
30 REM Program to evaluate a 2x2 or 3x3 determinant.
40 REM *********************************************
50 REM
100 DIM A(3,3) :REM Allow for largest array size
110 PRINT "This program will evaluate a 2x2 or 3x3 determinant"
120 INPUT "Enter size 2 or 3";N
130 IF N<2 OR N>3 THEN GOTO 120
140 PRINT "Enter values using the format (row,column)"
150 FOR I=1 TO N
160 PRINT
170 FOR J=1 TO N
180 PRINT "A(";I;",";J;")=";
190 INPUT A(I,J)
200 NEXT J
210 NEXT I
220 PRINT
230 ON N-1 GOSUB 340,370:REM Use subroutine for 2x2 or 3x3
240 PRINT "Determinant value is ";D
250 IF D=0 THEN PRINT "and determinant is singular"
260 PRINT
270 INPUT "More(YES or NO)";A$
280 IF A$="YES" THEN GOTO 120
290 PRINT "See you again"
300 END
340 REM Module to evaluate 2x2 determinant
350 D=A(1,1)*A(2,2)-A(2,1)*A(1,2)
360 RETURN
```

I / Loop　J / Loop

Calc 2 × 2

(Continued)

```
       ┌370 REM Module to evaluate 3x3 determinant
Calc.  │380 D1=A(1,1)*(A(2,2)*A(3,3)-A(3,2)*A(2,3))
3 × 3  │390 D2=A(2,1)*(A(1,2)*A(3,3)-A(3,2)*A(1,3))
       │400 D3=A(3,1)*(A(1,2)*A(2,3)-A(2,2)*A(1,3))
       │410 D=D1-D2+D3
       └420 RETURN
```

```
READY


RUN

This program will evaluate a 2x2 or 3x3 determinant
Enter size 2 or 3? 2
Enter values using the format (row,column)

A( 1 , 1 )=? 2
A( 1 , 2 )=? -1

A( 2 , 1 )=? -1
A( 2 , 2 )=? 3

Determinant value is  5

More(YES or NO)? YES
Enter size 2 or 3? 3
Enter values using the format (row,column)

A( 1 , 1 )=? 1
A( 1 , 2 )=? 0
A( 1 , 3 )=? 1

A( 2 , 1 )=? 0
A( 2 , 2 )=? 3
A( 2 , 3 )=? 2

A( 3 , 1 )=? -2
A( 3 , 2 )=? 1
A( 3 , 3 )=? 3

Determinant value is  13

More(YES or NO)? NO
See you again


READY
```

LOOP within a LOOP. In other words, I will remain equal to one for the entire range of J(1 to N) before the value of I will be incremented by one. For a 2×2 configuration, $A(1,1)$ and $A(1,2)$ will be requested for I equal 1 and $A(2,1)$ and $A(2,2)$ for I equal 2. Line 230 will choose the subroutine based on the value of N ($2 - 1 = 1$ for a 2×2 and $3 - 1 = 2$ for a 3×3) and then proceed to line 340 for a 2×2 or line 370 for 3×3 determinant. The equation defined by line 410 and the equations on lines 380 through 400 are essentially the same as those appearing in Program 8-1. However, it is important that you understand the matrix notation now because we will use it continuously in the following chapters. Note also the form of the

DIMENSION statement of line 100 for the nine elements of the 3 × 3 matrix.

```
    2 × 2
 ┌ ─ ─ ─ ─ ─ ─ ─ ┐
 │  A(1,1)   A(1,2) │  A(1,3)
 │  A(2,1)   A(2,2) │  A(2,3)
 └ ─ ─ ─ ─ ─ ─ ─ ┘
    A(3,1)   A(3,2)   A(3,3)
```

Program 8-3 is a test of whether the operations involved in calculating the third-order determinant are correctly understood. It is non-

sense to assume that since you now have a program that will do the job you no longer require an understanding of how it was performed.

Line 140 will randomly generate a number in the range of 1 through 5. Since the odds of a 3 being generated are one out of five, the chances of a negative sign in front of the number generated by line 160 are also one out of five. Note again the J loop within the I loop, so that all the

```
           10 REM *****  PROGRAM 8-3  *****
           20 REM ************************************************
           30 REM Test evaluation of the 3x3 determinant.
           40 REM ************************************************
           50 REM
          100 DIM A(3,3)
         ┌110 PRINT "For the following determinant values:"
          120 FOR I=1 TO 3
          130 FOR J=1 TO 3
 Random   140 S=INT(1+5*RND(X)) :REM Allow for one minus sign in 5 tries
 Gen.     150 IF S=3 THEN SG=-1 ELSE SG=+1
 of       160 A(I,J)=INT(1+10*RND(X))*SG :REM Select values from 1 to 10
 A(I,J)   170 PRINT "A(";I;",";J;")=";A(I,J),
          180 NEXT J
          190 PRINT
         └200 NEXT I
          210 PRINT
          220 GOSUB 370 :REM Evaluate determinant
 Input D  230 INPUT "The value of determinant D is";DA
 Test    ┌240 IF SGN(DA)<>SGN(D) THEN GOTO 260
 Calc.   └250 IF ABS(DA-D)<ABS(0.01*D) THEN PRINT "Correct": GOTO 270
          260 PRINT "No, the correct answer is";D
          270 PRINT
          280 INPUT "More(YES or NO)";A$
          290 IF A$="YES" THEN GOTO 110
          300 END
          370 REM Module to evaluate 3x3 determinant
          380 D1=A(1,1)*(A(2,2)*A(3,3)-A(3,2)*A(2,3))
          390 D2=A(2,1)*(A(1,2)*A(3,3)-A(3,2)*A(1,3))
          400 D3=A(3,1)*(A(1,2)*A(2,3)-A(2,2)*A(1,3))
          410 D=D1-D2+D3
          420 RETURN


 READY


 RUN


 For the following determinant values:
 A( 1 , 1 )= 5    A( 1 , 2 )= 9    A( 1 , 3 )= 9
 A( 2 , 1 )= 4    A( 2 , 2 )=-7    A( 2 , 3 )= 7
 A( 3 , 1 )=-2    A( 3 , 2 )= 1    A( 3 , 3 )= 9

 The value of determinant D is? -890
 Correct
```

```
More(YES or NO)? YES
For the following determinant values:
A( 1 , 1 )=-10   A( 1 , 2 )= 10   A( 1 , 3 )= 6
A( 2 , 1 )= 3    A( 2 , 2 )= 7    A( 2 , 3 )=-3
A( 3 , 1 )= 1    A( 3 , 2 )= 4    A( 3 , 3 )= 7

The value of determinant D is? 48
No, the correct answer is-820

More(YES or NO)? NO


READY
```

elements of a row will be generated before going on to the next row. The test commands of lines 240 and 250 are the same as used earlier, and lines 370 through 420 are the same as in Program 8-2.

8.3 LADDER NETWORKS

The first program devoted to *ladder* networks uses a sequence of equations defined for a particular length ladder. The solution to the three-section ladder network of Figure 8.1 is defined by lines 200 through 250 and lines 290 through

360 of Program 8-4. Note how the program progresses from the branch "most distant" from the source and then works back using the current-divider rule. Also compare the time required to do such a problem longhand versus using the computer, once the program is properly entered; this is especially true when you review the values chosen in the second run.

Program 8-5 will analyze ladder networks up to five sections, but, in fact, could be modified to analyze networks of any length without extending the length of the program. Five sections required 10 resistor values, as indicated by lines 160 through 200. Note that the absence

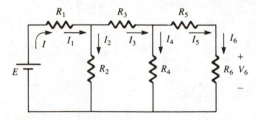

FIGURE 8.1

```
10 REM *****   PROGRAM 8-4   *****
20 REM *****************************************
30 REM Program to solve the 3-section ladder
40 REM network of Figure 8-1.
50 REM *****************************************
60 REM
100 PRINT "For the 3-section ladder network of Figure 8-1"
110 PRINT "enter the following network data:"
120 PRINT
```

```
       ┌ 130  INPUT  "R1=";R1
       │ 140  INPUT  "R2=";R2
Input  │ 150  INPUT  "R3=";R3
Data   │ 160  INPUT  "R4=";R4
       │ 170  INPUT  "R5=";R5
       │ 180  INPUT  "R6=";R6
       └ 190  INPUT  "and the supply voltage E=";E
       ┌ 200  REM Now calculate RT
       │ 210  RA=R5+R6
Calc.  │ 220  RB=R4*RA/(R4+RA)
RT     │ 230  RC=R3+RB
       │ 240  RD=R2*RC/(R2+RC)
       └ 250  RT=R1+RD
         260  PRINT
         270  PRINT "The circuit equivalent resistance is, RT=";RT;"ohms
         280  PRINT
       ┌ 290  REM Now calculate network currents
       │ 300  I=E/RT
       │ 310  I1=I
Calc.  │ 320  I2=RC*I/(RC+R2)
I      │ 330  I3=I-I2
       │ 340  I4=RA*I3/(RA+R4)
       │ 350  I5=I3-I4
       └ 360  I6=I5
       ┌ 370  PRINT "and the network currents are:"
       │ 380  PRINT "Supply current, I=";I;"amps"
       │ 390  PRINT "I1=";I1;"amps"
Output │ 400  PRINT "I2=";I2;"amps"
I      │ 410  PRINT "I3=";I3;"amps"
       │ 420  PRINT "I4=";I4;"amps"
       │ 430  PRINT "I5=";I5;"amps"
       └ 440  PRINT "I6=";I6;"amps"
V6     ┌ 450  PRINT :PRINT "V6=I6*R6=";I6*R6;"volts"
       └ 460  V6=I6*R6
P6       470  PRINT "P6=V6*I6=";V6*I6;"watts"
         480  END


READY


RUN
.
For the 3-section ladder network of Figure 8-1
enter the following network data:

R1=? 7
R2=? 6
R3=? 4
R4=? 6
R5=? 1
R6=? 2
and the supply voltage E=? 240

The circuit equivalent resistance is, RT= 10 ohms

and the network currents are:
Supply current, I= 24 amps
I1= 24 amps
I2= 12 amps
```

*Section 6.3, ICA

```
        I3= 12 amps
        I4= 4 amps
        I5= 8 amps
        I6= 8 amps

        V6=I6*R6= 16 volts
        P6=V6*I6= 128 watts


    READY

    RUN

    For the 3-section ladder network of Figure 8-1
    enter the following network data:

    R1=? 2.1E3
    R2=? 3.6E3
    R3=? 6.2E3
    R4=? 1E3
    R5=? 750
    R6=? 3.3E3
    and the supply voltage E=? 18.6E-3

    The circuit equivalent resistance is, RT= 4477.5869 ohms

    and the network currents are:
    Supply current, I= 4.154E-06 amps
    I1= 4.154E-06 amps
    I2= 2.7435E-06 amps
    I3= 1.4105E-06 amps
    I4= 1.1312E-06 amps
    I5= 2.7931E-07 amps
    I6= 2.7931E-07 amps

    V6=I6*R6= 9.2174E-04 volts
    P6=V6*I6= 2.5745E-10 watts


    READY



    10 REM *****  PROGRAM 8-5  *****
    20 REM **********************************
    30 REM Program analyzes a ladder network
    40 REM with up to 5 sections.
    50 REM **********************************
    60 REM
    100 PRINT "This program calculates RT and the supply current, I"
    110 PRINT "for a ladder network of up to 5 sections."
    120 PRINT:PRINT "Enter the ladder network resistor values"
    130 PRINT "Note: Input R=1E30 for the absence of a vertical element"
    140 PRINT "and 0 for the absence of a horizontal element"
    150 PRINT:PRINT
    160 DIM R(10) :REM Allow up to 5 sections with 10 resistors
    170 FOR I=1 TO 10
    180 PRINT "R(";I;")=";
    190 INPUT R(I)
    200 NEXT I
    210 PRINT
    220 INPUT "and value of E=";E
    230 REM Now calculate RT
    240 RT=1E30 :REM Initialize RT with an infinite(open-circuit) value
```

Loop Input R_x — (bracket annotation for lines 170–200)

```
 /     ┌250 FOR I=10 TO 2 STEP -2 :REM Work toward source 2 resistors at a time
Loop   │260 RV=R(I)*RT/(R(I)+RT)
(R_T)  │270 RT=RV+R(I-1)
       └280 NEXT I
        290 PRINT:PRINT "RT=";RT;"ohms"
        300 IT=E/RT
       ┌310 UN$="amperes"
Units  │320 IF IT<1E-3 THEN IT=IT*1E6:UN$="microamperes" :GOTO 340
       └330 IF IT<1 THEN IT=IT*1E3:UN$="milliamperes"
        340 PRINT "and the supply current I=";IT;UN$
        350 END
```

READY

RUN

This program calculates RT and the supply current, I
for a ladder network of up to 5 sections.

Enter the ladder network resistor values
Note: Input R=1E30 for the absence of a vertical element
and 0 for the absence of a horizontal element

R(1)=? 20
R(2)=? 20
R(3)=? 20
R(4)=? 20
R(5)=? 20
R(6)=? 20
R(7)=? 20
R(8)=? 20
R(9)=? 20
R(10)=? 20

and value of E=? 200

RT= 32.3636 ohms
and the supply current I= 6.1798 amperes

READY

RUN

This program calculates RT and the supply current, I
for a ladder network of up to 5 sections.

Enter the ladder network resistor values
Note: Input R=1E30 for the absence of a vertical element
and 0 for the absence of a horizontal element

R(1)=? 0
R(2)=? 120
R(3)=? 50
R(4)=? 1E30
R(5)=? 4E3
R(6)=? 36
R(7)=? 2.1E3
R(8)=? 1E3

```
R( 9 )=? 240
R( 10 )=? 120

and value of E=? 12.6

RT= 116.5759 ohms
and the supply current I= 108.0841 milliamperes


READY
```

of any horizontal element is limited to a short-circuit equivalent of zero ohms, because an open-circuit equivalent would separate the circuit. Similarly, the vertical elements are limited to the open-circuit equivalent, or they would short out the remainder of the network.

The choice of $R_T = 10^{30}$ ohms on line 240 can best be demonstrated by noting Figure 8.2. R_T is the total resistance determined for any combination of resistors to the right of R_{10}. The first calculation performed by line 260 is to determine the parallel combination of $R(10)$ and R_T. Since the result must be R_{10} the initial value of R_T must be equivalent to that of an open-circuit. Line 270 will then combine R_9 and R_{10} ($RT = RV + R(I-1) = R(10) + R(10-1) = R(10) + R(9)$) and return to line 250 for the next value of I. Note in the program that the value of I will decrease from 10 to 2 in steps of 2. During the second run, R_T is now the sum of $R(10)$ and $R(9)$ and will appear in parallel with $R(8)$. Line 260 will calculate the resulting parallel resistance and add it to $R(7)$. As noted earlier, this process of working toward the source could be easily expanded to any number of sections. It is one of the innovative approaches (compared to Program 8-4) that comes from developing your programming skills and continually

searching for the best routine. Eventually, the total resistance will be determined and the input current calculated (line 300). Lines 310 through 330 are interesting in that the printout of the resulting current will be in scientific notation. If the current is less than 1 mA, line 320 will multiply the result by 10^6 and apply the micro-ampere unit of measure. For instance, if $I = 0.5 \times 10^{-3} = 0.5E - 3$, then I would be multiplied by 10^6 to obtain $I = (0.5 \times 10^{-3})(10^6) = 0.5 \times 10^3 = 500\mu A$. If the test of line 320 is negative because $I = 0.2$ ampere, line 330 will multiply the result by 10^3 to obtain $(0.2)(10^3) = 200mA$. Of course, if the current is greater than or equal to 1 ampere, the label *amperes* will be applied. Note that the unit of measure increases from micro-ampere to milli-ampere to ampere. If 320 and 330 were reversed for the case of $I = 0.5 \times 10^{-3}$, what would be the result?

Note in the first run the equal values of R, as often used in analog-digital converters. In the second run, the network analyzed appears in Figure 8.3. Even though it is not a ladder

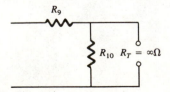

FIGURE 8.2

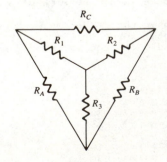

FIGURE 8.3

network in the strictest sense, the analysis is still appropriate and the correct results are obtained.

The last ladder program (8-6) will test your ability to properly analyze the three-section network of Figure 8.1. It will randomly select the resistor values from 10 to 1000 ohms and a supply voltage from 10 to 50 volts. Note in the test of the results for R_T that the sign check was not included because it assumes negative values will not be entertained. The check does appear in 490, however, since currents have a defined direction. Note in the first run the requirement for a level of accuracy within 1%.

```
10 REM *****  PROGRAM 8-6  *****
20 REM **************************************************
30 REM Program to test the solution of the 3-section
40 REM ladder network of Figure 8-1.
50 REM **************************************************
60 REM
100 PRINT "For the 3-section ladder network of"
110 PRINT "Figure 8-1 the network has the following components:"
120 PRINT
130 GOSUB 600 :REM Randomly select resistor value from 10 to 1000 ohms
140 R1=R :PRINT "R1=";R1;"ohms"
150 GOSUB 600 :R2=R
160 PRINT "R2=";R2;"ohms"
170 GOSUB 600 :R3=R
180 PRINT "R3=";R3;"ohms"
190 GOSUB 600 :R4=R
200 PRINT "R4=";R4;"ohms"
210 GOSUB 600 :R5=R
220 PRINT "R5=";R5;"ohms"
230 GOSUB 600 :R6=R
240 PRINT "R6=";R6;"ohms"
250 E=10*INT(1+5*RND(X)) :REM Randomly select E from 10 to 50 volts
260 PRINT "and supply voltage E=";E;"volts"
270 REM Now calculate RT
280 RA=R5+R6
290 RB=R4*RA/(R4+RA)
300 RC=R3+RB
310 RD=R2*RC/(R2+RC)
320 RT=R1+RD
330 PRINT
340 PRINT "What is the total resistance of the system";
350 INPUT RX
360 IF ABS(RT-RX)<ABS(0.01*RT) THEN PRINT "Correct" :GOTO 380
370 PRINT "No, it's";RT;"ohms"
380 REM Now calculate the network currents
390 I=E/RT
400 I1=I
410 I2=RC*I/(RC+R2)
420 I3=I-I2
430 I4=RA*I3/(RA+R4)
440 I5=I3-I4
450 I6=I5
460 REM Now test calculation of the current I6
470 PRINT
480 INPUT "I6=";IX
490 IF SGN(I6)<>SGN(IX) THEN GOTO 510
500 IF ABS(I6-IX)<ABS(0.01*I6) THEN PRINT "Correct" :GOTO 520
510 PRINT "No, it's";I6;"amps"
520 PRINT:INPUT "More(YES or NO)";A$
530 IF A$="YES" THEN PRINT :GOTO 100
540 END
```

(Continued)

```
600 REM Module to randomly select a resistor from 10 to 1000 ohms
610 R=10*INT(1+100*RND(X))
620 RETURN
```

RUN

```
For the 3-section ladder network of
Figure 8-1 the network has the following components:

R1= 780 ohms
R2= 220 ohms
R3= 1000 ohms
R4= 50 ohms
R5= 50 ohms
R6= 730 ohms
and supply voltage E= 40 volts

What is the total resistance of the system? 980
No, it's 961.7992 ohms

I6=? 4.35E-4
Correct

More(YES or NO)? YES

For the 3-section ladder network of
Figure 8-1 the network has the following components:

R1= 830 ohms
R2= 280 ohms
R3= 600 ohms
R4= 1000 ohms
R5= 650 ohms
R6= 120 ohms
and supply voltage E= 10 volts

What is the total resistance of the system? 1050
Correct

I6=? 1.15E-3
Correct

More(YES or NO)? NO
```

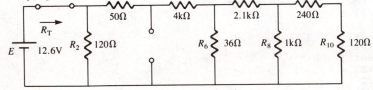

FIGURE 8.4

8.4 THE DELTA-WYE CONFIGURATION

The remaining programs of the chapter will perform the conversion operations between the delta (Δ) and the wye (Y) configuration and provide a general solution for the bridge network. In the first program (8-7), the resistors of the Y of Figure 8.4 are determined for the given Δ configuration. The program is relatively straightforward and should not require any additional comments.

```
10 REM ***** PROGRAM 8-7 *****
20 REM ************************************
30 REM Program converts from delta to wye.
40 REM ************************************
50 REM
100 PRINT "Enter the values of the resistances of"
110 PRINT "the delta connection of Figure 8-4."
120 PRINT
```

Input
(Δ)
```
130 INPUT "RA=";RA
140 INPUT "RB=";RB
150 INPUT "RC=";RC
```

Calc.
Δ → Y
```
160 REM Convert from delta to wye
170 D=RA+RB+RC
180 R1=RA*RC/D
190 R2=RB*RC/D
200 R3=RA*RB/D
210 PRINT
```

Output
(Y)
```
220 PRINT "The resistance values of a terminally equivalent"
230 PRINT "wye connection are:"
240 PRINT "R1=";R1;"ohms"
250 PRINT "R2=";R2;"ohms"
260 PRINT "R3=";R3;"ohms"
270 END
```

```
READY

.
RUN

Enter the values of the resistances of
the delta connection of Figure 8-4.

RA=? 20
RB=? 30
RC=? 10

The resistance values of a terminally equivalent
wye connection are:
R1= 3.3333 ohms
R2= 5 ohms
R3= 10 ohms


READY
```

*Example 7.25, ICA

In Program 8-8 you can choose either a
Y-Δ or Δ-Y conversion. For each conversion the
equations start at lines 1800 and 1900 respectively. Note the use of the colon on lines 1810,
1830, 1910, and 1930 to permit more than one
operation on each line. Note in the first run that

```
10 REM ***** PROGRAM 8-8 *****
20 REM ************************************
30 REM Program to perform the wye-delta or
40 REM delta-wye conversion.
50 REM ************************************
60 REM
```

Choice
```
100 REM Program will perform (1) Wye-delta or, (2) Delta-wye conversion
110 PRINT" Select conversion desired: (1) Wye-delta"
120 PRINT"                                        or, (2) Delta-wye"
130 INPUT "Choice=";C
```

(Continued)

```
140 IF C<>1 THEN IF C<>2 THEN GOTO 130
150 ON C GOSUB 1800,1900 :REM Perform conversion
160 ON C GOTO 200,300 :REM Print results
200 PRINT
210 PRINT"Results of the wye-delta conversion are:"
220 PRINT "RA=";RA
230 PRINT "RB=";RB
240 PRINT "RC=";RC
250 END
300 PRINT
310 PRINT "Results of the delta-wye conversion are:"
320 PRINT "R1=";R1
330 PRINT "R2=";R2
340 PRINT "R3=";R3
350 END
1800 REM Module to do wye-delta conversion
1810 INPUT "R1=";R1 :INPUT "R2=";R2 :INPUT "R3=";R3
1820 N=R1*R2+R2*R3+R1*R3
1830 RA=N/R2 :RB=N/R1 :RC=N/R3
1840 RETURN
1900 REM Module to do delta-wye conversion
1910 INPUT"RA=";RA :INPUT "RB=";RB :INPUT "RC=";RC
1920 D=RA+RB+RC
1930 R1=RA*RC/D :R2=RB*RC/D :R3=RA*RB/D
1940 RETURN
```

```
READY


RUN

 Select conversion desired: (1) Wye-delta
                      or, (2) Delta-wye
Choice=? 2
RA=? 90
RB=? 90
RC=? 90

Results of the delta-wye conversion are:
R1= 30
R2= 30
R3= 30


READY

RUN

 Select conversion desired: (1) Wye-delta
                      or, (2) Delta-wye
Choice=? 1
R1=? 20
R2=? 40
R3=? 60

Results of the wye-delta conversion are:
RA= 110
RB= 220
RC= 73.3333


READY
```

if all the values of one configuration are the same, the same must be true for the other, and they must be related by a factor of three. In the second run each value of the wye is different, resulting in a different value for the components of the delta. If two are the same then two of the other configuration must also be the same.

Program 8-9 randomly selects which conversion will be performed and the value of the resistor components. It will then test against each hand-calculated solution, since the value of each may be different. Keep in mind that the location of the resistors in Figure 8.4 is important because it affects the total resistance and, therefore, the currents of the network. Note in the runs provided that incorrect answers will result in an immediate printout of the correct answer for comparison purposes.

```
10 REM *****  PROGRAM 8-9  *****
20 REM ***********************************
30 REM Program to test the wye-delta or
40 REM delta-wye conversion.
50 REM ***********************************
60 REM
100 REM Randomly select either (1) wye-delta or, (2) delta-wye conversion
110 C=INT(1+2*RND(X)) :REM Select C=1 OR C=2
120 IF C=1 THEN PRINT "Do the following wye-delta conversion:"
130 IF C=2 THEN PRINT "do the following delta-wye conversion:"
140 ON C GOTO 150,340 :REM Determine test values and ask for conversion
150 REM Output values for wye-delta conversion
160 GOSUB 530 :REM Select value of R from 10 to 1000
170 R1=R
180 PRINT "R1=";R1;"ohms"
190 GOSUB 530 :R2=R
200 PRINT "R2=";R2;"ohms"
210 GOSUB 530 :R3=R
220 PRINT "R3=";R3;"ohms"
230 GOSUB 600 :REM Do delta-wye conversion
240 INPUT "RA=";RX
250 IF ABS(RA-RX)<ABS(0.01*RA) THEN PRINT "Correct" :GOTO 270
260 PRINT "No, RA=";RA;"ohms"
270 INPUT "RB=";RX
280 IF ABS(RB-RX)<ABS(0.01*RB) THEN PRINT "Correct" :GOTO 300
290 PRINT "No, RB=";RB;"ohms"
300 INPUT "RC=";RX
310 IF ABS(RC-RX)<ABS(0.01*RC) THEN PRINT "Correct" :GOTO 330
320 PRINT "No, RC=";RC;"ohms"
330 GOTO 560
340 REM Output values for delta-wye conversion
350 GOSUB 530 :REM Select value of R from 10 TO 1000
360 RA=R
370 PRINT "RA=";RA;"ohms"
380 GOSUB 530 :RB=R
390 PRINT "RB=";RB;"ohms"
400 GOSUB 530 :RC=R
410 PRINT "RC=";RC;"ohms"
420 GOSUB 650 :REM Do delta-wye conversion
430 INPUT "R1=";RX
440 IF ABS(R1-RX)<ABS(0.01*R1) THEN PRINT "Correct" :GOTO 460
450 PRINT "No, R1=";R1;"ohms"
460 INPUT "R2=";RX
470 IF ABS(R2-RX)<ABS(0.01*R2) THEN PRINT "Correct" :GOTO 490
480 PRINT "No, R2=";R2;"ohms"
490 INPUT "R3=";RX
500 IF ABS(R3-RX)<ABS(0.01*R3) THEN PRINT "Correct" :GOTO 520
510 PRINT "No. R3=";R3;"ohms"
```

(Continued)

```
520 GOTO 560
530 REM Select R from 10 TO 1000 in steps of 10 ohms
540 R=10*INT(1+100*RND(X))
550 RETURN
560 PRINT: INPUT "More(YES or NO)";A$
570 IF A$="YES" THEN PRINT"o.k." :GOTO 100
580 PRINT "So long for now"
590 END
600 REM Module to do wye-delta conversion
610 N=R1*R2+R2*R3+R1*R3
620 RA=N/R2 :RB=N/R1':RC=N/R3
630 RETURN
650 REM Module to perform delta-wye conversion
660 D=RA+RB+RC
670 R1=RA*RC/D :R2=RB*RC/D :R3=RA*RB/D
680 RETURN
```

READY

<u>RUN</u>


```
do the following delta-wye conversion:
RA= 480 ohms
RB= 920 ohms
RC= 420 ohms
R1=? 110
Correct
R2=? 212
Correct
R3=? 250
No. R3= 242.6374 ohms

More(YES or NO)? YES
o.k.
Do the following wye-delta conversion:
R1= 260 ohms
R2= 870 ohms
R3= 640 ohms
RA=? 1091
Correct
RB=? 3651.5
Correct
RC=? 1483.44
Correct

More(YES or NO)? NO
So long for now
```


READY

The last program (8-10) of the chapter will provide a general solution for the fairly complex bridge network of Figure 8.5. The solution will first find the terminally equivalent Y configuration for the lower delta and then solve for the various currents and the voltage. Following the Δ-Y conversion, the network will appear as shown in Figure 8.6. The values of R_1, R_2, and R_3 are determined by the module at 650. The values of R_x and R_y are determined by line 210, and the parallel combination is added to R_3 on line 230. The source current is then determined, followed by an application of the current-divider rule to determine I_4. Kirchhoff's current law will

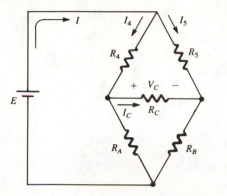

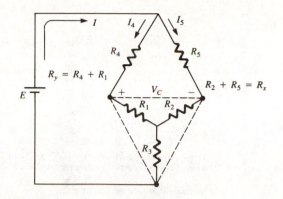

FIGURE 8.5   **FIGURE 8.6**

```
 10 REM *****  PROGRAM 8-10  *****
 20 REM *********************************************
 30 REM Program to provide the general solution
 40 REM for the bridge circuit of Figure 8-5.
 50 REM *********************************************
 60 REM
100 PRINT "To analyze the bridge network of Figure 8-5"
110 PRINT "enter the following data:"
120 PRINT
130 INPUT "RA=";RA
140 INPUT "RB=";RB
150 INPUT "RC=";RC
160 INPUT "R4=";R4
170 INPUT "R5=";R5
180 INPUT "E=";E
190 REM Now do network calculations
200 GOSUB 650 :REM Convert upper delta of bridge network into wye
210 RX=R2+R5 :RY=R1+R4
220 RP=RX*RY/(RX+RY)
230 RT=R3+RP
240 I=E/RT
250 I4=(R2+R5)*I/(R1+R2+R4+R5)
260 I5=I-I4
270 VC=I5*R5-I4*R4
280 IC=VC/RC
290 REM Print results
300 PRINT
310 PRINT "The input resistance, RT=";RT;"ohms"
320 PRINT "and the supply current I=";I;"amps"
330 PRINT
340 PRINT "Network currents are:"
350 PRINT "I4=";I4;"amps","and, I5=";I5;"amps"
360 PRINT "with the voltage across Rc, Vc=";VC;"volts"
370 PRINT "and current through Rc, Ic=";IC;"amps"
380 IF VC=0 OR IC=0 THEN PRINT "The bridge is balanced!"
390 END
650 REM Module to perform delta-wye conversion
660 REM Input of R's already done
670 D=RA+RB+RC
680 R1=RA*RC/D :R2=RB*RC/D :R3=RA*RB/D
690 RETURN
```

Input
Data — lines 130–180

Calc. — lines 200–280

Output
Control — lines 290–380

Calc.
$\Delta \rightarrow Y$ — lines 650–690

READY

```
RUN
*
To analyze the bridge network of Figure 8-5
enter the following data:

RA=? 3
RB=? 4
RC=? 1
R4=? 2
R5=? 2
E=? 20

The input resistance, RT= 2.7179 ohms
and the supply current I= 7.3585 amps

Network currents are:
I4= 3.7736 amps and, I5= 3.5849 amps
with the voltage across Rc, Vc=-.3774 volts
and current through Rc, Ic=-.3774 amps


READY

RUN

To analyze the bridge network of Figure 8-5
enter the following data:

RA=? 10
RB=? 20
RC=? 30
R4=? 40
R5=? 50
E=? 60

The input resistance, RT= 29.0476 ohms
and the supply current I= 2.0656 amps

Network currents are:
I4= 1.1803 amps and, I5= .8852 amps
with the voltage across Rc, Vc=-2.9508 volts
and current through Rc, Ic=-.098 amps


READY

*Problem 7.42a, ICA
```

provide I_5. Applying Kirchhoff's voltage law to the upper loop of Figure 8.5 will give the equation of line 270 for V_C and Ohm's law will provide I_C. If V_C and therefore I_C are zero, the bridge is balanced and line 380 will indicate such in the printout. In both runs, the balance condition is not satisfied. Note the negative answers in each run for V_C and I_C to demonstrate that the polarities and direction are the reverse of that appearing in Figure 8.5.

EXERCISES

Write a program to perform the following tasks.

1. Evaluate the following 2 × 2 and 3 × 3 determinants:

$$\begin{vmatrix} 22 & -40 \\ 6 & -.8 \end{vmatrix} \qquad \begin{vmatrix} 4 & 0 & 5 \\ -1 & 1.6 & 6 \\ 7.08 & 8 & 26 \end{vmatrix}$$

2. For the three-section ladder of Figure 8.1, tabulate I_T and the power to R_6 for R_6 in the range of 2 to 20 ohms, in increments of 2 ohms. Use $E = 30V$, $R_1 = R_2 = R_3 = R_4 = R_5 = 2$ ohms.

3. Expand Program 8-5 to include an analysis of up to N sections and tabulate R_T and N for N sections of resistors of equal value ($R_1 = R_2 = R_3 = \ldots$ etc.) For one run, choose $N = 10$ and each resistor $= 40$ ohms.

4. Plot the results of R_T versus N for problem 8-3. Label the N axis and use $N = 10$ and $R = 40$ ohms.

5. Develop a general solution for the network of Figure 8.7. Note that each value of the Y and the Δ is the same. Print out the magnitude of R_T, I_T, I_1 and I_2. For the first run, choose $E = 40V$, $R_1 = 60\Omega$ and $R_2 = 20\Omega$.

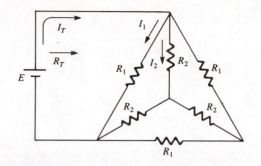

FIGURE 8.7

6. Given E, R_4, R_5, R_A and R_C, develop a routine for the bridge network of Figure 8.5 that will insure that the chosen resistor R_B will balance the bridge. Then calculate I, I_4, and I_5.

7. Develop a test routine that will test the solution for the current I of Figure 8.5, if $R_4 = R_5$ and $R_A = 2R_B$ but R_4, R_A, E and R_C are randomly generated.

Superposition, Thevenin's Theorem and Maximum Power Transfer

9

9.1 INTRODUCTION

This chapter will apply the Superposition, Thevenin, and Maximum Power Transfer theorems to networks of a fairly standard configuration. This will permit their application to more complex systems once the networks are reduced to the standard form. Special emphasis will be placed on providing the output results in the clearest form possible through tabulation and plotting routines.

9.2 SUPERPOSITION

The network of Figure 9.1 is the simplest possible using both a voltage and current source. It is, however, the resultant configuration for the network of Figure 9.2a when reduced to its simplest form, as shown in Figure 9.2b. Once the network of Figure 9.2b is analyzed using Program 9.1, the results can be obtained for the network of Figure 9.2a with little added difficulty.

Applying the superposition theorem will result in the network of Figure 9.3 when the current source is removed in Figure 9.1. Lines 180 through 230 of Program 9.1 will determine and output I_1 and I_2 for the voltage source E. The network of Figure 9.4 is the result if the effect of the voltage source E is removed. Lines 250 through 300 will determine and print out I_1 and I_2, due to the current source I using the current divider rule on lines 260 and 270. The total current is then the algebraic sum of the currents as determined by lines 320 and 330. To demonstrate that superposition is not applicable to power effects, the power to each element is calculated using the partial currents on lines 350 through 420, and the results compared to that obtained using the total current on lines 440 through 470. Note the significant differences in power levels in the printout of the first run.

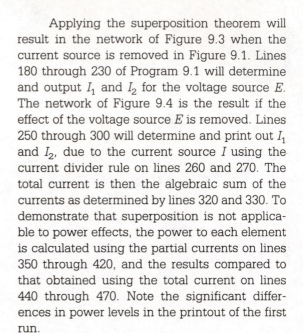

FIGURE 9.3

In the second run the voltage source is reversed, resulting in a negative sign for the partial currents I_1 and I_2. The reversal in voltage

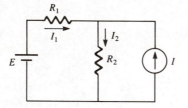

FIGURE 9.1

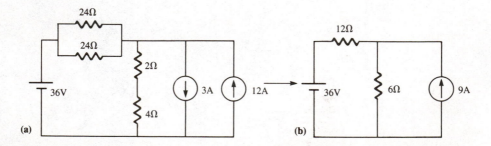

(a)   **(b)**

FIGURE 9.2

```
10 REM *****  PROGRAM 9-1  *****
20 REM *******************************************
30 REM Program to solve a standard two-loop
40 REM network using the Superposition Theorem.
50 REM *******************************************
60 REM
100 PRINT "For the standard two-loop network of"
110 PRINT "Figure 9-1 enter the following data:"
120 PRINT
```

Input Data
```
130 INPUT "R1=";R1
140 INPUT "R2=";R2
150 INPUT "Supply voltage, E=";E
160 INPUT "and current source, I=";I
170 PRINT
```

I due to *E*
```
180 REM Calculate partial currents due to E
190 I1=E/(R1+R2)
200 I2=I1
210 PRINT "Partial currents due to only E:"
220 PRINT "I1=";I1;"amps"
230 PRINT "I2=";I2;"amps"
240 PRINT
```

I due to I_S
```
250 REM Calculate partial currents due to I
260 J1=-R2*I/(R1+R2)
270 J2=R1*I/(R1+R2)
280 PRINT "Partial currents due to only I:"
290 PRINT "I1=";J1;"amps"
300 PRINT "I2=";J2;"amps"
310 PRINT
```

Total *I*
```
320 PRINT "Total current through R1 is";I1+J1;"amps"
330 PRINT "and the total current through R2 is";I2+J2;"amps"
340 PRINT
```

Incorrect *P*
```
350 REM Show power relations
360 P1=I1^2*R1 :REM Power using partial currents
370 P2=I2^2*R2 :REM due only to E
380 P3=J1^2*R1 :REM And power using only partial
390 P4=J2^2*R2 :REM currents due to I
400 PRINT
410 PRINT "Sum of powers to R1 due to both sources is";P1+P3;"watts"
420 PRINT "Sum of powers to R2 due to both sources is";P2+P4;"watts"
430 PRINT
```

P
```
440 P5=(I1+J1)^2*R1
450 P6=(I2+J2)^2*R2
460 PRINT "While the power to R1 due to total current is";P5;"watts"
470 PRINT "and the power to R2 due to total current is";P6;"watts"
480 PRINT
490 END
```

READY

<u>RUN</u>

For the standard two-loop network of
Figure 9-1 enter the following data:

R1=? <u>12</u>
R2=? <u>6</u>
Supply voltage, E=? <u>36</u>
and current source, I=? <u>9</u>

*Example 8.4, ICA

```
Partial currents due to only E:
I1= 2 amps
I2= 2 amps

Partial currents due to only I:
I1=-3 amps
I2= 6 amps

Total current through R1 is-1 amps
and the total current through R2 is 8 amps


Sum of powers to R1 due to both sources is 156 watts
Sum of powers to R2 due to both sources is 240 watts

While the power to R1 due to total current is 12 watts
and the power to R2 due to total current is 384 watts


READY

RUN

For the standard two-loop network of
Figure 9-1 enter the following data:

R1=? 2E3
R2=? 4E3
Supply voltage, E=? -22.8
and current source, I=? 4E-3

Partial currents due to only E:
I1=-3.8E-03 amps
I2=-3.8E-03 amps

Partial currents due to only I:
I1=-2.6667E-03 amps
I2= 1.3333E-03 amps

Total current through R1 is-6.4667E-03 amps
and the total current through R2 is-2.4667E-03 amps


Sum of powers to R1 due to both sources is .043 watts
Sum of powers to R2 due to both sources is .065 watts

While the power to R1 due to total current is .084 watts
and the power to R2 due to total current is .024 watts


READY
```

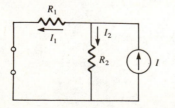

FIGURE 9.4

sources is indicated by a negative sign in response to the request for the voltage source level on line 150. The magnitudes resulting from the applied voltage and current of the second run are such that the total currents I_1 and I_2 are both negative and, therefore, opposite in direction to that shown in Figure 9.1.

9.3 THEVENIN'S THEOREM AND MAXIMUM POWER TRANSFER

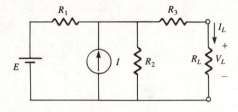

FIGURE 9.5

Program 9-2 will provide a general solution for the network of Figure 9.5 using Thevenin's theorem. The value of R_{Th} and E_{Th} for the network to the left of R_L will first be determined, fol-

```
        10 REM *****  PROGRAM 9-2  *****
        20 REM ****************************************
        30 REM Program to analyze the network of
        40 REM Figure 9-5 using Thevenin's theorem.
        50 REM ****************************************
        60 REM
       100 PRINT "For the network of Figure 9-5"
       110 PRINT "enter the following data:"
       120 PRINT
      ┌130 INPUT "R1=";R1 :REM Add 0 if resistor non-existant
       140 INPUT "R2=";R2 :REM Enter 1E30 if resistor non-existant
Input  150 INPUT "R3=";R3 :REM Enter 0 if resistor non-existant
Data   160 INPUT "RL=";RL
       170 INPUT "Supply voltage, E=";E
      └180 INPUT "and supply current, I=";I
       190 PRINT
      ┌200 REM Determine Rth
RTh   └210 RT=R3+R1*R2/(R1+R2)
      ┌220 REM Use superposition to determine Eth
       230 E1=R2*E/(R1+R2)
ETh    240 I2=R1*I/(R1+R2)
       250 E2=I2*R2
      └260 ET=E1+E2
IL &  ┌270 REM Determine IL and VL
VL     280 IL=ET/(RT+RL)
      └290 VL=IL*RL
      ┌300 PRINT "Using Thevenin's theorem:"
Output 310 PRINT "Rth=";RT;"ohms"
Control320 PRINT "and Eth=";ET;"volts"
       330 PRINT "with IL=";IL;"amps"
      └340 PRINT "and VL=";VL;"volts"
       350 END
```

```
READY

RUN
*
For the network of Figure 9-5
enter the following data:

R1=? 1E30
R2=? 4
R3=? 2
RL=? 7
Supply voltage, E=? 0
and supply current, I=? 12
```

*Example 8.7, ICA

```
Using Thevenin's theorem:
Rth= 6 ohms
and Eth= 48 volts
with IL= 3.6923 amps
and VL= 25.8462 volts


READY


RUN
.
For the network of Figure 9-5
enter the following data:

R1=? 3
R2=? 6
R3=? 0
RL=? 4
Supply voltage, E=? 9
and supply current, I=? 0

Using Thevenin's theorem:
Rth= 2 ohms
and Eth= 6 volts
with IL= 1 amps
and VL= 4 volts


READY
```

*Example 8.6, ICA

lowed by an application of Ohm's law to find the load current I_L and voltage V_L. Line 210 will determine R_{Th} as demonstrated by Figure 9.6.

For E_{Th}, the superposition theorem will be applied as shown in Figures 9.7 and 9.8. The calculation of E'_{Th} appears on line 230 while E''_{Th} is calculated by lines 240 and 250. E_{Th} is then determined by the sum of E'_{Th} and E''_{Th}, which is accomplished on line 260. An opposite polarity for E or a reversal in direction for I should be included as a negative sign when

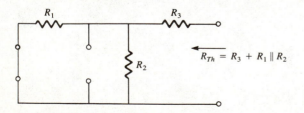

FIGURE 9.6

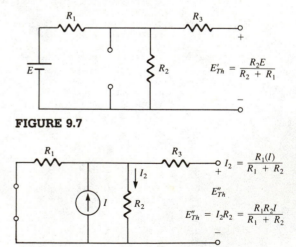

FIGURE 9.7

FIGURE 9.8

providing the values of each on lines 170 and 180. The defined polarities for I_L and V_L appear in Figure 9.5. A negative sign in the results indicates a polarity and direction opposite to that indicated in the figure.

The variations possible using the basic structure of Figure 9.5 are demonstrated by the two runs of the program. In the first run, the network of Figure 9.9 is analyzed, while in the second run the network of Figure 9.10 is examined.

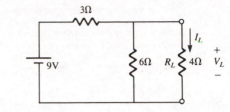

FIGURE 9.10

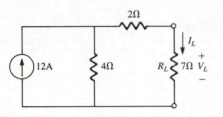

FIGURE 9.9

Program 9-3 tabulates a number of important quantities determined by Program 9-2 for a range of load values. Note that the program is similar to Program 9-2 through line 260. Lines 310 and 320 set the headings using various TAB values to separate the titles by 10 spaces. Note the semicolon at the end of line 310 to insure that all the headings are on the same line. On line 330 the range of load values is

```
        10 REM *****   PROGRAM 9-3  *****
        20 REM *******************************************
        30 REM Program to tabulate changes in load levels for
        40 REM a range of load values using Thevenin's theorem
        50 REM *******************************************
        60 REM
       100 PRINT "For the network of Figure 9-5"
       110 PRINT "enter the following data:"
       120 PRINT
      ┌130 INPUT "R1=";R1 :REM Enter 0 if resistor non-existant
       140 INPUT "R2=";R2 :REM Enter 1E30 if resistor non-existant
Input  150 INPUT "R3=";R3 :REM Enter 0 if resistor non-existant
Data   160 INPUT "RL=";RL
       170 INPUT "Supply voltage, E=";E
      └180 INPUT "and supply current, I=";I
       190 PRINT
      ┌200 REM Determine Rth
R_Th  └210 RT=R3+R1*R2/(R1+R2)
      ┌220 REM Use superposition to determine Eth
       230 E1=R2*E/(R1+R2)
E_Th   240 I2=R2*I/(R1+R2)
       250 E2=R1*R2*I/(R1+R2)
      └260 ET=E1+E2
Output┌270 PRINT "Using Thevenin's theorem:"
R_Th & 280 PRINT "Rth=";RT;"ohms"
E_Th  └290 PRINT "and Eth=";ET;"volts"
       300 PRINT
Table ┌310 PRINT TAB(7);"RL";TAB(15);"IL";TAB(25);"VL";
Heading└320 PRINT TAB(35);"PL";TAB(45);"PD";TAB(55);"n%"
      ┌330 FOR RL=RT/4 TO 4*RT STEP RT/4
       340 IL=ET/(RT+RL)
       350 VL=IL*RL
Calc.  360 PL=IL^2*RL
       370 PD=ET*IL
      └380 N=100*PL/PD
```

(Continued)

```
       ┌ 390  IF RL=RT THEN PRINT "Rth=";
Output │ 400  PRINT TAB(5);RL;TAB(13);IL;TAB(23);VL;
Control│ 410  PRINT TAB(33);PL;TAB(43);PD;TAB(53);N
       └ 420  NEXT RL
         430  END
```

READY

<u>RUN</u>

```
For the network of Figure 9-5
enter the following data:

R1=?  20
R2=?  1E30
R3=?  5
RL=?  25
Supply voltage, E=?  -10
and supply current,  I=?  4

Using Thevenin's theorem:
Rth= 25 ohms
and Eth= 70 volts
```

	RL	IL	VL	PL	PD	n%
	6.25	2.24	14	31.36	156.8	20
	12.5	1.8667	23.3333	43.5556	130.6667	33.3333
	18.75	1.6	30	48	112	42.8571
Rth=	25	1.4	35	49	98	50
	31.25	1.2444	38.8889	48.3951	87.1111	55.5556
	37.5	1.12	42	47.04	78.4	60
	43.75	1.0182	44.5455	45.3554	71.2727	63.6364
	50	.9333	46.6667	43.5556	65.3333	66.6667
	56.25	.8615	48.4615	41.7515	60.3077	69.2308
	62.5	.8	50	40	56	71.4286
	68.75	.7467	51.3333	38.3289	52.2667	73.3333
	75	.7	52.5	36.75	49	75
	81.25	.6588	53.5294	35.2664	46.1176	76.4706
	87.5	.6222	54.4444	33.8765	43.5556	77.7778
	93.75	.5895	55.2632	32.5762	41.2632	78.9474
	100	.56	56	31.36	39.2	80

READY

determined by the resultant Thevenin resistance. The minimum value is $R_{Th}/4$ and the maximum $4R_{Th}$, with steps of $R_{Th}/4$. Those values determined by the load level are then calculated and printed by lines 400 and 410. Note the printing of $R_{Th} =$ on line 390 to identify when the load equals the Thevenin resistance. Also note in the provided run that when $R_L = R_{Th} = 25$ ohms for the network of Figure 9.11, the power is a maximum value (49W) and the efficiency is 50% (as it must be

for maximum power conditions). In Figure 9.11 note the reversal of the source E, which will

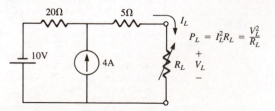

$$P_L = I_L^2 R_L = \frac{V_L^2}{R_L}$$

FIGURE 9.11

have a significant impact on the results for E_{Th}. The power P_D is that power supplied by the Thevenin equivalent source. The efficiency is determined by $\eta\% = P_o/P_i \times 100\% = P_L/P_D \times 100\%$. It is interesting to note that for any system the percent efficiency will always be 20% when $R_L = R_{Th}/4$ and 80% when $R_L = 4R_{Th}$. It clearly defines the range of efficiency we can expect for a range of load values.

The program content through line 290 of Program 9-4 is the same as in Program 9-3. In addition, the results will be plotted for the network of Figure 9.11 using the values appearing in the run of that program. If desired, a comparison can be made between the results of the two programs. The TAB (5) command of line 310 was chosen to insure that the data points do not appear on the same location as the labeling of

```
          10 REM *****  PROGRAM 9-4  *****
          20 REM *************************************************
          30 REM Program to plot changes in load levels for
          40 REM a range of load values using Thevenin's theorem.
          50 REM *************************************************
          60 REM
          100 PRINT "For the two-loop circuit of Figure 9-5"
          110 PRINT "enter the following data:"
          120 PRINT
          130 INPUT "R1=";R1  :REM Enter 0 if resistor non-existent
          140 INPUT "R2=";R2 :REM Enter 1E30 if resistor non-existent
          150 INPUT "R3=";R3 :REM Enter 0 if resistor non-existent
          160 INPUT "Supply voltage, E=";E
          170 INPUT "and supply current, I=";I
          180 PRINT
          190 REM Now do circuit calculations
  R_Th   ⌈200 REM Use Thevenin's theorem to determine Rth
         ⌊210 RT=R3+R1*R2/(R1+R2)
         ⌈220 REM Use superposition to determine Eth
          230 E1=R2*E/(R1+R2)
  E_Th    240 I2=R2*I/(R1+R2)
          250 E2=R1*R2*I/(R1+R2)
         ⌊260 ET=E1+E2
 Output  ⌈270 PRINT "Using Thevenin's theorem:"
 R_Th &   280 PRINT "Rth=";RT;"ohms"
 E_Th    ⌊290 PRINT "and Eth=";ET;"volts"
  Title  ⌈300 PRINT:PRINT "The following is a plot of PL vs RL:"
         ⌊310 PRINT:PRINT TAB(5);
  Calc.  ⌈320 PM=ET^2/(4*RT)
         ⌊330 SF=1
         ⌈340 FOR PO=INT(PM/10) TO PM STEP INT(PM/10)
  P_L     350 PRINT USING "####",PO;          ⌉
  axis    360 NEXT PO                          ⌡ Calc.
         ⌊370 PRINT "  PL(watts)"
         ⌈380 FOR RL=RT/4 TO 4*RT STEP RT/4
          390 IL=ET/(RT+RL)
          400 VL=IL*RL
  R_L     410 PL=IL^2*RL
  Loop    420 PO=INT(5+PL*SF+0.5) :REM Indent of 5  ⌉
          430 PRINT USING " ###",RL;←R_L axis       ⌡ Plot points
          440 PRINT TAB(PO);"*"
         ⌊450 NEXT RL
          460 PRINT "RL(ohms)" ←R_L axis
          470 PRINT
          480 END
```

READY

```
RUN

For the two-loop circuit of Figure 9-5
enter the following data:

R1=? 20
R2=? 1E30
R3=? 5
Supply voltage, E=? -10
and supply current, I=? 4

Using Thevenin's theorem:
Rth= 25 ohms
and Eth= 70 volts

The following is a plot of PL vs RL:

          4   8  12  16  20  24  28  32  36  40  44  48   PL(watts)
     6                                *
    13                                               *
    19                                                  *
    25                                                   *
    31                                                  *
    38                                                 *
    44                                               *
    50                                             *
    56                                          *
    63                                       *
    69                                      *
    75                                    *
    81                                 *
    88                               *
    94                              *
   100                          *
 RL(ohms)


    READY
```

the R_L axis. On the next line the maximum power level is determined by substituting the values of E_{Th} and R_{Th} into the general equation for maximum power. In this case, $P_M = 49$W is sufficiently less than the maximum number of divisions available to permit choosing a scale factor of one (SF = 1), which would allocate one watt per division of the horizontal scale. Of course, the plot is shifted by five before the count is initiated. In other words, a plot point of $P_L = 4$W is actually at an indent of $5 + 4 = 9$ spaces. On line 340 the axis of P_L, is printed in increments of $\text{INT}(P_{max}/10 = \text{INT}(49/10) = \text{INT}(4.9) = 4$ watts and extends from 4 watts to 48 watts. Even though the maximum power is greater than 48 watts ($=49$W) and appears on line 340, it is not an increment of 4 watts and

therefore will not appear as an axis value. The format statement of line 350 requires that a double-digit number such as 12 be printed with the 2 on the PO value and the one on the previous division. The number 16 will appear with the 6 on the $16 + 5 = 21$st division, and so on. At the completion of the LOOP defined by lines 340 through 360, the axis label, P_L (watts), will be printed two spaces (as defined by line 370) to the right of the last axis entry.

On lines 390 through 410, the required calculations are performed and the plot points defined by line 420. Note the presence of the required indent of 5 and the 0.5 to insure that the proper integer value is determined for PO on line 420. Line 430 will then label the resistance axis using the format indicated. Statement 440 will then print a plot point at each integer value of PO and continue the LOOP routine until the range of R_L is completed. The axis will be labeled by line 460. The resulting plot clearly indicates that maximum power is delivered to the load when $R_L = R_{Th} = 25$ ohms. However, the maximum power plot point appears to be at 50 watts rather than 49 watts. This is because the quote before the star on line 440 will introduce an additional space before the star is printed. Rather than correct the plot by adding a statement such as PO = PO-1 before line 440, it was left as is because a number of computer systems do not introduce the shift and this routine will result in an exact plot.

The last system to be analyzed using Thevenin's theorem appears in Figure 9.12. It is a general network that can be reduced to a multitude of standard forms by substituting open- and short-circuit equivalents for some of the resistor values.

The value of R_{Th} is determined in Program 9-5 on lines 300 through 340 using the configuration of Figure 9.13.

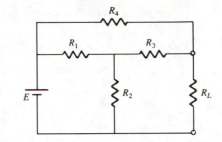

FIGURE 9.12

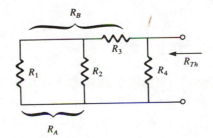

FIGURE 9.13

```
 10 REM *****  PROGRAM 9-5  *****
 20 REM ********************************************
 30 REM Program  to solve multi-loop network of
 40 REM Figure 9-12 using the Thevenin's theorem.
 50 REM ********************************************
 60 REM
100 PRINT "For the network of Figure 9-12"
110 PRINT "enter the following data:"
120 REM In this program a resistor can be replaced by
130 REM an open-circuit (1E30) or short-circuit (0) equivalent.
140 PRINT
150 INPUT "R1=";R1
160 INPUT "R2=";R2
170 INPUT "R3=";R3
180 INPUT "R4=";R4
190 INPUT "RL=";RL
200 INPUT "Supply voltage, E=";E
210 GOSUB 300 :REM Calculate Rth
220 GOSUB 350 :REM Calculate Eth
230 PRINT
240 PRINT "The Thevenin values are:"
250 PRINT "Rth=";RT;"ohms"
260 PRINT "and Eth=";ET;"volts"
270 PRINT "with IL=";ET/(RT+RL);"amps"
280 PRINT "and VL=";(ET*RL)/(RT+RL);"volts"
290 END
```

Input Data — lines 150–200
Calc. & Output Control — lines 240–280

(Continued)

```
     ┌ 300 REM Module to calculate Rth
     │ 310 RA=R1*R2/(R1+R2)
$R_{Th}$ │ 320 RB=R3+RA
     │ 330 RT=R4*RB/(R4+RB)
     └ 340 RETURN
     ┌ 350 REM Module to calculate Eth
     │ 360 RC=R1*(R3+R4)/(R1+R3+R4)
$E_{Th}$ │ 370 I2=E/(RC+R2)
     │ 380 I3=R1*I2/(R1+R3+R4)
     │ 390 ET=I3*R3+I2*R2
     └ 400 RETURN
```

READY

RUN

```
For the network of Figure 9-12
enter the following data:

R1=? 4E3
R2=? 8E3
R3=? 4E3
R4=? 1E3
RL=? 5E3
Supply voltage, E=? 16.8

The Thevenin values are:
Rth= 869.5652 ohms
and Eth= 16.0696 volts
with IL= 2.7378E-03 amps
and VL= 13.6889 volts
```

READY

RUN

```
For the network of Figure 9-12
enter the following data:

R1=? 6
R2=? 3
R3=? 8
R4=? 1E30
RL=? 12
Supply voltage, E=? 12

The Thevenin values are:
Rth= 10 ohms
and Eth= 4 volts
with IL= .1818 amps
and VL= 2.1818 volts
```

READY

The value E_{Th} is determined using the   and redrawn form of Figure 9.14, where

$$R_C = R_1 \| (R_3 + R_4),$$

with

$$I_2 = E/(R_C + R_2) \text{ [Ohm's law]}$$

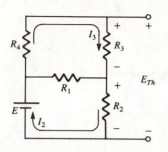

FIGURE 9.14

$$I_3 = R_1(I_2)/(R_1 + R_3 + R_4)$$
[Current Divider rule]

and finally

$$E_{Th} = I_3R_3 + I_2R_2$$

on line 390.

In the first run, the network is complete, while in the second run, the network analyzed

appears in Figure 9.15 because R_4 is approximated by an open-circuit. A number of other possible variations are possible, starting with the network of Figure 9.12.

The last program (9-6) of the section and chapter will test your ability to apply Thevenin's theorem to the network of Figure 9.12. The similarity with other testing programs introduced in this manual should make the program and resulting run fairly easy to understand.

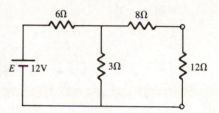

FIGURE 9.15

```
10 REM *****  PROGRAM 9-6  *****
20 REM **************************************
30 REM Program to test the application of
40 REM Thevenin's theorem to the network
50 REM of Figure 9-12.
60 REM **************************************
70 REM
100 PRINT "Determine Rth and Eth for the network"
110 PRINT "of Figure 9-12 for the following values:"
120 PRINT
130 GOSUB 450 :R1=R :REM Random selection of resistor value
140 PRINT "R1=";R1;"ohms"
150 GOSUB 450 :R2=R
160 PRINT "R2=";R2;"ohms"
170 GOSUB 450 :R3=R
180 PRINT "R3=";R3;"ohms"
190 GOSUB 450 :R4=R
200 PRINT "R4=";R4;"ohms"
210 E=10*INT(1+5*RND(X)) :REM Random selection of supply voltage
220 PRINT "and supply voltage E=";E;"volts"
230 PRINT
240 GOSUB 400 :REM calculate Rth
250 INPUT "Rth=";RX
260 IF ABS(RX-RT)<ABS(0.01*RT) THEN PRINT "Correct" :GOTO 280
270 PRINT "No, it's";RT;"ohms"
280 GOSUB 500 :REM Calculate Eth
290 INPUT "Eth=";EX
300 IF SGN(EX)<>SGN(ET) THEN GOTO 320
310 IF ABS(EX-ET)<ABS(0.01*ET) THEN PRINT "Correct" :GOTO 330
320 PRINT "No, it's";ET;"volts"
330 PRINT
340 INPUT "More(YES or NO)";A$
350 IF A$="YES" THEN GOTO 100
360 PRINT "Goodbye."
370 END
```

(Continued)

```
400 REM Module to calculate Rth
410 RA=R1*R2/(R1+R2)
420 RB=R3+RA
430 RT=R4*RB/(R4+RB)
440 RETURN
450 REM Module to pick R between 10 to 1000 ohms
460 R=10*INT(1+100*RND(X))
470 RETURN
500 REM Module to calculate Eth
510 RC=R1*(R3+R4)/(R1+R3+R4)
520 I2=E/(RC+R2)
530 I3=R1*I2/(R1+R3+R4)
540 ET=I3*R3+I2*R2
550 RETURN
```

```
READY


RUN

Determine Rth and Eth for the network
of Figure 9-12 for the following values:

R1= 710 ohms
R2= 530 ohms
R3= 360 ohms
R4= 390 ohms
and supply voltage E= 10 volts

Rth=? 249
No, it's 245.6197 ohms
Eth=? 7.88
Correct

More(YES or NO)? YES
Determine Rth and Eth for the network
of Figure 9-12 for the following values:

R1= 890 ohms
R2= 550 ohms
R3= 990 ohms
R4= 730 ohms
and supply voltage E= 20 volts

Rth=? 471.3
Correct
Eth=? 15.62
Correct

More(YES or NO)? NO
Goodbye.


READY
```

EXERCISES

Write a program to perform the following tasks:

1. Determine the current I_L and voltage V_L

for the network of Figure 9.16 using superposition. Print out the contributions of each source to both quantities.

2. Determine the Thevenin equivalent net-

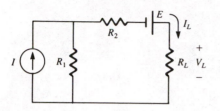

FIGURE 9.16

work for the components of Figure 9.16 external to R_L. Print out the resulting current I_L for a specified value of R_L.

3. Tabulate R_L, I_L, P_L and $\eta\%$ for R_L from $R_{Th}/5$ to $2\,R_{Th}$ in increments of $R_{Th}/5$ for the network of Figure 9.16.

4. Plot P_L vs. R_L for the range of problem 9-3. Label the R_L axis.

5. Plot

$$\eta\% = \frac{P_L}{P_i} \times 100\% = \frac{I_L{}^2 R_L}{E_{Th}I_L} \times 100\%$$

$$= \frac{R_L}{R_L + R_{Th}} \times 100\% \text{ versus } R_L$$

for the range of problem 9-3. Label the $\eta\%$ and R_L axis.

6. Determine the Thevenin equivalent circuit for the network external to the resistor R_1 of Figure 9.16. Use a fixed value for R_L.

7. Determine the Thevenin equivalent circuit for the network external to the resistor R_C of Figure 8.5.

8. Determine the Thevenin equivalent circuit for the network external to the resistor R_6 of Figure 8.1.

9. If a current source of 5A is added in parallel with the 50-ohm resistors of Figure 7.11, determine the current through R_1 using superposition. Feel free to reduce the network to its simplest form before applying the theorem. Print out the contribution due to each source.

10. Develop a testing routine for the application of Thevenin's theorem to the resistor R_6 of Figure 8.1 with the additional option that the resistors R_1, R_3, and R_5 can be randomly selected as 0 ohms, and the resistors R_2 and R_3 as 1E30 ohms.

Mesh and Nodal Analysis

10

10.1   INTRODUCTION

We will review the MESH (LOOP) and NODAL methods of analysis in the same chapter because of their obvious similarities. General solutions will be provided using the matrix notation and determinants introduced in Chapter 8. Finally, the Gaussian Elimination Method (Pivotal Reduction) will also be introduced as a method that can be applied to determinants of any order.

10.2   MESH ANALYSIS

The first program (10-1) will provide a general solution for the network of Figure 10.1. Applying the mesh analysis technique will result in the following equations:

$$I_1(R_1 + R_3) - I_2 R_3 = E_1$$
$$I_2(R_2 + R_3) - I_1 R_3 = -E_2$$

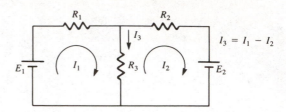

FIGURE 10.1

Applying determinants will result in:

$$I_1 = \frac{E_1(R_2 + R_3) - E_2 R_3}{D}$$

and $$I_2 = \frac{-E_2(R_1 + R_3) + E_1 R_3}{D}$$

with $$D = R_1 R_2 + R_1 R_3 + R_2 R_3$$

Note the equations for D, I_1, and I_2 on lines 200, 210, and 220 respectively. The results are particularly useful in that the sources can be reversed by introducing a negative sign, and more complex networks can be analyzed if you reduce them to the standard format of Figure

```
10 REM *****  PROGRAM 10-1  *****
20 REM *************************************************
30 REM Program to evaluate the loop currents for the
40 REM 2-loop network of Figure 10-1.
50 REM *************************************************
60 REM
100 PRINT "For the 2-loop network of Figure 10-1"
110 PRINT "enter the following data:"
120 PRINT
130 INPUT "R1=";R1
140 INPUT "R2=";R2
150 INPUT "R3=";R3
160 INPUT "Voltage, E1=";E1
170 INPUT "Voltage, E2=";E2
180 PRINT
190 REM Calculate I1 and I2
200 D=R1*R2+R1*R3+R2*R3
210 I1=(E1*(R2+R3)-E2*R3)/D
220 I2=(-E2*(R1+R3)+E1*R3)/D
230 PRINT "The loop currents are:"
240 PRINT "I1=";I1;"amps"
250 PRINT "I2=";I2;"amps"
260 END
```

Input — lines 130–170
Calc. — lines 190–220
Output — lines 230–250

READY

```
RUN
.
For the 2-loop network of Figure 10-1
enter the following data:

R1=? 2
R2=? 1
R3=? 4
Voltage, E1=? 2
Voltage, E2=? 6

The loop currents are:
I1=-1 amps
I2=-2 amps


READY


RUN

For the 2-loop network of Figure 10-1
enter the following data:

R1=? 1E3
R2=? 2.2E3
R3=? 3.3E3
Voltage, E1=? -5.4
Voltage, E2=? 8.6

The loop currents are:
I1=-4.5517E-03 amps
I2=-4.2947E-03 amps


READY

*Example 7.13, ICA
```

10.1. Note in each run that the resulting loop currents have directions opposite to that defined in Figure 10.1, due to the chosen component values. The current through R_3 can then be determined from $I_3 = I_1 - I_2$, which is $(-1 -(-2)) = (-1 + 2) = 1$ ampere in the direction indicated in Figure 10.1.

10.3 NODAL ANALYSIS

The second program is essentially the dual of Program 10-1 in that it will determine the two nodal voltages for the network of Figure 10.2. In fact, Program 10-2 can almost be determined

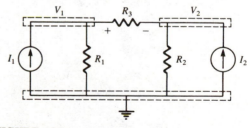

FIGURE 10.2

directly from Program 10-1 if the appropriate resistances and voltages sources were replaced by conductances and current sources. It is, of course, necessary to determine the conductance levels on line 200 and there are a few

```
         10 REM *****  PROGRAM 10-2  *****
         20 REM ******************************************
         30 REM Program to evaluate the nodal voltages
         40 REM of the two node network of Figure 10-2.
         50 REM ******************************************
         60 REM
         100 PRINT "For a two node network of Figure 10-2"
         110 PRINT "enter the following data:"
         120 PRINT
       ┌ 130 INPUT "R1=";R1
       │ 140 INPUT "R2=";R2
Input  │ 150 INPUT "R3=";R3
       │ 160 INPUT "Current, I1=";I1
       └ 170 INPUT "Current, I2=";I2
         180 PRINT
       ┌ 190 REM Calculate V1 and V2
       │ 200 G1=1/R1 :G2=1/R2 :G3=1/R3
Calc.  │ 210 D=G1*G2+G1*G3+G2*G3
       │ 220 V1=(I1*(G2+G3)+I2*G3)/D
       └ 230 V2=(I2*(G1+G3)+I1*G3)/D
       ┌ 240 PRINT "The nodal voltages are:"
Output │ 250 PRINT "V1=";V1;"volts"
       └ 260 PRINT "V2=";V2;"volts"
         270 END
```

READY

RUN

```
.
For a two node network of Figure 10-2
enter the following data:

R1=? 6
R2=? 4
R3=? 3
Current, I1=? -2
Current, I2=? 3

The nodal voltages are:
V1=-.9231 volts
V2= 4.6154 volts
```

READY

RUN

```
For a two node network of Figure 10-2
enter the following data:

R1=? 5.4E3
R2=? 3.9E3
R3=? 6.8E3
Current, I1=? 2E-3
Current, I2=? -0.5E-3

The nodal voltages are:
V1= 6.5236 volts
V2= 1.1385 volts
```

READY

*Example 7.21, ICA

important differences in sign on lines 220 and 230 as dictated by the following analysis of the network of Figure 10.2:

$$V_1[G_1 + G_3] - V_2 G_3 = I_1$$

$$V_2[G_2 + G_3] - V_1 G_3 = I_2$$

and using determinants:

$$V_1 = \frac{I_1[G_2 + G_3] + I_2 G_3}{D}$$

$$V_2 = \frac{I_2[G_1 + G_3] + I_1 G_3}{D}$$

with $D = G_1 G_2 + G_1 G_3 + G_2 G_3$

Negative results indicate that the nodal voltages are negative with respect to ground potential. The voltage V_{R_3} has the polarity defined by Figure 10.2 and a magnitude determined by $V_1 - V_2$. For the first run, it is $(-0.9231 - 4.6154) = -5.5385$ volts, while for the second run it is $6.5236 - 1.1385 = +5.3878$ volts.

10.4 MESH ANALYSIS— MATRIX NOTATION

We will now apply mesh analysis to the network of Figure 10.3 using matrix notation and mod-

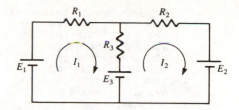

FIGURE 10.3

ules to perform the computation, input, and output routines.

The addition of the voltage source E_3 will add some complexity to the solution, but the resulting equations are very similar to those in Program 10-1. The resulting loop equations are:

$$(R_1 + R_3)I_1 + (-R_3)I_2 = E_1 - E_3$$

$$(-R_3)I_1 + (R_2 + R_3)I_2 = E_3 - E_2$$

which can be written using matrix notation as:

$$A(1,1)I_1 + A(1,2)I_2 = V(1)$$
$$A(2,1)I_1 + A(2,2)I_2 = V(2)$$

resulting in

$$A(1,1) = R_1 + R_3, A(1,2) = R_3, A(2,1) = -R_3$$

$$A(2,2) = R_2 + R_3, V(1) = E_1 - E_3$$
and $V(2) = E_3 - E_2$

as defined by lines 370 through 420 of Program 10-3.

```
10 REM *****   PROGRAM 10-3   *****
20 REM ***********************************
30 REM Program to determine the loop (mesh)
40 REM currents of Figure 10-3.
50 REM ***********************************
60 REM
100 DIM A(2,2),V(2)
110 PRINT "This program calculates the loop (mesh) currents"
120 PRINT "of a 2 loop network."
130 PRINT
140 GOSUB 300 :REM Get input data
150 GOSUB 500 :REM Calculate loop (mesh) currents
160 GOSUB 600 :REM Print output data
170 PRINT:PRINT
180 INPUT "More(YES or NO)";A$
190 IF A$="YES" THEN GOTO 130
200 END
```

(Continued)

```
          300 REM Module to accept input data
         ┌310 INPUT "E1=";E1
          320 INPUT "R1=";R1
          330 INPUT "E2=";E2
Input    │340 INPUT "R2=";R2
          350 INPUT "E3=";E3
         └360 INPUT "R3=";R3
         ┌370 A(1,1)=R1+R3
          380 A(1,2)=-R3
Define   │390 A(2,1)=-R3
Matrix    400 A(2,2)=R2+R3
          410 V(1)=E1-E3
         └420 V(2)=E3-E2
          430 RETURN
         ┌500 REM Module to calculate I1 & I2
          510 D=A(1,1)*A(2,2)-A(2,1)*A(1,2)
Calc.    │520 I1=(V(1)*A(2,2)-V(2)*A(1,2))/D
         └530 I2=(V(2)*A(1,1)-V(1)*A(1,2))/D
          540 RETURN
         ┌600 REM Module to output data
Output   │610 PRINT "Loop 1 current I1=";I1;"amps"
         └620 PRINT "Loop 2 current I2=";I2;"amps"
          630 RETURN


READY

RUN

This program calculates the loop (mesh) currents
of a 2 loop network.

E1=? 10
R1=? 20
E2=? 20
R2=? 5
E3=? 5
R3=? 1
Loop 1 current I1= .12 amps
Loop 2 current I2=-2.48 amps


More(YES or NO)? YES

E1=? -5.4
R1=? 1E3
E2=? 8.6
R2=? 2.2E3
E3=? 0
R3=? 3.3E3
Loop 1 current I1=-4.5517E-03 amps
Loop 2 current I2=-4.2947E-03 amps


More(YES or NO)? NO
```

In determinant form, the solution for I_1 is

$$I_1 = \frac{\begin{vmatrix} V(1) & A(1,2) \\ V(2) & A(2,2) \end{vmatrix}}{\begin{vmatrix} A(1,1) & A(1,2) \\ A(2,1) & A(2,2) \end{vmatrix}}$$

$$= \frac{V(1)A(2,2) - V(2)A(1,2)}{D}$$

with $D = A(1,1)A(2,2) - A(2,1)A(1,2)$

as appearing on lines 510 and 520. I_2 is similarly defined by line 530. The last subroutine of the sequence 600 through 620 will then print out the results. A comparison of this program with Program 10-1 reveals that the solution is not in terms of the network variables, but in terms of the matrix coefficients. In addition, the program is broken down into subroutines for additional control and clarity. The second run examines a network essentially identical in structure (due to $E_3 = 0$) to the network of Figure 10.1.

10.5 NODAL ANALYSIS— MATRIX NOTATION

The network of Figure 10.4 is the same as in section 10.2, except for the addition of the current source I_3—effectively the dual of the network of Figure 10.3.

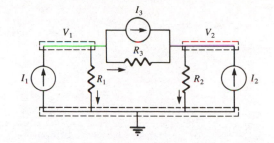

FIGURE 10.4

The current through R_1 and R_2 is defined by V_1/R_1 and V_2/R_2, while the current through R_3 is given by $(V_1 - V_2)/R_3$. In the first run I_{R_3} = (20 − 10)/5 = 2 amperes, while in the second run it is (2.5994 − [−7.7273])/6.8 × 10^3 = 1.5186mA. It would be relatively easy to add these calculations to Program 10-4.

```
10 REM *****  PROGRAM 10-4  *****
20 REM *********************************************
30 REM Program to determine the nodal voltages
40 REM for the network of Figure 10-4.
50 REM *********************************************
60 REM
100 DIM A(2,2),I(2)
110 PRINT "This program calculates the nodal voltages "
120 PRINT "for a two node network."
130 PRINT
140 GOSUB 300 :REM Get input data
150 GOSUB 500 :REM Calculate the nodal voltages
160 GOSUB 600 :REM Print output data
170 PRINT:PRINT
180 INPUT "More(YES or NO)";A$
190 IF A$="YES" THEN GOTO 130
200 END
```

Input
```
300 REM Module to accept input data
310 INPUT "I1=";I1
320 INPUT "R1=";R1 :G1=1/R1
330 INPUT "I2=";I2
340 INPUT "R2=";R2 :G2=1/R2
350 INPUT "I3=";I3
360 INPUT "R3=";R3 :G3=1/R3
```

Define Matrix
```
370 A(1,1)=G1+G3
380 A(1,2)=-G3
390 A(2,1)=-G3
400 A(2,2)=G2+G3
410 I(1)=I1-I3
420 I(2)=I2+I3
430 RETURN
```

Calc.
```
500 REM Module to calculate V1 & V2
510 D=A(1,1)*A(2,2)-A(2,1)*A(1,2)
520 V1=(I(1)*A(2,2)-I(2)*A(1,2))/D
530 V2=(I(2)*A(1,1)-I(1)*A(1,2))/D
540 RETURN
```

(Continued)

```
      ┌ 600 REM Module to output data
Output│ 610 PRINT "Node 1 voltage V1=";V1;"volts"
      │ 620 PRINT "Node 2 voltage V2=";V2;"volts"
      └ 630 RETURN


READY

RUN

This program calculates the nodal voltages
for a two node network.

I1=? 2
R1=? 5
I2=? 0
R2=? 2
I3=? 3
R3=? 5
Node 1 voltage V1= 20 volts
Node 2 voltage V2= 10 volts


More(YES or NO)? YES

I1=? 2E-3
R1=? 5.4E3
I2=? -3.5E-3
R2=? 3.9E3
I3=? 0
R3=? 6.8E3
Node 1 voltage V1= 2.5994 volts
Node 2 voltage V2=-7.7273 volts


More(YES or NO)? NO


READY
```

10.6 NODAL ANALYSIS— LADDER NETWORK

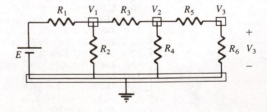

FIGURE 10.5

The next program will determine the voltage V_3 for the three-section network of Figure 10.5. The voltage source is first converted to a current source, as shown in Figure 10.6, and nodal analysis applied in the standard format. The resulting equations are

$$[G_1 + G_2 + G_3]V_1 \qquad\qquad [-G_3]V_2 \qquad\qquad 0 \qquad\quad = I$$
$$[-G_3]V_1 + [G_3 + G_4 + G_5]V_2 \qquad\quad [-G_5]V_3 = 0$$
$$0 \qquad\qquad\qquad [-G_5]V_2 [-G_5 - G_6]V_3 = 0$$

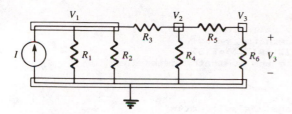

FIGURE 10.6

which result in the defined matrix coefficients appearing on line 280 through line 370 of Program 10-5. During the first pass through subroutine 500 (initiated by line 410), the value of the denominator determinant D is calculated using the equations first appearing in Program

8-1. Line 430 will then establish the values of A (1,3), A (2,3), and A (3,3) as equal to the supply current values (defined by line 370). The numerator determinant for V_3 can be determined using the equations of lines 510 through 540. Once V_3 is determined by line 450, it is printed out by line 480. Consider the enormous amount of time and energy saved by a program which performs all the calculations required here—especially when working with all the decimal values for the conductances and coming up with a solution accurate to the fourth-place, as appearing in the run of the program. The nodal voltages V_1 and V_2 could be determined directly by changing line 430 for each voltage or adding a few lines to the program.

```
10 REM *****   PROGRAM 10-5  *****
20 REM *********************************************
30 REM Program to determine the nodal voltage V3
40 REM for the general three-node network
50 REM of Figure 10-5.
60 REM *********************************************
70 REM
100 DIM A(3,3),I(3)
110 PRINT "This program calculates the nodal voltage V3 "
120 PRINT "for the network of Figure 10-5."
130 PRINT
140 GOSUB 200 :REM Get input data
150 GOSUB 400 :REM Calculate the node voltage V3
160 GOSUB 470 :REM PRINT Output data
170 PRINT:PRINT:INPUT "More (YES or NO)";A$
180 IF A$="YES" THEN GOTO 130
190 END
200 REM Module to accept input data
210 INPUT "E=";E
220 INPUT "R1=";R1 :G1=1/R1
230 INPUT "R2=";R2 :G2=1/R2
240 INPUT "R3=";R3 :G3=1/R3
250 INPUT "R4=";R4 :G4=1/R4
260 INPUT "R5=";R5 :G5=1/R5
270 INPUT "R6=";R6 :G6=1/R6
280 A(1,1)=G1+G2+G3
290 A(1,2)=-G3
300 A(1,3)=0
310 A(2,1)=-G3
320 A(2,2)=G3+G4+G5
330 A(2,3)=-G5
340 A(3,1)=0
350 A(3,2)=-G5
360 A(3,3)=G5+G6
370 I(1)=E/R1 :I(2)=0 :I(3)=0
380 RETURN
```

Input Data — lines 210–270
Define Matrix — lines 280–370

(Continued)

```
           400 REM Module to calculate V3
           410 GOSUB 500 :REM Evaluate determinant
           420 DE=D :REM Save the denominator value
           430 A(1,3)=I(1) :A(2,3)=I(2) :A(3,3)=I(3)
           440 GOSUB 500 :REM Calculate the numerator determinant
      V3   450 V3=D/DE
           460 RETURN
  Output ⎡ 470 REM Module to output data
         ⎣ 480 PRINT "The node 3 voltage V3=";V3;"volts"
           490 RETURN
         ⎡ 500 REM Module to evaluate a 3x3 determinant
           510 D1=A(1,1)*(A(2,2)*A(3,3)-A(3,2)*A(2,3))
  Calc.    520 D2=A(2,1)*(A(1,2)*A(3,3)-A(3,2)*A(1,3))
  D        530 D3=A(3,1)*(A(3,1)*A(1,2)-A(2,3)*A(1,3))
         ⎣ 540 D=D1-D2+D3
           550 RETURN


  READY


  RUN

  *
  This program calculates the nodal voltage V3
  for the network of Figure 10-5.

  E=? 240
  R1=? 12
  R2=? 6
  R3=? 4
  R4=? 6
  R5=? 1
  R6=? 2
  The node 3 voltage V3= 10.6667 volts


  More (YES or NO)? NO


  READY


  *Example 7.24, ICA
```

10.7   MESH ANALYSIS— FOUR LOOP SYSTEM

A module for evaluating determinants of any order will be introduced in this section to permit a complete analysis of a network with four-loop currents, as shown in Figure 10.7. The resulting equations require that the fourth-order determinant be evaluated before the various loop currents can be determined. The module of Program 10-6, starting at 500 and extending through 910, will perform this task using the Gaussian Elimination Method (or Pivotal Re-

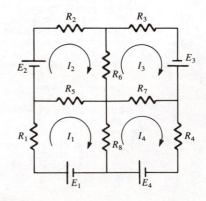

FIGURE 10.7

```
 10 REM *****  PROGRAM 10-6  *****
 20 REM *********************************************
 30 REM Program to determine the mesh currents for the
 40 REM general four-loop network of Figure 10-7.
 50 REM *********************************************
 60 REM
100 DIM A(4,4),E(4)
110 INPUT "E(1)=";E(1)
120 INPUT "E(2)=";E(2)
130 INPUT "E(3)=";E(3)
140 INPUT "E(4)=";E(4)
150 PRINT
160 FOR KK=1 TO 4  :REM Repeat for each loop current
170 RESTORE :REM Prepare to re-read resistance coefficients
180 IF KK=1 THEN PRINT "For the following resistance coefficients:"
190 FOR I=1 TO 4
200 FOR J=1 TO 4
210 READ A(I,J)
220 IF KK=1 THEN PRINT A(I,J); :REM Print resistance coefficients
230 NEXT J
240 IF KK=1 THEN PRINT
250 NEXT I
260 GOSUB 500
270 IF KK=1 THEN PRINT "Denominator determinant is";D
280 PRINT
290 DE=D :REM Save D as denominator
300 REM Now read voltage values to solve for the mesh currents
310 PRINT "For the following inserted voltage coefficients:"
320 FOR I=1 TO 4
330 A(I,KK)=E(I)
340 PRINT "A(";I;",";KK;")=";A(I,KK)
350 NEXT I
360 GOSUB 500 :REM Now calculate numerator determinant
370 PRINT "Numerator determinant is";D
380 PRINT
390 PRINT "Value of I";KK;"is";D/DE;"amps"
400 NEXT KK
410 END
420 DATA 14,-5,0,-8
430 DATA -5,13,-6,0
440 DATA 0,-6,16,-7
450 DATA -8,0,-7,19
500 REM Module to evaluate 4X4 determinant
510 REM using the Gaussian elimination method
520 S=1 :REM Sign, S is positive
530 N=4  :REM Order of determinant (4x4)
540 L=N-1 :REM Number of rows to be reduced
550 REM Start overall loops for (N-1) pivots
560 FOR I=1 TO L
570 REM Find the largest term in I-th column for pivot
580 B=0 :REM Initialize B=0
590 FOR K=I TO N
600 T=ABS(A(K,I)) :REM T=value of each term in I-th column
610 IF T>B THEN B=T:L=K :REM Determine magnitude and row of largest term
620 NEXT K
630 REM Check whether a non-zero term has been found
640 IF B=0 THEN PRINT "A non-zero term was not found" :END
650 REM Compare row with largest term (L) to row being reduced (I)
660 IF I<>L THEN GOTO 680
670 IF I=L THEN GOTO 750 :REM Largest term in row being reduced
680 REM Since I is not equal to L, switch rows I and L
690 S=-S :REM Required change when rows interchanged
```

(Continued)

```
700 FOR J=1 TO N
710 TM=A(I,J)
720 A(I,J)=A(L,J)
730 A(L,J)=TM
740 NEXT J
750 REM Start pivotal reduction
760 P=A(I,I) :REM P=Pivot term
770 NR=I+1 :REM NR=Next Row
780 REM For each row below the pivot row do the following:
790 FOR J=NR TO N :REM J=row
800 C=A(J,I)/P :REM C=Constant
810 FOR K=I TO N :REM For each column including the I-th column
820 A(J,K)=A(J,K)-C*A(I,K)
830 NEXT K
840 NEXT J
850 NEXT I
860 REM Pivotal reduction complete- now compute determinant
870 D=S :REM Initially set the determinant D=+1 or -1
880 FOR I=1 TO N
890 D=D*A(I,I) :REM Determines product of all diagnol terms
900 NEXT I
910 RETURN
```

READY

<u>RUN</u>

```
E(1)=? 10
E(2)=? 20
E(3)=? 30
E(4)=? 40


For the following resistance coefficients:
 14 -5  0 -8
 -5 13 -6  0
  0 -6 16 -7
 -8  0 -7 19
Denominator determinant is 1.6091E+04


For the following inserted voltage coefficients:
A( 1 , 1 )= 10
A( 2 , 1 )= 20
A( 3 , 1 )= 30
A( 4 , 1 )= 40
Numerator determinant is 9.2395E+04


Value of I 1 is 5.742 amps


For the following inserted voltage coefficients:
A( 1 , 2 )= 10
A( 2 , 2 )= 20
A( 3 , 2 )= 30
A( 4 , 2 )= 40
Numerator determinant is 9.8011E+04


Value of I 2 is 6.0911 amps


For the following inserted voltage coefficients:
A( 1 , 3 )= 10
A( 2 , 3 )= 20
```

```
A( 3 , 3 )= 30
A( 4 , 3 )= 40
Numerator determinant is 9.1303E+04

Value of I 3 is 5.6742 amps

For the following inserted voltage coefficients:
A( 1 , 4 )= 10
A( 2 , 4 )= 20
A( 3 , 4 )= 30
A( 4 , 4 )= 40
Numerator determinant is 8.032E+04

Value of I 4 is 4.9916 amps


READY
```

duction Method, as it is sometimes called). We do not have the space for a detailed description of this common technique, but there are a number of excellent texts available that cover it. The method essentially takes a determinant of the form appearing in Figure 10.8 and converts it to one having the parameters indicated in Figure 10.9. The magnitude of the solution of a determinant having the format of Figure 10.9 is then equal to the product of the diagonal terms, namely

$$D = S(a'_{11} \times a''_{22} \times a'''_{33} \times a''''_{44})$$

where S is a sign determined by the number of rows that are interchanged.

It is possible that the parameters of the first row may remain the parameters of the first row of Figure 10.9, but such an occurence requires that a_{11} of Figure 10.8 be the largest parameter in the first column as defined by lines 570 through 620. If not, the prime values of the first row of Figure 10.9 must be determined.

$$\begin{vmatrix} a_{11} & a_{12} & a_{13} & a_{14} \\ a_{21} & a_{22} & a_{23} & a_{24} \\ a_{31} & a_{32} & a_{33} & a_{34} \\ a_{41} & a_{42} & a_{43} & a_{44} \end{vmatrix}$$

FIGURE 10.8

$$\begin{vmatrix} a'_{11} & a'_{12} & a'_{13} & a'_{14} \\ 0 & a''_{22} & a''_{23} & a''_{24} \\ 0 & 0 & a'''_{33} & a'''_{34} \\ 0 & 0 & 0 & a''''_{44} \end{vmatrix}$$

FIGURE 10.9

The term a_{11}, and any diagonal term that results from a reduction to the format defined by Figure 10.9, is called a pivot, since multiples of its magnitudes are employed in setting all the parameters below the pivot to zero. This is where the often-used title, Pivotal-Reduction, comes from. An expanded number of REM statements were included to describe the process to the degree possible within the program itself. Note that the I loop extends from 560 to 850, with both a K and J loop within. In sequence, the largest value in a column is first determined by an exchange of rows if the pivotal element is not the largest. The new magnitude of the rows below the pivotal row are then determined using the equation appearing on line 820. The same process is repeated for each row until all the terms below the pivot term are zero. The I loop will then increase by one, and the next pivotal row and pivotal term chosen. The Gaussian Elimination Method can also be applied to 2×2, 3×3, 5×5 or larger determinants by changing the number N of line 530. If you feel you have a recurring need for such a program or subroutine, you should carefully read a detailed description in any one of the many available texts, so that you understand the reason for particular maneuvers.

Note the dimension statement of line 100 defining a number of subscripted variables to appear in the program, and note the unique way the data appearing on lines 420 through 450 is read, using the read commands of lines 190 through 250. The search for the data value will automatically move from line 210 to read $A(1,1) = 14$, $A(1,2) = -5$, $A(1,3) = 0$, and $A(1,4) = -8$ for the initial value of $I = 1$ and the J LOOP passing through the sequence 1 through 4. The value of I will then be incremented, resulting in $A(2,1)$, $A(2,2)$, etc.

Of course, the data values must be typed in on lines 420 through 450 before the program is RUN, or the values of $A(I, J)$ will all be zero. In addition, the data points are read in sequence, so there is no need to place them in four rows, except that it clearly defines the value read for each $A(I, J)$ value.

Note that the KK LOOP extends from 160 all the way through 400. Each pass will determine one of the loop currents for the network of Figure 10.7. Each run requires that the data points be reread on line 170 to be sure that the program either starts at the first data point or does not read beyond the last data point and pick up erroneous data.

Line 220 states that the data points will be printed out during the first pass only as indicated by the provided run of the program. In addition, the value of the denominator determinant will only be printed out for the first pass

as dictated by line 270. Lines 300 through 350 will list the values of supply voltage using a LOOP routine to define the row and column, rather than listing each separately. Line 360 will initiate the calculation of the numerator determinant (with the source values properly included in the determinant) and then print out the results for the numerator determinant and the desired loop current.

As noted in the provided run, the matrix location of the voltage sources is provided to demonstrate how they are relocated for each loop current. Note the excellent level of accuracy obtained for such a lengthy, demanding set of calculations. It is likely that performing all the calculations by hand would result in an incorrect answer for the loop current more than 50% of the time. However, once the program is properly written and entered, the user can assume 100% accuracy if the data is not particularly unusual and entered correctly.

Note that the equations to solve the network were not included in the program. The resulting coefficients were entered as data points on lines 420 through 450, and the voltage source values on line 110 through 140. For completeness, the general equations for the system of Figure 10-7 are provided below.

Note the symmetry of the coefficients about the diagonal axis. That is, $A_{21} = A_{12}$, $A_{31} = A_{13}$, etc. For the provided run, $R_1 = 1\Omega$, $R_2 = 2\,\Omega$, $R_3 = 3\Omega$, $R_4 = 4\Omega$, $R_5 = 5\Omega$, $R_6 = 6\Omega$, $R_7 = 7\Omega$, $R_8 = 8\Omega$, $E_1 = 10V$, $E_2 = 20V$, $E_3 = 30V$, and $E_4 = 40V$. The result is that $A_{11} = (R_1 + R_5 + R_8) = 14$, $A_{22} = (R_2 + R_5 + R_6) = 13$, $A_{33} = (R_3 + R_6 + R_7) = 16$ and $A_{44} = (R_4 + R_7 + R_8) = 19$.

$$
\begin{aligned}
[R_1 + R_5 + R_8]I_1 + & [-R_5]I_2 + & 0 & + & [-R_8]I_4 = E_1 \\
[-R_5]I_1 + [R_2 + R_5 + R_6]I_2 + & [-R_6]I_3 + & 0 & = E_2 \\
0 + & [-R_6]I_2 + [R_3 + R_6 + R_7]I_3 + & [-R_7]I_4 = E_3 \\
[-R_8]I_1 + & 0 + & [-R_7]I_3 + [R_4 + R_7 + R_8]I_4 = E_4
\end{aligned}
$$

10.8 TEST ROUTINE— NODAL ANALYSIS

The last program (10-7) of the chapter will test your skills in determining the nodal voltages for the network of Figure 10.4 using randomly generated values for the resistors and current sources. A negative sign for the current sources will be generated one out of three times (on an average basis) by line 910 through 930 of the subroutine at 900. The program itself is the same as Program 10-2 with the addition of the random generation modules and testing routines. Even though a minus sign appeared in each run, you can expect that a minus sign will appear in only one out of three runs (on an average basis).

```
10 REM *****  PROGRAM 10-7  *****
20 REM *************************************************
30 REM Program to test calculation of the nodal voltages
40 REM of the two-node network of Figure 10-4.
50 REM *************************************************
60 REM
100 DIM A(2,2),I(2)
110 PRINT "This program tests your calculations of the nodal"
120 PRINT "voltages V1 and V2 for the network of"
130 PRINT "Figure 10-4."
140 PRINT:PRINT "Randomly selected resistors are:"
150 GOSUB 800 :R1=R :G1=1/R1
160 PRINT "R1=";R1;"ohms"
170 GOSUB 800 :R2=R :G2=1/R2
180 PRINT "R2=";R2;"ohms"
190 GOSUB 800 :R3=R :G3=1/R3
200 PRINT "R3=";R3;"ohms"
210 PRINT
220 PRINT "and supply currents are:"
230 GOSUB 900 :I1=I
240 PRINT "I1=";I1;"amps"
250 GOSUB 900 :I2=I
260 PRINT "I2=";I2;"amps"
270 GOSUB 900 :I3=I
280 PRINT "I3=";I3;"amps"
290 PRINT
300 REM Determine array A(I,J)
310 A(1,1)=G1+G3
320 A(1,2)=-G3
330 A(2,1)=-G3
340 A(2,2)=G2+G3
350 I(1)=I1-I3
360 I(2)=I2+I3
370 REM Calculate nodal voltages V1 & V2
380 GOSUB 500
390 INPUT "Value of V1=";VX
395 IF SGN(V1)<>SGN(VX) THEN GOTO 410
400 IF ABS(V1-VX)<ABS(0.01*V1) THEN PRINT"Correct" :GOTO 420
410 PRINT "No,it's";V1;"volts"
420 PRINT
430 INPUT "Value of V2=";VX
435 IF SGN(V2)<>SGN(VX) THEN GOTO 450
440 IF ABS(V2-VX)<ABS(0.01*V2) THEN PRINT "Correct" :GOTO 460
450 PRINT "No,it's";V2;"volts"
460 PRINT
470 END
```

(Continued)

```
500 REM Module to calculate V1 & V2
510 D=A(1,1)*A(2,2)-A(2,1)*A(1,2)
520 V1=(I(1)*A(2,2)-I(2)*A(1,2))/D
530 V2=(I(2)*A(1,1)-I(1)*A(1,2))/D
540 RETURN
800 REM Module to randomly select a resistor value from 1 to 10
810 R=INT(1+10*RND(X))
820 RETURN
900 REM Module to randomly select a value of supply current
910 S=INT (1+3*RND(X)) :REM Select negative sign in 1 of 3 times
920 IF S=2 THEN SG=-1 ELSE SG=+1
930 I=SG*INT(1+5*RND(X)) :REM Range of magnitude = 1 to 5 amps
940 RETURN
```

READY

RUN

This program tests your calculations of the nodal
voltages V1 and V2 for the network of
Figure 10-4.

Randomly selected resistors are:
R1= 7 ohms
R2= 4 ohms
R3= 5 ohms

and supply currents are:
I1=-4 amps
I2= 3 amps
I3= 2 amps

Value of V1=? -14.8
Correct

Value of V2=? 4.5
Correct

READY

RUN

This program tests your calculations of the nodal
voltages V1 and V2 for the network of
Figure 10-4.

Randomly selected resistors are:
R1= 4 ohms
R2= 10 ohms
R3= 3 ohms

and supply currents are:
I1= 2 amps
I2=-1 amps
I3= 2 amps

Value of V1=? 2.8
No,it's 2.3529 volts

Value of V2=? 4.12
Correct

READY

EXERCISES

Write a program to perform each of the follow-ing tasks:

1. Determine the mesh currents I_1 and I_2 for the network of Figure 10.10.

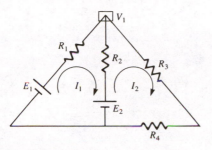

FIGURE 10.10

2. Convert the voltage sources of Figure 10.10 to current sources and determine the nodal voltage V_1. Then, print out the magnitude of the currents I_{R_1}, I_{R_2}, and I_{R_3}

3. Determine the source current for the three-section ladder network of Figure 8.1 using the mesh analysis approach and matrix nota-tion.

4. Convert the voltage source of Figure 8.1 to a current source and determine the nodal volt-age across the last element of the ladder using nodal analysis and matrix notation.

5. Find the source current for the bridge net-work of Figure 8.5 using mesh analysis and matrix notation.

6. Replace the voltage source of Figure 8.5 with a current source and determine the volt-age across the source using nodal analysis and matrix notation.

7. Convert each of the voltage sources of Figure 10.7 to a current source and determine all the nodal voltages using nodal analysis and matrix notation.

8. Develop a testing routine for applying mesh analysis to the network of Figure 10.1. Randomly generate the components of the net-work and generate a negative sign for each source on one out of three occasions (on an average basis). Limit each R to less than 1000 ohms but more than 10 ohms, and make the magnitude of E more than 5 volts but less than 50 volts.

9. Develop a testing routine for applying nodal analysis to the network of Figure 10.2. Randomly generate each R in the range indi-cated in problem 10-8 and limit I to the range $1A < I < 10A$.

10. Convert the 3×3 matrix of Figure 10.11a to one having the format of Figure 10.11b using your own set of commands. Feel free to borrow from the Pivotal Reduction method described in the chapter.

$$\begin{vmatrix} a_{11} & a_{12} & a_{13} \\ a_{21} & a_{22} & a_{23} \\ a_{31} & a_{32} & a_{33} \end{vmatrix} \qquad \begin{vmatrix} a'_{11} & a'_{12} & a'_{13} \\ 0 & a''_{22} & a''_{23} \\ 0 & 0 & a'''_{33} \end{vmatrix}$$

$$\text{(a)} \qquad\qquad \text{(b)}$$

FIGURE 10.11

Magnetism

11

11.1   INTRODUCTION

The study of magnetism can result in some interesting areas of application for computer techniques. A conversion routine will be developed to insure that the proper magnitude of a quantity is substituted into the equations appearing in the programs to follow. An interpolation routine will be used to permit transferring a non-linear continuous curve to discrete data points in a computer program. Finally, a "cut and try" approach, often applied when working with non-linear relationships, will be introduced to establish a foundation for analyzing such systems in the future.

11.2   MAGNETIC UNIT CONVERSIONS

Program 11-1 will perform the conversion process between a number of units frequently employed in the analysis of magnetic circuits. The provided run of the program reveals that, once the conversion to be performed is chosen,

the magnitude of the quantity to be converted is provided and the program will print out the magnitude of that quantity in the preferred system of units.

Note on line 100 that there will be 16 subscripted string variables labeled UN$ and 8 subscripted variables labeled CF. Lines 150 through 220 will identify the 16 values of UN$ as UN$(1) = inches, UN$(2) = square inches, etc. It is similar to the data read in for Program 10-6, except that we are now reading in string variables that happen to spell out a unit of measure. Line 270 holds the eight conversion factors necessary to convert from one unit of measure to the other. Only eight are required, since a conversion process in one direction will involve multiplying the given magnitude by a conversion factor, while the reverse process will involve dividing the given quantity by the same conversion factor. Lines 300 through 390 simply provide a listing of the conversions that can be performed. Note that, even though the list is typed in sequential order, the TAB statements will line up the possibilities in two straight columns in the printout.

```
10 REM *****  PROGRAM 11-1  *****
20 REM ******************************************************
30 REM Conversion between units in magnetic calculations.
40 REM ******************************************************
50 REM
100 DIM UN$(16),CF(8)
110 PRINT "This program will perform conversions"
120 PRINT "between a number of standard units"
130 PRINT "used in magnetic calculations."
140 PRINT
150 REM Read in unit and conversion factor data
160 FOR I=1 TO 16
170 READ UN$(I)
180 NEXT I
190 DATA inches,sq.inches,cm,sq.cm,lines
200 DATA Maxwells,lines/sq.inch,Maxwells/sq.cm,
210 DATA meters,sq.meters,meters,sq.meters,webers
220 DATA webers,Wb/sq.meter,Wb/sq.meter
230 REM Read in conversion factors
240 FOR I=1 TO 8
250 READ CF(I)
260 NEXT I
270 DATA 0.0254,6.452E-4,0.01,1E-4,1E-8,1E-8,0.155E-4,1E-4
280 PRINT "Which of the following conversions do you wish to perform?"
290 PRINT
```

Define UN$(I) (units) — lines 190–220
Conversion Factors — lines 230–270

```
 ┌─300 PRINT "(1) inches to meters";TAB(36);"( 9) meters to inches"
 │  310 PRINT "(2) sq.inches to sq.meters";TAB(36);"(10) sq.meters to inches"
 │  320 PRINT "(3) cm to meters";TAB(36);"(11) meters to cm"
Unit│  330 PRINT "(4) sq.cm to sq.meters";TAB(36);"(12) sq.meters to sq.cm"
Conversions│  340 PRINT "(5) lines to webers";TAB(36);"(13) webers to lines"
 │  350 PRINT "(6) Maxwells to webers";TAB(36);"(14) webers to Maxwells"
 │  360 PRINT "(7) lines/sq.inch to Wb/sq.meter";
 │  370 PRINT TAB(36);"(15) Wb/sq.meter to lines/sq.inch"
 │  380 PRINT "(8) Maxwell/sq.cm to Wb/sq.meter";
 └─390 PRINT TAB(36);"(16) Wb/sq.meter to Maxwells/sq.cm"
    400 PRINT
 ┌─410 PRINT TAB(15);"Choice (Enter 0 to stop)";
Choice│  420 INPUT C
 │  430 IF C=0 THEN END
 └─440 IF C<1 OR C>16 THEN GOTO 410
    450 PRINT
Input 460 INPUT "Value to convert";V
    470 IF C<=8 THEN A=V*CF(C) ELSE A=V*(1/CF(C-8))
Output┌─480 PRINT "Which converts to";A;
 └─490 IF C<=8 THEN PRINT UN$(C+8) ELSE PRINT UN$(C-8)
    500 PRINT
    510 GOTO 280
```

READY

RUN

This program will perform conversions
between a number of standard units
used in magnetic calculations.

Which of the following conversions do you wish to perform?

(1) inches to meters   (9) meters to inches
(2) sq.inches to sq.meters   (10) sq.meters to inches
(3) cm to meters   (11) meters to cm
(4) sq.cm to sq.meters   (12) sq.meters to sq.cm
(5) lines to webers   (13) webers to lines
(6) Maxwells to webers   (14) webers to Maxwells
(7) lines/sq.inch to Wb/sq.meter   (15) Wb/sq.meter to lines/sq.inch
(8) Maxwell/sq.cm to Wb/sq.meter   (16) Wb/sq.meter to Maxwells/sq.cm

                Choice (Enter 0 to stop)? 13

Value to convert? 50
Which converts to 5E+09 lines

Which of the following conversions do you wish to perform?

(1) inches to meters   (9) meters to inches
(2) sq.inches to sq.meters   (10) sq.meters to inches
(3) cm to meters   (11) meters to cm
(4) sq.cm to sq.meters   (12) sq.meters to sq.cm
(5) lines to webers   (13) webers to lines
(6) Maxwells to webers   (14) webers to Maxwells
(7) lines/sq.inch to Wb/sq.meter   (15) Wb/sq.meter to lines/sq.inch
(8) Maxwell/sq.cm to Wb/sq.meter   (16) Wb/sq.meter to Maxwells/sq.cm

                Choice (Enter 0 to stop)? 1

Value to convert? 100
Which converts to 2.54 meters

```
Which of the following conversions do you wish to perform?

 (1) inches to meters         ( 9) meters to inches
 (2) sq.inches to sq.meters   (10) sq.meters to inches
 (3) cm to meters             (11) meters to cm
 (4) sq.cm to sq.meters       (12) sq.meters to sq.cm
 (5) lines to webers          (13) webers to lines
 (6) Maxwells to webers       (14) webers to Maxwells
 (7) lines/sq.inch to Wb/sq.meter  (15) Wb/sq.meter to lines/sq.inch
 (8) Maxwell/sq.cm to Wb/sq.meter  (16) Wb/sq.meter to Maxwells/sq.cm

            Choice (Enter 0 to stop)? 0


  READY
```

Once line 410 makes the conversion choice, it will request the value to be converted on line 460 and then decide whether the operation requires a multiplication or division of the factors involved. If C is less than or equal to 8, the conversion operation involves multiplying the value provided by the conversion factor at the location in the data chain of line 270, defined by C. If C is greater than 8, the given value is divided by the conversion factor at location $C - 8$. For instance, if the conversion is from webers to lines, as in the first run, the value of C is 13, which is greater than 8. The given value of 50 is therefore divided by the conversion factor at $(C - 8) = (13 - 8) = 5$, which is $1E - 8 = 10^{-8}$. If the process were to convert from lines to webers, the conversion factor at $(C) = 5 = 1E - 8$ would have been used, but it would have been multiplied by 50 to determine the result in webers. Note the use of $UN\$(C + 8)$ and $UN\$(C - 8)$ on line 490 to insure that the proper unit of measure is chosen from the data list of lines 190 through 220.

11.3 *B-H* CURVES

The analysis of magnetic circuits requires that the non-linear *B-H* curve be used to determine the flux density B for a given magnetizing force H or vice versa. Program 11-2 will provide the value of H for a given value of B using specific data points (equally spaced) off the non-linear cast steel curve of Figure 11.1. Note on line 150 that the value of B (0.1 for the first non-zero entry) is followed by the corresponding value of H (50 for cast-steel), as determined from Figure 11.1. The process then continues through 1T for equal increments in T (1T = 1 Tesla

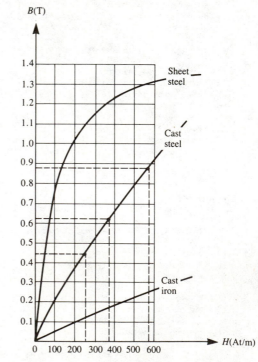

FIGURE 11.1

```
           10 REM *****   PROGRAM 11-2   *****
           20 REM ***********************************
           30 REM Lookup of H for input value of B
           40 REM for cast steel for B<1T.
           50 REM ***********************************
           60 REM
          100 DIM B(10),H(10)
          110 REM Read table data of B,H
Read    ┌120 FOR I=1 TO 10
Data    │130 READ B(I),H(I)
Points  │140 NEXT I
        │150 DATA 0,0,0.1,50,0.2,100,0.3,160,0.4,220,0.5,290,0.6,360,0.7,430
        └160 DATA 0.8,510,0.9,590,1,700
          170 REM Input B value for table lookup
          180 PRINT
Input B   190 INPUT "Flux density, B=";B
          200 IF B>1 THEN PRINT "Data only for B<1T - re-enter B" :GOTO 180
          210 REM Do lookup
Locate  ┌220 FOR I=1 TO 9
Region  └230 IF B(I)<B THEN NEXT I
        ┌240 REM Interpolate
        │250 IS=(H(I+1)-H(I))/(B(I+1)-B(I))
        └260 H=H(I)+IS*(B-B(I))
Output H 270 PRINT "For B=";B;"Wb/sq.m, H=";H;"At/m"
          280 END
```

READY

<u>RUN</u>


```
Flux density, B=? 0.88
For B= .88 Wb/sq.m, H= 574 At/m
```

READY


```
Flux density, B=? 0.45
For B= .45 Wb/sq.m, H= 255 At/m
```

READY


```
Flux density, B=? 1.3
Data only for B<1T - re-enter B

Flux density, B=? 0.63
For B= .63 Wb/sq.m, H= 374 At/m
```

READY

= 1Wb/m^2). The significance of line 250 is best described using Figure 11.2. Note the straight line approximation for the curve between values of B. The slope of the line between intersections is given by

$$\text{slope} = m = \frac{\Delta y}{\Delta x} = \frac{B(I + 1) - B(I)}{H(I + 1) - H(I)}$$

and the inverse slope by:

$$IS = \frac{H(I + 1) - H(I)}{B(I + 1) - B(I)}$$

as appearing on line 250.

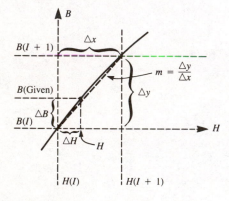

FIGURE 11.2

For a given value of B as shown in the figure, ΔB and ΔH are then related by the slope $= S = \Delta B/\Delta H$ or $IS = \Delta H/\Delta B$, which results in $\Delta H = IS \times \Delta B$. Returning to Figure 11.2, we then find that the value of H for a particular value of B is given by

$$H = H(I) + \Delta H = H(I) + IS \times \Delta B$$

and

$$H = H(I) + IS (B - B(I))$$

which is line 260 of the program. The smaller the interval Δy between chosen data points, the better the accuracy. However, in this case, approximating the curve by straight line segments between values of B separated by 0.1T

should provide excellent results. The actual curve is certainly very close to the straight line segments between the chosen data points.

The best test for such a program is to choose a few values of B and note how closely they come to the values we would determine graphically using a straight edge. The results obtained for the three runs of the program appear in Figure 11.1, and the match appears to be excellent. Keep in mind that the maximum value of B for cast steel in the range of the graph is 1T and any input value of B greater than 1T will result in the output appearing in the third run of the program.

For a range of B, Program 11-3 will look up the value of H and print out the results in tabular form, as shown in the provided run of the program. Line 140 defines the increment in B at 0.05T, which results in 20 data points for the range of 0.05 to 1. The format for the output appears within the quotes of line 160, with the order of the data defined by B, H at the end of the line. Note that the value of H at the data points of lines 230 and 240 corresponds with the results obtained in the table for the same values of B, and that the spacing between values of H is not the same, due to the non-linear curve. If the B-H curve were a straight line, all the resulting increments in H would be equal.

In Program 11-4 the user can choose the material and ask for the value of H for a particular flux density B. Naturally, the data of lines 1270 through 1330 must reference each material. Line 1250 will read the corresponding values of B for each material for the value of H appearing in the first column. For cast iron (denoted $CI(I)$) the values are 0, 0.05, 0.1 etc., for cast steel ($CS(I)$) the values are 0, 0.20, 0.37, etc., and for sheet steel ($SS(I)$), the values are 0, 0.73, 1.03, etc. Note that the values of H are incremented equally and the value of B determined. If the value of B had been chosen for cast iron, there would have been only three data points at $B = 0$, 0.1T, and 0.2T, since the curve

```
              10 REM *****   PROGRAM 11-3  *****
              20 REM ********************************
              30 REM Table of B vs H
              40 REM for cast steel for B<1T.
              50 REM ********************************
              60 REM
             100 DIM B(10),H(10)
Table        110 GOSUB 190 :REM Read table data
Heading ┌    120 PRINT TAB(5);"B";TAB(14);"H"
        └    130 PRINT "  (Wb/sq.m)      (At/m)"
Range   ┌    140 FOR B=0.05 TO 1 STEP 0.05
of B    │    150 GOSUB 260 :REM Obtain value of H
&       │    160 PRINT USING "    #.##        ###.#",B,H
Table   └    170 NEXT B
Format       180 END
             190 REM Read table of data points
        ┌    200 FOR I=0 TO 10
Read    │    210 READ B(I),H(I)
Data    │    220 NEXT I
Points  │    230 DATA 0,0,0.1,50,0.2,100,0.3,160,0.4,220,0.5,290,0.6,360,0.7,430
        └    240 DATA 0.8,510,0.9,590,1,700
             250 RETURN
             260 REM Interpolate value of H
Locate  ┌    270 FOR I=1 TO 9
Region  └    280 IF B(I)<B THEN NEXT I
        ┌    290 REM Interpolate
        │    300 IS=(H(I+1)-H(I))/(B(I+1)-B(I))
        └    310 H=H(I)+IS*(B-B(I))
             320 RETURN
```

READY

RUN

B	H
(Wb/sq.m)	(At/m)
0.05	25.0
0.10	50.0
0.15	70.0
0.20	100.0
0.25	130.0
0.30	160.0
0.35	185.0
0.40	220.0
0.45	255.0
0.50	290.0
0.55	325.0
0.60	360.0
0.65	390.0
0.70	430.0
0.75	470.0
0.80	480.0
0.85	535.0
0.90	590.0
0.95	645.0
1.00	700.0

READY

```
 10 REM *****   PROGRAM 11-4   *****
 20 REM *********************************************************
 30 REM Program to determine H for various types of metals.
 40 REM *********************************************************
 50 REM
100 REM Program determines H for a particular core type
110 REM and value of B.
120 DIM H(6),B(6),CI(6),CS(6),SS(6)
130 GOSUB 1200 :REM Read table data
140 PRINT:PRINT "Enter core material: (1) Cast iron"
150 PRINT TAB(21);"(2) Cast steel"
160 PRINT TAB(21);"(3) Sheet steel"
170 INPUT CM
180 GOSUB 1400 :REM Obtain flux density data for selected core material
190 INPUT "B=";B
200 REM Now determine H
210 GOSUB 1600
220 PRINT
230 PRINT "For B of";B;"Wb/sq.m, H is";H;"At/m"
240 PRINT:PRINT
250 INPUT "More data (YES or NO)";A$
260 IF A$="YES" THEN GOTO 140
270 END
1200 REM Read stored data:
1210 REM CI( ) holds flux density for cast iron
1220 REM CS( ) holds flux density for cast steel
1230 REM SS( ) holds flux density for sheet steel
1240 FOR I=0 TO 6
1250 READ H(I),CI(I),CS(I),SS(I)
1260 NEXT I
1270 DATA 0,0,0,0
1280 DATA 100,0.05,0.20,0.73
1290 DATA 200,0.10,0.37,1.03
1300 DATA 300,0.15,0.51,1.15
1310 DATA 400,0.19,0.65,1.23
1320 DATA 500,0.23,0.79,1.29
1330 DATA 600,0.27,0.91,1.32
1340 RETURN
1400 REM Module to select flux density data
1410 IF CM<0 OR CM>3 THEN END
1420 ON CM GOSUB 1440,1500,1550
1430 RETURN
1440 REM Flux for cast iron
1450 FOR I=0 TO 6
1460 B(I)=CI(I)
1470 NEXT I
1480 RETURN
1500 REM Flux for cast steel
1510 FOR I=0 TO 6
1520 B(I)=CS(I)
1530 NEXT I
1540 RETURN
1550 REM Flux for sheet steel
1560 FOR I=0 TO 6
1570 B(I)=SS(I)
1580 NEXT I
1590 RETURN
1600 REM Module to determine H given B
1610 FOR I=0 TO 6
1620 IF B(I)<B THEN NEXT I
1630 REM Now interpolate
1640 IS=(H(I)-H(I-1))/(B(I)-B(I-1))
1650 H=H(I-1)+IS*(B-B(I-1))
1660 RETURN
```

Input *B* 190 (annotation for line 190)

Output *H* 230 (annotation for line 230)

Read Data Points for each metal (annotation for lines 1240–1330)

CI (annotation for lines 1440–1480)

CS (annotation for lines 1500–1540)

SS (annotation for lines 1550–1590)

Determine *H* (annotation for lines 1600–1660)

```
READY

RUN


Enter core material:  (1) Cast iron
                      (2) Cast steel
                      (3) Sheet steel
? 3
B=? 0.3

For B of .3 Wb/sq.m, H is 41.0959 At/m


More data (YES or NO)? YES

Enter core material:  (1) Cast iron
                      (2) Cast steel
                      (3) Sheet steel
? 1
B=? 0.25

For B of .25 Wb/sq.m, H is 550 At/m


More data (YES or NO)? NO


READY
```

goes off the graph when H is greater than 600 At/m. This procedure will result in six data points. Note that line 1270 is needed so that values of H less than 100 or values of B less than 0.05 have two data points to determine the desired value of B or H. Once the value of CM is known, line 180 will send the program to the subroutine at 1400 to set $B(I)$ equal to the values determined by the chosen material. For example, if cast steel were chosen, $CM = 2$ and line 1420 would send us to the subroutine at line 1500. Each value of $B(I)$ on line 1520 will then equal the corresponding value of $CS(I)$ resulting in $B(1) = CS(1) = 0$ and $B(2) = CS(2) = 0.20$, with $B(3) = CS(3) = 0.37$ and so on. The computer will then have the appropriate values of B to apply to the equations of lines 1640 and 1650. Even though a different set of data points was used, the results appearing in the two runs of the program are excellent.

11.4   THE TRANSFORMER

For the transformer of Figure 11.3, the secondary current I_2 will be determined for a particular magnetomotive force at the primary and N_2 turns in the secondary. In order to use the proper B-H curve, Program 11-5 must first ask for the core material on line 180 and then move on to the proper subroutine, as in Program 11-4, to establish the appropriate values of $B(I)$. Once the values of $B(I)$ are established, the program

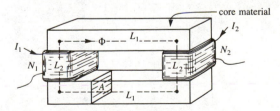

FIGURE 11.3

```
       10 REM *****   PROGRAM 11-5   *****
       20 REM ********************************************
       30 REM Program to determine the secondary current
       40 REM of a transformer.
       50 REM ********************************************
       60 REM
      100 DIM H(6),B(6),CI(6),CS(6),SS(6)
      110 GOSUB 1200 :REM Read table data
      120 PRINT "For the magnetic circuit of Figure 11-3"
      130 PRINT "enter the following data:"
      140 PRINT
      150 PRINT:PRINT "Enter core material: (1) Cast iron"
      160 PRINT TAB(21);"(2) Cast steel"
      170 PRINT TAB(21);"(3) Sheet steel"
      180 INPUT CM
      190 GOSUB 1400 :REM Obtain flux density data for selected core material
      200 INPUT "Value of flux (Wb)=";FL
      210 INPUT "Cross sectional area, A (in sq. meters)=";A
      220 B=FL/A
      230 REM Determine H
      240 GOSUB 1600
      250 PRINT
      260 PRINT "For B of ";B;"Wb/sq.m, H is";H;"At/m"
      270 PRINT
      280 INPUT "Length 1(meters)=";L1
      290 INPUT "Length 2(meters)=";L2
      300 NI=2*H*L1+2*H*L2
      310 PRINT:PRINT "Total ampere-turns=";NI;"At"
      320 PRINT:INPUT "Number of primary turns, Np=";N1
      330 INPUT "Number of secondary turns, Ns=";N2
      340 INPUT "Primary current, Ip=";I1
      350 I2=(N1*I1-NI)/N2
      360 PRINT:PRINT "Secondary current, Is=";I2;"amperes"
      370 END
     1200 REM Read stored data:
     1210 REM CI( ) holds flux density for cast iron
     1220 REM CS( ) holds flux density for cast steel
     1230 REM SS( ) holds flux density for sheet steel
     1240 FOR I=0 TO 6
     1250 READ H(I),CI(I),CS(I),SS(I)
     1260 NEXT I
     1270 DATA 0,0,0,0
     1280 DATA 100,0.05,0.20,0.73
     1290 DATA 200,0.10,0.37,1.03
     1300 DATA 300,0.15,0.51,1.15
     1310 DATA 400,0.19,0.65,1.23
     1320 DATA 500,0.23,0.79,1.29
     1330 DATA 600,0.27,0.91,1.32
     1340 RETURN
     1400 REM Module to select flux density data
     1410 IF CM<0 OR CM>3 THEN END
     1420 ON CM GOSUB 1440,1500,1550
     1430 RETURN
     1440 REM Flux for cast iron
     1450 FOR I=0 TO 6
     1460 B(I)=CI(I)
     1470 NEXT I
     1480 RETURN
     1500 REM Flux for cast steel
     1510 FOR I=0 TO 6
     1520 B(I)=CS(I)
     1530 NEXT I
     1540 RETURN
```

Margin annotations:
- Determine Data Points (lines 150–190)
- B (lines 200–220)
- H (lines 230–260)
- ℓ (lines 280–290)
- NI (lines 300–310)
- N_P, N_S
- I_P
- I_S (lines 320–360)
- As in Prog. 11-4 (lines 1200–1260)

```
1550 REM Flux for sheet steel
1560 FOR I=0 TO 6
1570 B(I)=SS(I)
1580 NEXT I
1590 RETURN
1600 REM Module to determine H given B
1610 FOR I=0 TO 6
1620 IF B(I)<B THEN NEXT I
1630 REM Interpolate
1640 IS=(H(I)-H(I-1))/(B(I)-B(I-1))
1650 H=H(I-1)+IS*(B-B(I-1))
1660 RETURN
```

READY

RUN
.

```
For the magnetic circuit of Figure 11-3
enter the following data:


Enter core material: (1) Cast iron
                     (2) Cast steel
                     (3) Sheet steel
? 3
Value of flux (Wb)=? 1.5E-5
Cross sectional area, A (in sq. meters)=? 0.15E-3

For B of  .1 Wb/sq.m, H is 13.6986 At/m

Length 1(meters)=? 0.05
Length 2(meters)=? 0.03

Total ampere-turns= 2.1918 At

Number of primary turns, Np=? 60
Number of secondary turns, Ns=? 30
Primary current, Ip=? 2

Secondary current, Is= 3.9269 amperes
```

READY

*Example 10.7, ICA

returns to line 200 to request the parameters of the system.

The value of B is determined first, followed by a run of the interpolation routine to calculate the resulting value of H for the given core material. The remaining parameters are then provided and the total required magnetomotive force determined from $NI_T = 2HL_1 + 2HL_2$ (line 300). The current I_2 is determined from $NI_T = N_1 I_1 - N_2 I_2$, written in the following form: $I_2 = (N_1 I_1 - NI_T)/N_2$ (line 350). Note that the module extending from 1400 through 1660 is the same as in Program 11-4. In fact, the major difference between the programs is that the calculations appear on lines 200 through 360. In the first run of the program, the resulting value of B is quite low. Note, however, that the resulting value of H matches the graphical value beautifully and that the accuracy of the answer is maintained.

11.5 DETERMINING Φ

The analysis of magnetic circuits in which the magnetomotive force is given and the flux desired requires the use of a "cut and try" approach, due to the non-linear relationship between B and H. As its name implies, the "cut and try" approach is one which applies approximations that will result in a solution greater than the actual value. The result will then be reduced (*cut*) and retested (*try* again) against the given value, until the percent difference between the results is acceptable.

For the system of Figure 11.4, Ohm's law for magnetic circuits states that $\Phi = NI/\mathscr{R}$, where NI is the applied magnetomotive force and $\mathscr{R}$ is the reluctance of the circuit. If we ignore the sheet steel section altogether, the net reluctance of the remaining cast steel is obviously less than the actual value, resulting in a flux greater than that desired—the first objective of this cut and try approach has been achieved.

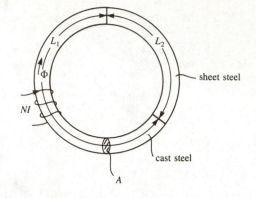

FIGURE 11.4

Applying the above to the following:

$$NI = H_{\text{sheet-steel}} L_1 + H_{\text{cast-steel}} L_2$$
$$\text{(Ampere's circuital law)}$$

will result in

$$H_{\text{cast steel}} = \frac{NI}{L_2} \text{ (Given)}$$

as appearing on line 250 of Program 11-6.

```
 10 REM *****  PROGRAM 11-6  *****
 20 REM **********************************
 30 REM Program to help determine the flux
 40 REM of a two-material core.
 50 REM **********************************
 60 REM
100 DIM H(6),B(6),CI(6),CS(6),SS(6)
110 GOSUB 1200 :REM Read table data
120 PRINT "This program will determine the flux for the"
130 PRINT "two-metal core of Figure 11-4."
140 PRINT
150 PRINT "Enter the following data:"
160 INPUT "NI=";NI
170 INPUT "Length of material 1, L1=";L1
180 INPUT "Length of meterial 2, L2=";L2
190 INPUT "Cross-sectional area of core, A=";A
200 PRINT
210 PRINT "Select type of material (1) Cast iron, (2) Cast steel, or"
220 PRINT "(3) Sheet steel"
230 INPUT "Material 1: ";C1
240 INPUT "Material 2: ";C2
250 H2=NI/L2 :REM Calculate approximate H for material 2
260 CM=C2 :GOSUB 1400 :REM Read flux density data for selected material
270 H=H2 :GOSUB 1800 :REM Determine B given H
280 CM=C1 :GOSUB 1400 :GOSUB 1600 :REM Determine H given B
290 H1=H
300 REM Calculate total NT
310 NT=H1*L1+H2*L2
```

Input Data (bracket for lines 150–240)

Determine NI_T (bracket for lines 250–310)

```
Test   ┌320 IF NT<NI THEN PRINT "Data out of range - retry." :C=0 :GOTO 140
NIₜ    └330 IF ABS(NT-NI)>ABS(0.05*NI) THEN GOTO 400
        340 C=C+1 :PRINT:PRINT "After";C;"tries solution is:"
        350 PRINT
Print  ┌360 PRINT "Flux=";B*A;"webers"
Results│370 PRINT "H1=";H1;" H2=";H2;" B=";B
       └380 PRINT "and, NT=";NT
        390 END
        400 C=C+1 :REM Count number of tries
ΔB      410 B=0.95*B :REM Reduce B by 5%
New H₂  420 CM=C2 :GOSUB 1400 :GOSUB 1600 :REM Determine H for given B
New H₁  430 GOTO 280
        1200 REM Read stored data:
        1210 REM CI( ) holds flux density for cast iron
        1220 REM CS( ) holds flux density for cast steel
        1230 REM SS( ) holds flux density for sheet steel
        1240 FOR I=0 TO 6
        1250 READ H(I),CI(I),CS(I),SS(I)
        1260 NEXT I
        1270 DATA 0,0,0,0
        1280 DATA 100,0.05,0.20,0.73
        1290 DATA 200,0.10,0.37,1.03
        1300 DATA 300,0.15,0.51,1.15
        1310 DATA 400,0.19,0.65,1.23
        1320 DATA 500,0.23,0.79,1.29
        1330 DATA 600,0.27,0.91,1.32
        1340 RETURN
        1400 REM Module to select flux density data
        1410 IF CM<0 OR CM>3 THEN END
        1420 ON CM GOSUB 1440,1500,1550
        1430 RETURN
        1440 REM Flux for cast iron
        1450 FOR I=0 TO 6
        1460 B(I)=CI(I)
        1470 NEXT I
        1480 RETURN
        1500 REM Flux for cast steel
        1510 FOR I=0 TO 6
        1520 B(I)=CS(I)
        1530 NEXT I
        1540 RETURN
        1550 REM Flux for sheet steel
        1560 FOR I=0 TO 6
        1570 B(I)=SS(I)
        1580 NEXT I
        1590 RETURN
        1600 REM Module to determine H given B
        1610 FOR I=0 TO 6
        1620 IF B(I)<B THEN NEXT I
        1630 REM Interpolate
        1640 IS=(H(I)-H(I-1))/(B(I)-B(I-1))
        1650 H=H(I-1)+IS*(B-B(I-1))
        1660 RETURN
        1800 REM Module to determine B given H
        1810 FOR I=0 TO 6
        1820 IF H(I)<H THEN NEXT I
        1830 REM Interpolate
        1840 S=(B(I)-B(I-1))/(H(I)-H(I-1))
        1850 B=B(I-1)+S*(H-H(I-1))
        1860 RETURN

READY
```

```
RUN

This program will determine the flux for the
two-metal core of Figure 11-4.

Enter the following data:
NI=? 50
Length of material 1, L1=? 0.20
Length of meterial 2, L2=? 0.12
Cross-sectional area of core, A=? 0.003

Select type of material (1) Cast iron, (2) Cast steel, or
(3) Sheet steel
Material 1: ? 3
Material 2: ? 2

After 40 tries solution is:

Flux= 2.7326E-04 webers
H1= 12.4775   H2= 416.6667   B= .091
and, NT= 52.4955


READY

RUN

This program will determine the flux for the
two-metal core of Figure 11-4.

Enter the following data:
NI=? 75
Length of material 1, L1=? .04
Length of meterial 2, L2=? .12
Cross-sectional area of core, A=? .003

Select type of material (1) Cast iron, (2) Cast steel, or
(3) Sheet steel
Material 1: ? 3
Material 2: ? 2

After 8 tries solution is:

Flux= 1.9693E-03 webers
H1= 89.9229   H2= 625   B= .6564
and, NT= 78.5969


READY
```

Referring to Figure 11.1, if H is too large, as it is in the above calculations, the resulting value of B, and therefore Φ, will also be too large.

On line 270 the resulting value of B is determined for the cast steel and then used to determine the value of H for sheet steel on line 280. Note the slight change in the equation of

the subroutine of 1800 to determine the value of B for the given H value (a reversal of previous programs). The total NI is then calculated using the equation of line 310, followed by a test of whether the calculated NI is within 5% of the *given NI*. Once the difference between the two is less than 5%, the calculated value of NI is considered close enough and the resulting flux determined and printed out by line 360. If the 5% criteria is not satisfied, line 330 will send the program to line 400 to print out the number of tries and the results thus far. Remember that *NT* is the calculated value and *NI* the given value. Line 410 will reduce the flux density (and therefore flux Φ, since it is related by a constant) to 95% of the value determined by line 270, and the value of *H* for each material will be determined by line 420 and 280, followed by a new calculation for *NI*. The cutting and testing will continue until the result is within the 5% guideline.

Note in the first run the requirement for 40 passes to converge on the 5% criteria while the second run only required 8. Note also that the resulting differences between *NI* and *NT* is less than 5% of *NI*.

11.6 TESTING THE USE OF THE *B-H* CURVE

The last program (11-7) of the chapter will test your skills reading the *B-H* curve of Figure 11.1. The boundary of $H = 600At/m$ limits the ran-

```
 10 REM ***** PROGRAM 11-7 *****
 20 REM *******************************************
 30 REM Program to test the use of the B-H curve.
 40 REM *******************************************
 50 REM
100 DIM H(6),B(6),CI(6),CS(6),SS(6)
110 GOSUB 1200 :REM Read table data
120 CM=INT(1+3*RND(X)) :REM Select material type
130 IF CM=1 THEN PRINT "For cast iron";
140 IF CM=2 THEN PRINT "For cast steel";
150 IF CM=3 THEN PRINT "For sheet steel";
160 GOSUB 1400 :REM Obtain flux density data for selected core material
170 PRINT " what is the value of H (At/m) for B=";
180 IF CM=1 THEN B=INT(1+27*RND(X))/100 :REM B up to 0.27
190 IF CM=2 THEN B=INT(1+91*RND(X))/100  :REM B up to 0.91
200 IF CM=3 THEN B=INT(1+132*RND(X))/100 :REM B up to 1.32
210 PRINT B;"T"
220 GOSUB 1600  :REM Determine value of H
230 INPUT "H=";HX
240 IF ABS(H-HX)<ABS(0.01*H) THEN PRINT "Correct" :GOTO 260
250 PRINT "No, it's";H;"At"
260 END
1200 REM Read stored data:
1210 REM CI( ) holds flux density for cast iron
1220 REM CS( ) holds flux density for cast steel
1230 REM SS( ) holds flux density for sheet steel
1240 FOR I=0 TO 6
1250 READ H(I),CI(I),CS(I),SS(I)
1260 NEXT I
1270 DATA 0,0,0,0
1280 DATA 100,0.05,0.20,0.73
1290 DATA 200,0.10,0.37,1.03
1300 DATA 300,0.15,0.51,1.15
1310 DATA 400,0.19,0.65,1.23
1320 DATA 500,0.23,0.79,1.29
1330 DATA 600,0.27,0.91,1.32
1340 RETURN
1400 REM Module to select flux density data
1410 IF CM<0 OR CM>3 THEN END
1420 ON CM GOSUB 1440,1500,1550
1430 RETURN
```

(Continued)

```
1440 REM Flux for cast iron
1450 FOR I=0 TO 6
1460 B(I)=CI(I)
1470 NEXT I
1480 RETURN
1500 REM Flux for cast steel
1510 FOR I=0 TO 6
1520 B(I)=CS(I)
1530 NEXT I
1540 RETURN
1550 REM Flux for sheet steel
1560 FOR I=0 TO 6
1570 B(I)=SS(I)
1580 NEXT I
1590 RETURN
1600 REM Module to determine H given B
1610 FOR I=0 TO 6
1620 IF B(I)<B THEN NEXT I
1630 REM Interpolate
1640 IS=(H(I)-H(I-1))/(B(I)-B(I-1))
1650 H=H(I-1)+IS*(B-B(I-1))
1660 RETURN
```

```
READY

RUN

For sheet steel what is the value of H (At/m) for B= 1.27 T
H=? 467
Correct


READY

RUN

For cast steel what is the value of H (At/m) for B= .68 T
H=? 241
No, it's 421.4286 At


READY

RUN

For cast iron what is the value of H (At/m) for B= .16 T
H=? 325
Correct


READY
```

domly generated value of B to those appearing on lines 180, 190, and 200. For $CM = 1$, representing cast iron, the range of B is .01 to 0.27. The provided value of H is then tested against the computer-generated value on line 240 and, if it falls within 1% you will receive a correct answer indication.

EXERCISES

Write a program to perform the following tasks:

1. Provide the value of H for cast steel for $B < 1T$, using an interval of $B = 0.05T$ between data points rather than the 0.1T employed in

Program 11-2. Compare the values of H obtained at $B = 0.35$ and 0.67 with each program and comment on whether there is a significant increase in the level of accuracy.

2. Repeat program 11-1, but change the interval to 0.2T and compare the results with those obtained for both intervals in problem 11-1.

3. Provide the value of B for cast steel for $H < 700$ At/m, using an interval of $H = 50$At/m between data points.

4. Use the data points of Program 11-3 to develop a technique for determining B from H, using the slope $= S$ defined in section 11-3.

5. Tabulate the results of problem 11-4 for H from 50 to 700At/m in increments of 50At/m and compare the results to that obtained by Program 11-3. The data points will be different, but you can compare the values.

6. Determine I for the magnetic circuit of Figure 11.5. For the first run, use $\Phi = 3 \times 10^{-4}$ Wb, $N = 50$, $A = 6 \times 10^{-4}$ m^2, $\ell_{abcd} = 0.3$m, $\ell_{defa} = 0.1$m, $M_1 =$ sheet steel and $M_2 =$ cast steel.

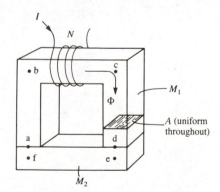

FIGURE 11.5

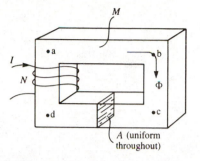

FIGURE 11.6

7. Determine I for the magnetic circuit of Figure 11.6. For the first run, use $M =$ sheet steel, $\Phi = 3.5 \times 10^{-4}$ Wb, $A = 8 \times 10^{-4}$ m^2, $\ell_{ab} = \ell_{cd} = 0.4$m, $\ell_{bc} = \ell_{da} = 0.3$m, and $N = 100$.

8. Given NI for the circuit of Figure 11.6, determine the flux Φ. For the first run, use the same data for A, ℓ and M as in problem 11-7, and use $NI = 150$ At.

9. Given NI for the circuit of Figure 11.5, determine the flux Φ. For the first run, use the same data for A, ℓ and M as in problem 11-6, but use $NI = 300$ At.

10. For the ideal transformer, $N_1 I_1 = N_2 I_2$ and $I_2 = N_1 I_1 / N_2$. Run Program 11-5 for the provided values of the first run and calculate the percent difference between values of I_2 determined by Program 11-5 and the above equation.

11. Develop a testing routine for the system of problem 11-6. Limit the component values to the following: 6×10^{-4}m$^2 < A < 12 \times 10^{-4}$m^2, 1×10^{-4} Wb $< \Phi < 5 \times 10^{-4}$Wb, and $50 < N < 200$. Permit the choice of any of the three materials but limit the lengths to those appearing in Figure 11.5.

Capacitors and Inductors

12

12.1 INTRODUCTION

The similarities between the various equations and plots that result from the analysis of capacitive and inductive networks permit an examination of each in the same chapter. The first few programs will use some of the basic equations associated with each element to develop a series of useful routines. The transient behavior of R-L and R-C circuits will then be examined using a tabulating or plotting routine.

12.2 SELF-INDUCTANCE

The first program (12-1) will determine the self-inductance of a coil using the equation:

$$L = \frac{N^2 \mu A}{\ell}$$

with ℓ measured in meters, A in m² and μ defined by the core material. The program is fairly straightforward except for the sequence starting at line 200 to determine the appropriate units. Line 200 states that any result for L less than 99 μH should have μH as a unit of measure. For a result of 2000 pF, for instance, which can also be written as $2000 \times 10^{-12} = 0.002 \times 10^{-6}$, the result would be multiplied by 10^6 and the unit μH applied. The result for this example is $(0.002 \times 10^{-6})(10^6) = 0.002 \ \mu$H. If greater than 99 μH but less than 99mH, the result would be multiplied by 10^3 and mH applied, while for L greater than 99 mH, the unit Henry (H) would be appropriate. Note in the first run that the unit μH is appropriate, since the result was less than 99 μH. The second run demonstrates the impact of the N^2 term in the equation by doubling the turns and leaving the other parameters fixed.

```
            10 REM *****   PROGRAM 12-1  *****
            20 REM ***********************************
            30 REM Program to calculate self-inductance
            40 REM ***********************************
            50 REM
            100 PRINT "To calculate the self-inductance of a coil"
            110 PRINT "enter the following data:"
            120 PRINT
          ┌ 130 INPUT "Number of turns, N=";N
          │ 140 INPUT "Core permeability, mu=";MU
  Input   │ 150 INPUT "Core diameter, d (in meters)";D
          │ 160 A=3.14159*D^2/4
          └ 170 INPUT "Mean length of core, l (in meters)=";LE
            180 PRINT
  Calc.     190 L=N^2*MU*A/LE
          ┌ 200 IF L<99E-6 THEN L=L*1E6 :UN$="uH" :GOTO 230
  Units   │ 210 IF L<99E-3 THEN L=L*1E3 :UN$="mH" :GOTO 230
          └ 220 UN$="H"
  Output   230 PRINT "Inductance, L=";L;UN$
            240 END


            READY

            *
            RUN

            To calculate the self-inductance of a coil
            enter the following data:

            Number of turns, N=? 100
            Core permeability, mu=? 12.56E-7
```

*Example 11.1, ICA

```
Core diameter, d (in meters)? .004
Mean length of core, l (in meters)=? 0.08

Inductance, L= 1.9729 uH


READY


To calculate the self-inductance of a coil
enter the following data:

Number of turns, N=? 200
Core permeability, mu=? 12.56E-7
Core diameter, d (in meters)? 0.004
Mean length of core, l (in meters)=? 0.08

Inductance, L= 7.8917 uH


READY
```

12.3 ENERGY STORED BY A CAPACITOR OR INDUCTOR

The following program (12-2) will permit a choice of an inductive or capacitive element and then calculate the energy stored by the chosen unit. The equations appearing on lines 440 and 640 are relatively simple, but some effort was required to provide a clear format for the printout and to apply the proper unit of measurement. Note that even though two elements are involved, and the energy calculated using two different equations, lines 500 through 530 are the same for each, reducing the program length by at least four lines.

```
       10 REM *****  PROGRAM 12-2  *****
       20 REM ******************************************
       30 REM Program to calculate the energy stored
       40 REM in a capacitor or inductor.
       50 REM ******************************************
       60 REM
      ┌100 PRINT"Calculate the energy stored in:(1)a Capacitor,or (2)an Inductor"
Choice│110 INPUT "Choice=";C
      │120 IF C=1 OR C=2 THEN GOTO 130 ELSE GOTO 110
      └130 ON C GOSUB 400,600
       140 PRINT
      ┌150 INPUT "More (YES or NO)";A$
Continue?└160 IF A$="YES" THEN PRINT:PRINT:GOTO 100
       170 END
      ┌400 REM Module to calculate the energy stored by a capacitor
      │410 PRINT:PRINT "To calculate the energy stored by a capacitor, enter:"
    C │420 INPUT "Capacitor value, C=";C
      │430 INPUT "Voltage across capacitor, E=";E
      │440 WC=C*E^2/2
      └450 W=WC :GOTO 500
      ┌500 IF W<99E-6 THEN W=W*1E6 :UN$="uJ" :GOTO 530
      │510 IF W<99E-3 THEN W=W*1E3 :UN$="mJ" :GOTO 530
Output│520 UN$="J"
      │530 PRINT:PRINT "Energy stored is";W;UN$
      └540 RETURN
```

(Continued)

```
600 REM Module to calculate the energy stored by an inductor
610 PRINT:PRINT "To calculate the energy stored in an inductor, enter:"
620 INPUT "Inductor value, L=";L
630 INPUT "Current through inductor, I=";I
640 WL=0.5*L*I^2
650 W=WL :GOTO 500
```

READY

RUN

Calculate the energy stored in:(1)a Capacitor,or (2)an Inductor
Choice=? 2

To calculate the energy stored in an inductor, enter:
Inductor value, L=? 25E-3
Current through inductor, I=? 5

Energy stored is .3125 J

More (YES or NO)? YES

Calculate the energy stored in:(1)a Capacitor,or (2)an Inductor
Choice=? 1

To calculate the energy stored by a capacitor, enter:
Capacitor value, C=? 50E-6
Voltage across capacitor, E=? 25

Energy stored is 15.625 mJ

More (YES or NO)? NO

READY

12.4 SERIES AND PARALLEL CAPACITORS AND INDUCTORS

In Program 12-3 any number of series or parallel capacitors or inductors can be combined. The fact that the total inductance or capacitance of series inductors or parallel capacitors is simply the sum of the elements permits using the routine beginning at line 700 to determine the result for each situation. In the same way, the routine at 800 can be used to find the total inductance or capacitance of parallel inductors or series capacitors. Note in each routine the need to set G and S to zero before totaling the input values. The GO TO 720 command on line 750 reveals that the input value for each element in a train of series inductors is provided before returning to line 620. The program has essentially six components. The first is the heading that extends through line 60. Next, lines 100 through 130 permit a choice of inductive or capacitive values. Lines 300 through 490 are reserved for capacitive elements, while lines 500 through 680 are limited to inductive components. The last two ''sections'' provide the required mathematical calculations.

Do not become confused with the role of the S and the P in the printout statements of lines 420 and 470. The S and P variables refer to the process performed by the routines at 700

```
        10 REM *****   PROGRAM 12-3   *****
        20 REM ********************************************
        30 REM Calculating the series or parallel equivalent
        40 REM of C's or L's
        50 REM ********************************************
        60 REM
Choose ┌100 PRINT:PRINT "Which element will be combined, C or L";
C or L │110 INPUT C$
       │120 IF C$="C" THEN GOTO 300
       └130 IF C$="L" THEN GOTO 500
        140 GOTO 100
        300 REM Combining capacitors in series or parallel
Choose  310 PRINT
Series ┌320 INPUT "Combine in series (S) or parallel (P)";C$
or     │330 IF C$="S" THEN GOTO 400
Parallel└340 IF C$="P" THEN GOTO 450
C       350 GOTO 310
        400 PRINT "Enter values (0 when done)"
        410 GOSUB 800
C_T −   420 PRINT:PRINT "Equivalent series capacitance is";P;"farads"
(Series)430 GOTO 480
        450 PRINT "Enter values (0 when done)"
        460 GOSUB 700
C_T −   470 PRINT:PRINT "Equivalent parallel capacitance is";S;"farads"
(Parallel)480 PRINT :INPUT "More (YES or NO)";A$
        490 IF A$="YES" THEN GOTO 100 ELSE END
        500 REM Combining inductors in series or parallel
Choose  510 PRINT
Series ┌520 INPUT "Combine in series (S) or parallel (P)";C$
or     │530 IF C$="S" THEN GOTO 600
Parallel└540 IF C$="P" THEN GOTO 650
L       550 GOTO 510
        600 PRINT "Enter values (0 when done)"
        610 GOSUB 700
L_T −   620 PRINT:PRINT "Equivalent series inductance is";S;"henry"
(Series)630 GOTO 480
        650 PRINT "Enter values (0 when done)"
        660 GOSUB 800
L_T −   670 PRINT:PRINT "Equivalent parallel inductance is";P;"henry"
(Parallel)680 GOTO 480
       ┌700 REM Module to obtain sum of input values
       │710 S=0
Sum    │720 INPUT "Value=";V
       │730 IF V=0 THEN RETURN
       └740 S=S+V
        750 GOTO 720
       ┌800 REM Module to obtain inverse sum of input values
Inverse│810 G=0
Sum    │820 INPUT "Value";V
       │830 IF V=0 THEN P=1/G:RETURN
       └840 G=G+1/V
        850 GOTO 820


    READY

    RUN


    Which element will be combined, C or L? C

    Combine in series (S) or parallel (P)? P
    Enter values (0 when done)
```

```
Value=? 100
Value=? 200
Value=? 200
Value=? 300
Value=? 0

Equivalent parallel capacitance is 800 farads

More (YES or NO)? YES

Which element will be combined, C or L? L

Combine in series (S) or parallel (P)? P
Enter values (0 when done)
Value? 4E-3
Value? 3E-3
Value? 5E-3
Value? 0

Equivalent parallel inductance is 1.2766E-03 henry

More (YES or NO)? YES

Which element will be combined, C or L? C

Combine in series (S) or parallel (P)? S
Enter values (0 when done)
Value? 22E-6
Value? 300E-6
Value? 0

Equivalent series capacitance is 2.0497E-05 farads

More (YES or NO)? NO


READY
```

and 800 respectively and not to whether the elements are in series (S) or parallel (P), as appearing on line 320. Note in the provided run how nicely the program computes the series or parallel combination for either capacitors or inductors and for any number of elements.

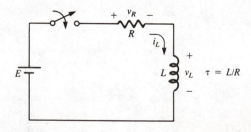

FIGURE 12.1

12.5 TRANSIENT ANALYSIS, *R-L* CIRCUIT

The transient behavior of the voltage or current for the *R-L* circuit of Figure 12.1 establishes a relationship with time that involves the factor e^x. The result is a mathematical relationship that can often be confusing and difficult. In the

analysis of Program 12-4, the important elements of the general equations are calculated in sequential order and then substituted when required. The time constant is a useful quantity in that it defines the transition period (5τ) and provides the basis for a range and increment on line 270. Since e^x appears in the library of the

BASIC language, it appears simply as EXP(X) on lines 280 and 300. The major part of the calculations and the printout of the results appear entirely on lines 270 through 330. Lines 100 through 260 request the input data and title the columns of the output.

For any input data, the number of lines in the printout will always be 10, as defined by line 270. Note how the various quantities level off after the fourth time constant (at 0.08 seconds) in the provided printout.

```
10 REM *****  PROGRAM 12-4  *****
20 REM ******************************************************
30 REM Program to provide a table of values for
40 REM The RL circuit of Figure 12-1.
50 REM ******************************************************
60 REM
100 PRINT "For the RL circuit of Figure 12-1."
110 PRINT "enter the following data:"
120 PRINT
130 INPUT "Supply voltage, E=";E
140 INPUT "Resistor, R=";R
150 INPUT "Inductor, L=";L
160 TA=L/R
170 PRINT
180 PRINT "Time constant, TAU=";TA;"seconds"
190 PRINT
200 PRINT "For a period of 5 time constants the voltages,"
210 PRINT "current, and energy stored are:"
220 PRINT
230 PRINT TAB(5);"T";TAB(15);"VL";TAB(25);"VR";TAB(35);"IL";TAB(45);"WL"
240 PRINT TAB(2);"(seconds)";TAB(13);"(volts)";TAB(23);"(volts)";
250 PRINT TAB(33);"(amps)";TAB(43);"(joules)"
260 PRINT
270 FOR T=TA/4 TO 5*TA STEP TA/4
280 IL=(E/R)*(1-EXP(-T/TA))
290 VR=IL*R
300 VL=E*EXP(-T/TA)
310 WL=0.5*L*IL^2
320 PRINT TAB(3);T;TAB(13);VL;TAB(23);VR;TAB(33);IL;TAB(43);WL
330 NEXT T
340 END
```

Table Heading → lines 230–250

T Loop → lines 270–330

Calc. → lines 280–310

```
READY


RUN
*
For the RL circuit of Figure 12-1.
enter the following data:

Supply voltage, E=? 50
Resistor, R=? 2E3
Inductor, L=? 4.0

Time constant, TAU= 2E-03 seconds

For a period of 5 time constants the voltages,
current, and energy stored are:
```

*Example 11.4, ICA

T (seconds)	VL (volts)	VR (volts)	IL (amps)	WL (joules)
5E-04	38.94	11.06	5.53E-03	6.1161E-05
1E-03	30.3265	19.6735	9.8367E-03	1.9352E-04
1.5E-03	23.6183	26.3817	.013	3.48E-04
2E-03	18.394	31.606	.016	4.9947E-04
2.5E-03	14.3252	35.6748	.018	6.3634E-04
3E-03	11.1565	38.8435	.019	7.5441E-04
3.5E-03	8.6887	41.3113	.021	8.5331E-04
4E-03	6.7668	43.2332	.022	9.3456E-04
4.5E-03	5.27	44.73	.022	1.0004E-03
5E-03	4.1042	45.8958	.023	1.0532E-03
5.5E-03	3.1964	46.8036	.023	1.0953E-03
6E-03	2.4894	47.5106	.024	1.1286E-03
6.5E-03	1.9387	48.0613	.024	1.1549E-03
7E-03	1.5099	48.4901	.024	1.1756E-03
7.5E-03	1.1759	48.8241	.024	1.1919E-03
8E-03	.9158	49.0842	.025	1.2046E-03
8.5E-03	.7132	49.2868	.025	1.2146E-03
9E-03	.5554	49.4446	.025	1.2224E-03
9.5E-03	.4326	49.5674	.025	1.2285E-03
.01	.3369	49.6631	.025	1.2332E-03

READY

12.6 TRANSIENT ANALYSIS, *R-C* CIRCUIT

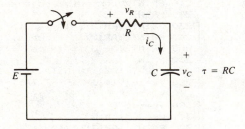

FIGURE 12.2

The analysis of a circuit such as Figure 12.2 is normally limited to a particular set of calculations. In Program 12-5 the user can choose the quantity to be determined and the program will request the values required to perform the necessary calculations. For instance, if the time required to reach a particular voltage V_C during the transient rise toward E volts is requested, the operation described by line 170 is applicable and $C = 5$.

If $C = 5$, line 220 will send the program to 800, where a module will request the information necessary to perform the required calculations. The general equation for the voltage

```
10 REM *****  PROGRAM 12-5  *****
20 REM *************************************************
30 REM Program to calculate various quantities
40 REM of an RC circuit.
50 REM *************************************************
60 REM
100 PRINT "For the RC circuit of Figure 12-2"
110 PRINT "choose the quantity you wish to know:"
120 PRINT
```

```
      130 PRINT TAB(10);"(1) Value of Tau"
      140 PRINT TAB(10);"(2) Value of R"
      150 PRINT TAB(10);"(3) Value of C"
Choice 160 PRINT TAB(10);"(4) Value of Vc at desired t"
      170 PRINT TAB(10);"(5) Value of t at desired Vc"
      180 PRINT:PRINT TAB(15);"Choice";
      190 INPUT C
      200 IF C<1 OR C>5 THEN 180
      210 PRINT
Continue? 220 ON C GOSUB 300,400,500,600,800 :REM Perform specified calculation
      230 PRINT:INPUT "More(YES or NO)";A$
      240 IF A$="YES" THEN GOTO 120
      250 END
      300 REM Module to compute Tau
Calc.  310 INPUT "R=";R
τ      320 INPUT "C=";C
       330 TA=R*C
      340 PRINT:PRINT "Tau=";TA;"seconds"
      350 RETURN
      400 REM Module to compute R
Calc.  410 INPUT "C=";C
R      420 INPUT "TAU=";TA
       430 R=TA/C
      440 PRINT:PRINT "R=";R;"ohms"
      450 RETURN
      500 REM Module to compute C
Calc.  510 INPUT "R=";R
C      520 INPUT "TAU=";TA
       530 C=TA/R
      540 PRINT:PRINT "C=";C;"farads"
      550 RETURN
      600 REM Module to compute Vc
       610 INPUT "R=";R
       620 INPUT "C=";C
       630 TA=R*C
       640 INPUT "Supply voltage, E=";E
Calc.  650 PRINT:PRINT "Tau=";TA;"seconds"
Vc     660 PRINT "Voltage Vc is desired at (enter 0 when done) t=";
       670 INPUT T
       680 IF T=0 THEN RETURN
       690 IF T>=10*TA THEN VC=E :GOTO 710
       700 VC=E*(1-EXP(-T/TA))
      710 PRINT "Vc=";VC;"volts"
      720 PRINT
      730 GOTO 660
      800 REM Module to compute time, t
       810 INPUT "R=";R
       820 INPUT "C=";C
       830 TA=R*C
       840 INPUT "Supply voltage, E=";E
       850 PRINT:PRINT "Tau=";TA;"seconds"
Calc.  860 PRINT "The time t is desired when (enter 0 when done) Vc=";
t      870 INPUT VC
       880 IF VC=0 THEN RETURN
       890 IF VC>E THEN PRINT "Sorry, not possible - retype" :GOTO 860
       900 T=-TA*LOG(1-VC/E)
      910 PRINT "Time, t=";T;"seconds"
      920 PRINT
      930 GOTO 860
      940 RETURN

      READY
```

```
RUN

For the RC circuit of Figure 12-2
choose the quantity you wish to know:

            (1) Value of Tau
            (2) Value of R
            (3) Value of C
            (4) Value of Vc at desired t
            (5) Value of t at desired Vc

            Choice? 2

C=? 0.001
TAU=? 20

R= 2E+04 ohms

More(YES or NO)? YES

            (1) Value of Tau
            (2) Value of R
            (3) Value of C
            (4) Value of Vc at desired t
            (5) Value of t at desired Vc

            Choice? 1

R=? 2000
C=? 10E-6

Tau= .02 seconds

More(YES or NO)? YES

            (1) Value of Tau
            (2) Value of R
            (3) Value of C
            (4) Value of Vc at desired t
            (5) Value of t at desired Vc

            Choice? 5

R=? 2.7E3
C=? 50E-6
Supply voltage, E=? 12

Tau= .135 seconds
The time t is desired when (enter 0 when done) Vc=? 11
Time, t= .3355 seconds

The time t is desired when (enter 0 when done) Vc=? 6
Time, t= .094 seconds

The time t is desired when (enter 0 when done) Vc=? 1
Time, t= .012 seconds

The time t is desired when (enter 0 when done) Vc=? 0

More(YES or NO)? NO


READY
```

across the capacitor during the charging phase is given by

$$v_c = E(1 - e^{-t/\tau})$$

rearranging: $\quad \dfrac{V_C}{E} = 1 - e^{-t/\tau}$

and $\quad e^{-t/\tau} = 1 - \dfrac{V_C}{E}$

Using logorithms,

$$-t/\tau = \log\left(1 - \frac{V_C}{E}\right)$$

and finally,

$$t = -\tau \log\left(1 - \frac{V_C}{E}\right)$$

which is the equation appearing on line 900. Once *t* is printed out by line 910, line 930 will send the program back to line 860 to ask for another value of V_C. If $V_C = 0$, the program will return to line 230 to determine if there are any other calculations to be performed. For each choice of lines 130 through 170, there is a corresponding subroutine starting at 300, 400, 500, 600, and 800. Although the total length of the program could be reduced by isolating those calculations that require *R* and *C* and

simply asking for each before entering the appropriate subroutine, the format of Program 12-5 clearly defines the various subroutines and permits adding or deleting a subroutine without adversely affecting the rest of the program. Note in the printout that the operations defined by lines 160 and 170 are the only two that received multiple runs because all the parameters remained the same—only the desired time *t* or the desired voltage V_C changed. Note in the last choice of the printout how quickly the voltage V_C reached one-half of its maximum value of 12 volts and how long it took to reach 11 volts. In the next program, a plot of V_C versus *t* will be generated using some of the plotting statements introduced in earlier programs.

Program 12-6 will plot the transient behavior of the voltage V_C of an *R-C* network as it approaches its steady-state value of *E* volts. Although steps 130 through 160 define 26 data points for V_C, the dimension statement of V(100) was chosen to permit data points to be added by simply changing the step level.

The equation to be plotted is

$$v_C = E(1 - e^{-t/\tau})$$

```
 10 REM *****   PROGRAM 12-6   *****
 20 REM ************************************
 30 REM Program to plot the voltage Vc
 40 REM for the network of Figure 12-2.
 50 REM ************************************
 60 REM
100 DIM V(100),T(100),D(100)
110 INPUT "Supply voltage, E=";E
120 INPUT "Time constant, Tau=";TA
130 TF=5*TA
140 TS=TA/5
150 I=0
160 FOR T=0 TO TF STEP TS
170 I=I+1
180 V(I)=E*(1-EXP(-T/TA))
190 D(I)=V(I)
200 T(I)=T
210 NEXT T
220 N=I
230 FS=40 :REM Full scale=40
240 CP=10 :REM Sets center position at 10 divisions from the left border
250 GOSUB 3300
260 PRINT:PRINT TAB(10);"Voltage scale is";E/(FS-CP);"volts/division"
270 END
```

Annotation on lines 160–210: T Loop Calc. $v_C(t)$

```
              3300 REM Module to plot data in array D(I) with N entries
             ⎡3310 FM=ABS(D(I))
Determine    │3320 FOR I=2 TO N
             │3330 IF ABS(D(I))>FM THEN FM=ABS(D(I))
 V_Cmax      │3340 NEXT I
             ⎣3350 REM FM has largest value in array
  v_C        ⎡3360 FOR I=1 TO FS
  axis       │3370 PRINT "-";
             ⎣3380 NEXT I
 V_Cmax       3390 PRINT INT(FM+0.5)
             ⎡3400 FOR I= 1 TO N
             │3410 P=INT(CP+((FS-CP)*D(I))/FM+0.5)
  Plot       │3415 PRINT USING "##.## ",T(I);
  v_C        │3420 IF P=CP THEN PRINT TAB(CP);"*"
             │3430 IF P<CP THEN PRINT TAB(P);"*";TAB(CP);"I"
             │3440 IF P>CP THEN PRINT TAB(CP);"I";TAB(P);"*"
             ⎣3450 NEXT I
              3460 PRINT "Time(seconds)"
              3470 RETURN
```

READY

RUN

Supply voltage, E=? <u>10</u>
Time constant, Tau=? <u>0.4</u>
```
------------------------------------------- 10
    0.00      *
    0.08      I       *
    0.16      I             *
    0.24      I                *
    0.32      I              *
    0.40      I             *
    0.48      I                *
    0.56      I                  *
    0.64      I                   *
    0.72      I                    *
    0.80      I                    *
    0.88      I                     *
    0.96      I                     *
    1.04      I                      *
    1.12      I                      *
    1.20      I                       *
    1.28      I                       *
    1.36      I                       *
    1.44      I                       *
    1.52      I                        *
    1.60      I                        *
    1.68      I                        *
    1.76      I                        *
    1.84      I                        *
    1.92      I                        *
    2.00      I                        *
Time(seconds)
```

Voltage scale is .3333 volts/division

READY

```
RUN

Supply voltage, E=? 10
Time constant, Tau=? 0.2
------------------------------------------------ 10
  0.00      *
  0.04      I      *
  0.08      I          *
  0.12      I              *
  0.16      I                  *
  0.20      I                     *
  0.24      I                       *
  0.28      I                        *
  0.32      I                         *
  0.36      I                          *
  0.40      I                           *
  0.44      I                            *
  0.48      I                            *
  0.52      I                             *
  0.56      I                             *
  0.60      I                              *
  0.64      I                              *
  0.68      I                              *
  0.72      I                              *
  0.76      I                              *
  0.80      I                              *
  0.84      I                              *
  0.88      I                              *
  0.92      I                              *
  0.96      I                              *
  1.00      I                              *
Time(seconds)

Voltage scale is .3333 volts/division


READY
```

Since the plot is that of v_C versus t, the only required quantities are E and τ (lines 110 and 120). For t equal to the values defined by line 160, the value of v_C will be determined by line 180 and stored as $D(I)$ by line 190. Subscripted values of t will also be stored as $T(I)$ by line 200. As an example, if $T = 0$, $v_c = E(1 - e^{-0}) = E(1 - 1) = 0$ volts and $D(1) = V(1) = 0$ and $T(1) = T = 0$. For $T = \tau/5$, $v_c = E(1 - e^{-\tau/5/\tau}) = E(1 - e^{-1/5}) = E(1 - 0.8187) = 0.1813E$ and $D(2) = V(2) = 0.1813E$ with $T(2) = T = \tau/5$ and so on. The number of data points is stored as N by line 220. A moment's examination of lines 130, 140 and 160 should reveal in this case that $N = 26$ (don't forget to include $T = 0$ as a data point).

Lines 230 and 240 reveal that the plot will be limited to 40 divisions, with the count beginning at 10 divisions. Since all values of v_C will be positive, CP is the zero axis for the graph. Subroutine 3300 is very similar to those applied in earlier plotting programs. Lines 3310 through 3350 determine the largest value for $D(I)$. (We know from experience that E is the largest possible value.) Lines 3360 through 3380 will print a dashed line that will extend from the left-hand margin of the page to three divisions less than the full-scale value of 40. Line 3390 will then add 0.5 to FM to be sure that the integer value determined is correct. For instance, if E equals 10 volts and FM equals 9.995, then INT(9.995 + 0.5) = INT(10.495) = 10, as it should be. The resulting value will then be printed at the end of the dashed line. The full implication of line 3410 can best be demonstrated with another example. If $E = 10$ volts and $FS = 40$ divisions, the number of divisions available for the plot is $FS - CP = 30$ divisions, resulting in $E/(FS\text{-}CP) = 10/30 = 0.333$ volts per division. If $D(I) = 3.0$ volts, then the plot point should be $3.0/0.333 = 9$ divisions from the center position of 10, or 19 divisions from the left-hand margin. Substituting into line 3410 we have:

$$P = INT(10 + ((40 - 10) \times 3.0)/10 + 0.5)$$
$$= INT(10 + 9 + 0.5)$$
$$= INT(19.5)$$
$$= 19$$

Remember that the 0.5 is added to insure an integer value of 10 for the case of INT(9.999) rather than 9. The next line and those to follow provide a time scale, plot the data points, and establish the reference axis.

Since the time scale is linked to the value of τ, a change in τ will not change the resulting graph, as shown in the second run. In fact, a change in τ or E will not change the appearance of the curve since both are included in determining the scale for the corresponding axis. To obtain varying plots the time scale or voltage axis must remain fixed and not linked to the circuit values.

The last program (12-7) is a testing routine developed as an extension of Program 12-5. In this case, however, since some quantities are not required (or in fact are the quantities to be determined), the printout statements of lines 1450 through 1480 are controlled by the calculation to be performed. If C is desired as determined by a random generation of $CH = 3$, then the randomly generated value of C would not be printed out by line 1460. Note that V_C is limited by the randomly generated value of E on line 1240 and the value of t is limited in range by the calculated value of τ on line 1260. The operation

```
10 REM ***** PROGRAM 12-7 *****
20 REM *********************************************
30 REM Program to test the calculation of various
40 REM quantities for the RC circuit of Figure 12-2.
50 REM *********************************************
60 REM
100 PRINT "This program tests the calculation of various"
110 PRINT "quantities for the RC circuit of Figure 12-2."
120 PRINT
130 FOR J=1 TO 5 :REM Allow up to 5 questions
140 GOSUB 1000 :REM Choose one of four calculations
150 GOSUB 1200 :REM Obtain values for calculation
160 GOSUB 1400 :REM Print circuit values and request answer
170 GOSUB 1600 :REM Do calculation and test input
180 PRINT
190 PRINT "So far you have answered";RA;"question(s) correctly"
200 PRINT
210 NEXT J
220 PRINT
230 PRINT "You have answered";RA;"correctly out of";J
240 END
1000 REM Select one of four calculations to test
1010 REM CH=1 solve for t
1020 REM CH=2 solve for R
1030 REM CH=3 solve for C
1040 REM CH=4 solve for Vc
1050 CH=INT(1+4*RND(X))
1060 RETURN
```

```
1200 REM Obtain values for calculations
1210 E=10*INT(1+5*RND(X))  :REM 10<E<50
1220 R=1000*INT(1+100*RND(X))  :REM 1000<R<100,000
1230 C=1E-6*INT(1+1000*RND(X))  :REM 1E-6<C<1E-3
1240 VC=INT(E*RND(X))  :REM 0<Vc<E
1250 TA=R*C
1260 T=0.1*TA*INT(1+100*RND(X))  :REM 0.1*TA<T<10*TA
1270 RETURN
1400 REM List selected data
1410 PRINT
1420 PRINT J;". For the values below:"
1430 PRINT
1440 PRINT "Supply voltage E=";E;"volts"
1450 IF CH=1 OR CH=3 OR CH=4 THEN PRINT "R=";R;"ohms"
1460 IF CH=1 OR CH=2 OR CH=4 THEN PRINT "C=";C;"farads"
1470 IF CH<4 THEN PRINT "Vc=";VC;"volts"
1480 IF CH>1 THEN PRINT "t=";T;"seconds"
1490 PRINT "What is the value of ";
1500 IF CH=1 THEN PRINT "t";
1510 IF CH=2 THEN PRINT "R";
1520 IF CH=3 THEN PRINT "C";
1530 IF CH=4 THEN PRINT "Vc";
1540 INPUT XA
1550 RETURN
1600 REM Do calculations and test input
1610 IF CH=1 THEN A=-R*C*LOG(1-VC/E)
1620 IF CH=2 THEN A=-T/(C*LOG(1-VC/E))
1630 IF CH=3 THEN A=-T/(R*LOG(1-VC/E))
1640 IF CH=4 THEN A=E*(1-EXP(-T/(R*C)))
1650 IF ABS(A-XA)<ABS(0.01*A) THEN PRINT:PRINT "Correct"  :RA=RA+1 :GOTO 1670
1660 PRINT:PRINT "No, it's";A
1670 RETURN
```

```
READY

RUN
```

This program tests the calculation of various
quantities for the RC circuit of Figure 12-2.


```
 1 . For the values below:

Supply voltage E= 30 volts
R= 3.1E+04 ohms
C= 4.28E-04 farads
Vc= 14 volts
What is the value of t? 8.3

Correct

So far you have answered 1 question(s) correctly


 2 . For the values below:

Supply voltage E= 40 volts
R= 6.1E+04 ohms
C= 4.7E-05 farads
Vc= 21 volts
What is the value of t? 25.3
```

```
No, it's 2.1343

So far you have answered 1 question(s) correctly


 3 . For the values below:

Supply voltage E= 50 volts
R= 5000 ohms
C= 1.4E-05 farads
t= .14 seconds
What is the value of Vc? 43.2

Correct

So far you have answered 2 question(s) correctly


 4 . For the values below:

Supply voltage E= 30 volts
C= 9.64E-04 farads
Vc= 19 volts
t= 132.8392 seconds
What is the value of R? 137E3

Correct

So far you have answered 3 question(s) correctly


 5 . For the values below:

Supply voltage E= 30 volts
R= 8.9E+04 ohms
C= 3.37E-04 farads
Vc= 25 volts
What is the value of t? 53.7

Correct

So far you have answered 4 question(s) correctly


You have answered 4 correctly out of 5

READY
```

to be performed is controlled by the variable CH on lines 1600 through 1640.

Note in the printout that the random generation of CH by line 1050 did not result in a request for the value of C, but asked for the value of t three times and R and V_C once.

EXERCISES

Write a program to perform the following tasks.

1. Using the general equation for the self-inductance of a coil, determine the number of

turns N, the area A, or the length ℓ, given the other necessary quantities. Choose which is to be calculated and provide the necessary input data. For the first runs of the program, use $N = 100$, $d = 0.004$m, $\ell = 0.08$m, $\mu = 4\pi \times 10^{-7}$, and $L = 1.974$ μH as required, and also as a check against your results.

2.   Determine the capacitance of a capacitor given the area of the plates, the distance between the plates, and a choice of whether the dielectric is air ($\epsilon_r = 1.0006$), paraffined paper ($\epsilon_r = 2.5$), oil ($\epsilon_r = 4.0$), mica ($\epsilon_r = 5$), porcelain ($\epsilon_r = 6$) or bakelite ($\epsilon_r = 7$). In addition, determine the electric field strength between the plates and the charge on the plates as determined by the applied voltage. For the first run, choose mica and use $A = 0.01$m^2, $d = 1.5 \times 10^{-3}$m, and $V = 450$ volts.

3.   For the configuration of Figure 12.3, determine C_1, C_2, or C_3 given C_T and the other two capacitive values.

4.   Repeat Program 12-4 for an R-C circuit. For the first run use $E = 20$V, $R = 4$kΩ, and $C = 0.1\mu$F.

5.   Repeat Program 12-5 for an R-L network. For the first run use $E = 60$V, $R = 1.5$ kΩ, and $L = 9$mH.

6.   Repeat Program 12-6 for the voltage v_L of an R-L circuit. Use the values of problem 12-5.

7.   Repeat Program 12-6 for the current i_L of an R-L circuit. Use the values of problem 12-5.

8.   Generate an interpolation routine that will provide the magnitude of e^{-x} for a range of x from 0 to 5. Use an interval for x of 0.2 and any curve or data listing you can find.

9.   Use the results of problem 12-8 to determine the values of the voltage across the capacitor of an R-C network as it charges toward E volts. Determine v_C at 1т, 2т, 3т, 4т, and 5т and compare to the results obtained using the internal e^{-x} function. For the first run, use $E = 20$V, $R = 4$kΩ and $C = 0.1\mu$F.

10.   Given the following voltage (Figure 12.4) across a 2 μF capacitor, determine the magnitude of the average current for each time interval of the same slope. Be sure the printout indicates the time interval and the sign of the resulting current.

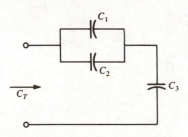

FIGURE 12.3

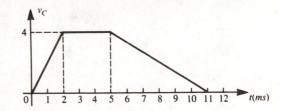

FIGURE 12.4

Sinusoidal-AC

13

13.1 INTRODUCTION

Chapters 4 through 12 were an application of the basic laws, theorems, and methods of analysis to dc steady-state systems. We will now enter the mathematically more complex realm of ac networks through a sequence of programs designed to gradually prepare us to use complex algebra in the following analysis.

Initially, a number of subroutines will be developed that will play an integral part in the more lengthy programs to follow. Once the content of each subroutine is clearly understood, we will not need to describe that portion of the program and can emphasize the analysis particular to that program. We will be progressing through a building process that will permit the review of a lengthy complex program in a reduced period of time.

13.2 DEGREES AND RADIANS

The first program (13-1) will perform the basic calculations associated with converting degrees to radians and vice versa. The calculations are not difficult to perform, but developing the layout of the printout requires some thought and development. Note the provided value of π on line 130. Some systems have a stored value of a very high degree of accuracy that can be

```
10 REM ***** PROGRAM 13-1  *****
20 REM *****************************************
30 REM Program to perform conversions between
40 REM degrees and radians.
50 REM *****************************************
60 REM
100 PRINT "This program performs conversions between"
110 PRINT "degrees and radians"
120 PRINT
130 P=3.14159 :REM P is the value of pi
140 PRINT TAB(10);"(1) Convert degrees to radians"
150 PRINT TAB(10);"(2) Convert radians to degrees"
160 PRINT
170 PRINT TAB(5);"Which conversion will be performed(enter 0 to stop)";
180 PRINT
190 INPUT CH
200 IF CH=0 THEN END
210 IF CH>2 THEN 170
220 ON CH GOSUB 300,400
230 GOTO 120
300 REM Convert degrees to radians
310 INPUT "Angle, in degrees=";A
320 PRINT
330 R=(P/180)*A
340 PRINT "Angle is";R;"radians"
350 PRINT
360 RETURN
400 REM Convert radians to degrees
410 INPUT "Angle, in radians=";R
420 PRINT
430 A=(180/P)*R
440 PRINT "Angle is";A;"degrees"
450 PRINT
460 RETURN
```

READY

```
RUN

This program performs conversions between
degrees and radians

        (1) Convert degrees to radians
        (2) Convert radians to degrees

    Which conversion will be performed(enter 0 to stop)
? 2
Angle, in radians=? 1.5

Angle is 85.9437 degrees

        (1) Convert degrees to radians
        (2) Convert radians to degrees

    Which conversion will be performed(enter 0 to stop)
? 1
Angle, in degrees=? 48

Angle is .8378 radians

        (1) Convert degrees to radians
        (2) Convert radians to degrees

    Which conversion will be performed(enter 0 to stop)
? 0


    READY
```

incorporated in the program with a simple reference to &PI, PI, or whatever that system uses.

Note on lines 200 and 210 that a choice of operation of 0 or a number greater than 2 will cause the program to either END or present the question on line 170 again. Although straightforward, the program can be helpful to have if a number of conversions must be performed. Always keep in mind that once the program is properly entered and stored, the only printout you will see after the RUN statement is the request for input data and the result of the conversion.

13.3 ANGULAR VELOCITY AND FREQUENCY

Program 13-2 is essentially an extended version of Program 13-1 that permits the choice of six operations involving the frequency and angular velocity of a sinusoidal waveform. If the fre-

```
 10 REM *****  PROGRAM 13-2  *****
 20 REM ***************************************************
 30 REM Program performs calculations involving f,w, and T.
 40 REM ***************************************************
 50 REM
100 PRINT "This program performs calculations involving f,w, and T"
110 P=3.14159 :REM Some BASIC interpreters have predefined pi: &PI, PI, etc.
120 PRINT
130 PRINT "Choose the desired operation (enter 0 to stop):"
140 PRINT
```

(Continued)

```
      ┌ 150 PRINT TAB(5);"(1) f to T";TAB(25);"(4) T to f"
    Choice
      │ 160 PRINT TAB(5);"(2) f to w";TAB(25);"(5) w to f"
      │ 170 PRINT TAB(5);"(3) T to w";TAB(25);"(6) w to T"
      │ 180 PRINT TAB(15);"Choice";
      └ 190 INPUT CH
        200 IF CH=0 THEN END
        210 IF CH>6 THEN 180
        220 PRINT
        230 ON CH GOSUB 250,300,350,400,450,500
        240 GOTO 120
      ┌ 250 REM (1) Convert f to T
   T  │ 260 INPUT "Frequency, f=";F
   ↑  │ 270 T=1/F
   f  └ 280 PRINT "Period, T=";T;"seconds"
        290 RETURN
      ┌ 300 REM (2) Convert f to w
   3  │ 310 INPUT "Frequency, f=";F
   ↑  │ 320 W=2*P*F
   f  └ 330 PRINT "Angular velocity, w=";W;"radians per second"
        340 RETURN
      ┌ 350 REM (3) Convert T to w
   3  │ 360 INPUT "Period, T=";T
   ↑  │ 370 W=2*P/T
   T  └ 380 PRINT "Angular velocity, w=";W;"radians per second"
        390 RETURN
      ┌ 400 REM (4) Convert T to f
   f  │ 410 INPUT "Period, T=";T
   ↑  │ 420 F=1/T
   T  └ 430 PRINT "Frequency, f=";F;"Hertz"
        440 RETURN
      ┌ 450 REM Convert w to f
   f  │ 460 INPUT "Angular velocity, w=";W
   ↑  │ 470 F=W/(2*P)
   3  └ 480 PRINT "Frequency, f=";F;"Hertz"
        490 RETURN
      ┌ 500 REM (6) Convert w to T
   T  │ 510 INPUT "Angular velocity, w=";W
   ↑  │ 520 T=2*P/W
   3  └ 530 PRINT "Period, T=";T;"seconds"
        540 RETURN
```

READY

RUN

This program performs calculations involving f,w, and T

Choose the desired operation (enter 0 to stop):

```
    (1) f to T        (4) T to f
    (2) f to w        (5) w to f
    (3) T to w        (6) w to T
            Choice? 3
```

Period, T=? 0.05
Angular velocity, w= 125.6636 radians per second

Choose the desired operation (enter 0 to stop):

```
    (1) f to T        (4) T to f
    (2) f to w        (5) w to f
    (3) T to w        (6) w to T
            Choice? 5
```

```
Angular velocity, w=? 10
Frequency, f= 1.5916 Hertz

Choose the desired operation (enter 0 to stop):

     (1) f to T          (4) T to f
     (2) f to w          (5) w to f
     (3) T to w          (6) w to T
             Choice? 1

Frequency, f=? 250
Period, T= 4E-03 seconds

Choose the desired operation (enter 0 to stop):

     (1) f to T          (4) T to f
     (2) f to w          (5) w to f
     (3) T to w          (6) w to T
             Choice? 0


READY
```

quency were to be determined from the angular velocity using the equation $f = \omega/2\pi$, then $CH = 5$ on line 190 and line 230 would send the program to the subroutine at 450. The value of ω would then be provided in response to the question of line 460 and the frequency determined by line 470 and printed out by line 480. Line 490 would then RETURN the program to 240, where it would be directed back to 120 to choose the next operation to be performed. Choosing $CH = 0$ would end the program at line 200. Each subroutine at 250, 300, 350, 400,

450, and 500 is very similar in format, demonstrating that the program, although long, is not difficult to understand or write.

13.4 AVERAGE AND EFFECTIVE VALUE

The next program (13-3) is quite interesting in that it is short but still sufficiently powerful to determine the average and effective value of a

```
10 REM ***** PROGRAM 13-3 *****
20 REM ************************************************
30 REM Program to calculate the average and effective
40 REM voltages of a multiple pulse signal.
50 REM ************************************************
60 REM
100 DIM V(10),T(10) :REM Allow for up to 10 pulse cycles
110 PRINT "Enter voltage and time information for a"
120 PRINT "multiple pulse waveform (enter 0 when done)"
130 PRINT
140 INPUT "Total period of pulse train=";TP
150 FOR I=1 TO 10
160 PRINT "Voltage";I;"=";
170 INPUT V(I)
180 INPUT "for period,T=";T(I)
190 IF T(I)<=0 THEN PRINT "No!!" :GOTO 180
200 IF T+T(I)>TP THEN PRINT "last interval too big - redo" :GOTO 180
210 T=T+T(I)
220 IF T=TP THEN N=I :GOTO 240
230 NEXT I
240 PRINT
```

(Continued)

```
        250 REM Perform calculations
        260 FOR I=1 TO N
 A      270 NU=NU+V(I)*T(I)
 T      280 D=D+T(I)
 A²     290 N2=N2+V(I)^2*T(I)
        300 NEXT I
Av.     310 AV=NU/D
Eff.    320 EF=SQR(N2/D)
        330 PRINT "Average voltage V=";AV;"volts"
        340 PRINT "and effective voltage V=";EF;"volts"
        350 END
```


READY

RUN

Enter voltage and time information for a
multiple pulse waveform (enter 0 when done)

Total period of pulse train=? 8
Voltage 1 =? 2
for period,T=? 1
Voltage 2 =? -4
for period,T=? 2
Voltage 3 =? 0
for period,T=? 3
Voltage 4 =? 8
for period,T=? 2

Average voltage V= 1.25 volts
and effective voltage V= 4.5277 volts

READY

RUN

Enter voltage and time information for a
multiple pulse waveform (enter 0 when done)

Total period of pulse train=? 6
Voltage 1 =? 5
for period,T=? 1
Voltage 2 =? -5
for period,T=? 1
Voltage 3 =? 5
for period,T=? 1
Voltage 4 =? -5
for period,T=? 1
Voltage 5 =? 5
for period,T=? 1
Voltage 6 =? -5
for period,T=? 1

Average voltage V= 0 volts
and effective voltage V= 5 volts

READY

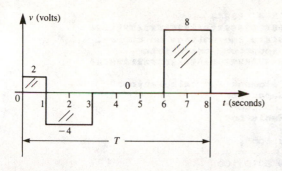

FIGURE 13.1

train of one to ten pulses. In Figure 13.1, for example, there are five pulse cycles even though two have zero magnitude. A pulse cycle is therefore defined by a provided magnitude and a time interval. Note on line 160 how the value of *I* will label each pulse without having to repeat the statement 10 times. Each input value is defined by $V(1)$, $V(2)$, etc. For the example above, $V(1) = 2$ with $T(1) = 1$. $T(1)$ cannot be less than zero as defined by line 190, and the sum of all the pulse time intervals cannot be greater than the period of the waveform. Although line 200 may seem unnecessary, it checks whether the correct values for the pulse intervals have been provided. Once the total *T* is equal to the period of the entire waveform, the variable *N* is set equal to *I* and the program leaves the routine to move on to line 240. Note that the *I* loop extends from 150 all the way down to 230 to request the description of the pulse train.

The variable *NU* of line 270 is initially zero and the first pass with $I = 1$ results in $NU = 0 + V(1) \times T(1) = 0 + 2 \times 1 = 2$ for the example of Figure 13.1. The variable *NU* will store the net area under the pulse train as we consider each pulse. Similarly, *D* is initially zero, with a first pass resulting in $D = 0 + T(1) = 1$ on line 280. Line 280, therefore, is storing the time interval between zero and the end of the pulse defined by *I*. Line 290 will result in $N2 = 0 + V(1)^2 \times T(1) = (2)^2 \times 1 = 4$, which is

the area under the squared value of the pulse, in preparation for the effective value to be calculated by line 320. During the second run with $I = 2$, $NU = 2 + V(2) \cdot T(2) = 2 + (-4)(2) = 2 - 8 = -6$, $D = 1 + T(2) = 1 + 2 = 3$ and $N2 = 4 + V(2)^2 \cdot T(2) = 4 + (-4)^2 \cdot 2 = 4 + 32 = 36$.

Once the *I* loop is completed after five passes, the program will move on to line 310 to calculate the average value using the net algebraic area (*NU*) and the total period *D*. The effective value will then be determined by $\sqrt{N_2/D}$ and the results printed out by lines 330 and 340. The first run of the program examines the pulse train of Figure 13.1, while the second run examines the square wave of Figure 13.2. The second run confirms that the average value of a square wave is zero and the effective value equal to the peak value of the pulse train.

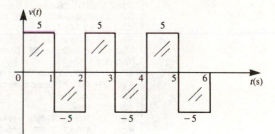

FIGURE 13.2

13.5 *R, L*, AND *C* ELEMENTS

The impact of the resistor, inductor, or capacitor on the phase angle of the voltage across or current through the element will be considered in Program 13-4. In fact, the program permits a choice of elements and input quantity and will respond with the sinusoidal expression for the desired voltage or current. The individual steps are not difficult to follow, but the result obtained suggests we have attained a higher level of sophistication.

```
         10 REM *****  PROGRAM 13-4  *****
         20 REM ********************************************
         30 REM Program to calculate the voltage or current
         40 REM of a capacitor, inductor, or resistor.
         50 REM ********************************************
         60 REM
        ┌100 PRINT:PRINT "Select element for calculation:"
         110 PRINT TAB(15);"(1) Capacitor"
         120 PRINT TAB(15);"(2) Inductor"
         130 PRINT TAB(15);"(3) Resistor"
Options  140 PRINT
         150 PRINT TAB(10);"Selection";
         160 INPUT CH
         170 IF CH<1 OR CH>3 THEN GOTO 100
         180 PRINT
        └190 INPUT "Calculate (1) Voltage, or (2) Current  ";CA
         200 PRINT
         210 IF CA<1 OR CA>2 THEN GOTO 170
         220 ON CH GOSUB 300,400,500
         230 PRINT
         240 IF CA=1 THEN M=VM :PH=VA :PRINT "v(t)=";
         250 IF CA=2 THEN M=IM :PH=IA :PRINT "i(t)=";
         260 GOSUB 560
         270 PRINT:INPUT "More(YES or NO)",A$
         280 IF A$="YES" THEN GOTO 100
         290 END
        ┌300 REM Routine for capacitor element
         310 INPUT "Capacitance=";C
         320 GOSUB 640 :REM input signal values
C        330 XC=1/(W*C)
         340 IF CA=1 THEN VM=MA*XC :VA=PH-90
         350 IF CA=2 THEN IM=MA/XC :IA=PH+90
        └360 RETURN
        ┌400 REM Routine for inductor element
         410 INPUT "Inductance=";L
         420 GOSUB 640
L        430 XL=W*L
         440 IF CA=1 THEN VM=MA*XL :VA=PH+90
         450 IF CA=2 THEN IM=MA/XL :IA=PH-90
        └460 RETURN
        ┌500 REM Routine for resistor element
         510 INPUT "Resistance=";R
         520 GOSUB 640
R        530 IF CA=1 THEN VM=MA*R :VA=PH
         540 IF CA=2 THEN IM=MA/R :IA=PH
        └550 RETURN
        ┌560 REM Module to print sinusoidal equation
         570 PRINT USING "###.###",M;
Sinusoidal 580 PRINT "sin(";W;
Function  590 PRINT "t";
         600 IF PH=0 THEN PRINT ")"; :GOTO 680
         610 IF PH>0 THEN PRINT "+";
         620 PRINT PH;")"
        └630 RETURN
        ┌640 REM Routine to accept input signal data values
         650 INPUT "Magnitude of input=";MA
Input    660 INPUT "at angle (degrees) ";PH
Data     670 INPUT "and signal frequency (Hertz)=";F
         680 W=2*3.14159*F
        └690 RETURN


         READY
```

```
RUN


Select element for calculation:
                (1) Capacitor
                (2) Inductor
                (3) Resistor

        Selection? 1

Calculate (1) Voltage, or (2) Current  ? 2

Capacitance=? 10E-6
Magnitude of input=? 10
at angle (degrees) ? 25
and signal frequency (Hertz)=? 60

i(t)=  0.038sin( 376.9908 t+ 115 )

More(YES or NO)? YES

Select element for calculation:
                (1) Capacitor
                (2) Inductor
                (3) Resistor

        Selection? 3

Calculate (1) Voltage, or (2) Current  ? 1

Resistance=? 1E3
Magnitude of input=? 10E-3
at angle (degrees) ? -45
and signal frequency (Hertz)=? 1E3

v(t)= 10.000sin( 6283.18 t-45 )

More(YES or NO)? YES

Select element for calculation:
                (1) Capacitor
                (2) Inductor
                (3) Resistor

        Selection? 2

Calculate (1) Voltage, or (2) Current  ? 2

Inductance=? 5E-3
Magnitude of input=? 100
at angle (degrees) ? 30
and signal frequency (Hertz)=? 100

i(t)= 31.831sin( 628.318 t-60 )

More(YES or NO)? NO


READY
```

Before examining the program in detail, note the separation by modules starting at 300, 400, 500, 560, and 640. Each has its singular function in the program and is easily understood by itself. If a resistor is chosen, CH of line 160 is 3 and if the input quantity is a sinusoidal current $CA = 1$ since the voltage will be determined. Since $CH = 3$, lines 220 will send the program to the resistor subroutine at 500, where the value of R will be requested. Once the element value is known, subroutines 300, 400, and 500 will direct the program to line 640, where the input data regarding the applied voltage or current is requested. The program will then return to line 530 where $CA = 1$ will use Ohm's law in the form $V = IR$ and state that the phase angle of the voltage equals that of the current. Line 550 will send the program back to line 230, where a line space will be provided before $CA = 1$ defines the magnitude and angle of the input quantity and starts to print out $v(t)$. It then proceeds to line 560 for the output sinusoidal expression. Note on line 600 that, if $PH = 0$, the bracket will be printed immediately rather than have $+0°$ in the solution. Otherwise, the phase angle PH will be added by lines 610 and 620. Remember that PH need not be zero for a resistor if the input current or voltage has a phase angle associated with it. In the case of a capacitor or inductor, the phase shift is incorporated with a plus or minus 90° when Ohm's law is applied. Note the change in sign for each device, as determined by whether the voltage or current is being determined. Consider also that the program is calculating X_L and X_C and not requiring their magnitude as input quantities.

The provided runs of the program demonstrate how well the proper phase angle is calculated, no matter what phase angle is associated with the input quantity. The program clearly performs a number of simple but important calculations and provides an appropriate format for the desired voltage or current.

Program 13-5 will not only list the impedances of the R, L, and C element for a chosen range of frequencies, but also calculate and tabulate the current and power for each frequency. The heart of the program extends from lines 290 to 400. Lines 100 through 280 simply request the input data necessary to perform the calculations and title the program.

```
10 REM *****  PROGRAM 13-5  *****
20 REM *******************************************
30 REM Program to tabulate frequency, impedance,
40 REM and power for an R, L, or C element.
50 REM *******************************************
60 REM
100 PRINT "This program will tabulate the impedance and power"
110 PRINT "of an R, L, or C element over a range of frequencies."
120 PRINT
130 PRINT "Select element for calculation:"
140 PRINT TAB(15);"(1) Capacitor"
150 PRINT TAB(15);"(2) Inductor"
160 PRINT TAB(15);"(3) Resistor"
170 PRINT
180 PRINT TAB(10);"Selection";
190 INPUT CH
200 IF CH<1 OR CH>3 THEN GOTO 120
210 PRINT
220 PRINT "Value of ";
230 IF CH=1 THEN INPUT "capacitor";C
240 IF CH=2 THEN INPUT"inductor";L
250 IF CH=3 THEN INPUT "resistor";R
260 PRINT
270 INPUT "Voltage magnitude(rms)";E
280 PRINT
```

```
Heading ⎡290 PRINT "Frequency      Impedance      Current      Power(real)"
        ⎣300 PRINT "  (Hz)          (ohms)        (amps)       (watts)"
         310 PRINT
        ⎡320 FOR F=100 TO 1000 STEP 100
         330 W=2*3.14159*F
         340 IF CH=1 THEN X=1/(W*C)⎤
         350 IF CH=2 THEN X=W*L     ⎥ Calc.
  f      360 IF CH=3 THEN X=R      ⎦
  Loop   370 IE=E/X
         380 IF CH=3 THEN P=E^2/R ELSE P=0
         390 PRINT TAB(1);F;TAB(13);X;TAB(27);IE;TAB(39);P
        ⎣400 NEXT F
         410 END
```

READY

<u>RUN</u>

This program will tabulate the impedance and power
of an R, L, or C element over a range of frequencies.

Select element for calculation:
                    (1) Capacitor
                    (2) Inductor
                    (3) Resistor

          Selection? <u>2</u>

Value of inductor? <u>2E-3</u>

Voltage magnitude(rms)? <u>25</u>

Frequency (Hz)	Impedance (ohms)	Current (amps)	Power(real) (watts)
100	1.2566	19.8944	0
200	2.5133	9.9472	0
300	3.7699	6.6315	0
400	5.0265	4.9736	0
500	6.2832	3.9789	0
600	7.5398	3.3157	0
700	8.7965	2.8421	0
800	10.0531	2.4868	0
900	11.3097	2.2105	0
1000	12.5664	1.9894	0

READY

This program will tabulate the impedance and power
of an R, L, or C element over a range of frequencies.

Select element for calculation:
                    (1) Capacitor
                    (2) Inductor
                    (3) Resistor

          Selection? <u>3</u>

Value of resistor? <u>120</u>

(Continued)

```
Voltage magnitude(rms)? 22
```

Frequency (Hz)	Impedance (ohms)	Current (amps)	Power(real) (watts)
100	120	.1833	4.0333
200	120	.1833	4.0333
300	120	.1833	4.0333
400	120	.1833	4.0333
500	120	.1833	4.0333
600	120	.1833	4.0333
700	120	.1833	4.0333
800	120	.1833	4.0333
900	120	.1833	4.0333
1000	120	.1833	4.0333

```
READY
```

The frequency range and increment is fixed by line 320, but a different set of limits could be defined by calling for line 320 before running the program and changing the values. The F loop proceeds from line 320 to 400. During the first pass, $f = 100$ Hz, resulting in $\omega = 2\pi f = 2\pi(100) = 628.318$ for line 330, and if $CH = 2$, $X_L = \omega L = (628.318)(2 \times 10^{-3}) = 1.2566$ ohms for an inductor of 2mH. The program calculations are limited to finding the current through a device for a particular input voltage. For R, the *rms* value of the current is determined by line 370 using Ohm's law, followed by the power calculation of line 380. Obviously, $P = 0$ for pure inductive and capacitive elements. The calculated value will then be printed out as defined by the TAB statements of line 390, which are sensitive to the table headings of lines 290 and 300. The LOOP will then repeat itself with $f = 200$ Hz, 300 Hz, and so on until $f = 1000$ Hz.

Note for the inductor in the first run how the magnitude of X_L increases with frequency, resulting in a decreasing level of current, while, for the resistor, the impedance remains fixed with frequency, resulting in an unchanging level of current. In fact, note that, for the resistor, the power is also independent of frequency.

13.6 PLOTTING THE SINUSOIDAL WAVEFORM

The next two programs will introduce routines for plotting the sinusoidal function. As indicated by lines 180 and 200 of Program 13-6, the BASIC language deals solely with radian measure rather than degrees. Recall that the conversion from degrees to radians is simply Degrees/57.296, since 1 radian = 57.296 degrees. Line 180 states that the plot will extend from 0(degrees or radians) to 360° or 2 π radians, with increments of 360°/50 = 7.2 degrees or 6.283/50 = 0.126 radians. Although 51 data points will result, the dimension statement was set at 100 on line 100 in case the increment was reduced to provide additional data points. On line 200 the magnitude of the sinusoidal function is determined at each value of ωt. For $M = 20$ and $PH = 30$ degrees, the first pass would result in $D(1) = 20 \sin (0 + 30°/57.296) = 10$, as indicated in the printout of the first run, and $D(2) = 20 \sin(6.283/50 + 30°/57.296) = 20 \sin(0.649$ radians$) = 20 \sin(37.2°) = 12.09$, and so on. The resulting value of $N = I$ on line 220 is determined by the increment on line 180 and not by some specified range of I. On line 3310,

FM is initially set equal to $D(1) = 10$ for the above example, FS is set equal to 40, with the center position CP at $FS/2 = 20$ on line 240. The largest value of FM is then determined by the I loop so that the scale factor can be set against the maximum swing of the signal. In this case, the maximum swing is naturally defined by the peak value of the sinusoidal voltage, but remember that the routine from 3300 through 3470 is a general purpose module that is simply recalled from memory when needed. For divisions 1 to $FS - 3 = 40 - 3 = 37$, the dash of line 3370 is printed, followed by a print-

ing of the maximum value on line 3390. Remember that the 0.5 on line 3390 is there only to insure that the value of 20 will result from 19.998 because $INT(19.998 + 0.5) = 20$. For this routine, line 3410 is:

$$P = INT(20 + ((40 - 20) \cdot D(I))/FM + 0.5)$$
$$= INT(20 + (20/FM)D(I) + 0.5)$$

The role of the 0.5 is the same as on line 3390, while the 20 is setting the center of zero line at 20 divisions from the left-hand margin. The 20/FM is the scale factor. It is spreading the 40 divisions over the full peak-to-peak value

```
 10 REM *****  PROGRAM 13-6  *****
 20 REM *******************************************
 30 REM Program to plot a single sinusoidal waveform.
 40 REM *******************************************
 50 REM
100 DIM D(100) :REM Allow for up up to 100 data values
110 PRINT "Input the following data for plotting"
120 PRINT "a sinusoidal waveform"
130 PRINT
140 INPUT "Signal amplitude is ";M
150 PRINT
160 INPUT "Phase angle(degrees) =",PH
170 PRINT
180 FOR WT=0 TO 6.283 STEP 6.283/50
190 I=I+1
200 D(I)=M*SIN(WT+PH/57.296)
210 NEXT WT
220 N=I
230 PRINT "Data stored for";N;"points"
240 FS=40 :CP=FS/2
250 GOSUB 3300
260 END
3300 REM Module to plot data in array D(I) with N entries
3310 FM=ABS(D(I))
3320 FOR I=2 TO N
3330 IF ABS(D(I))>FM THEN FM=ABS(D(I))
3340 NEXT I
3350 REM FM has largest value in array
3360 FOR I=1 TO FS
3370 PRINT "-";
3380 NEXT I
3390 PRINT INT(FM+0.5)
3400 FOR I= 1 TO N
3410 P=INT(CP+((FS-CP)*D(I))/FM+0.5)
3420 IF P=CP THEN PRINT TAB(CP);"*"
3430 IF P<CP THEN PRINT TAB(P);"*";TAB(CP);"I"
3440 IF P>CP THEN PRINT TAB(CP);"I";TAB(P);"*"
3450 NEXT I
3470 RETURN
```

Annotations to the left of the code:
- θ = ωt Loop (lines 180–210)
- Plotting Routine (lines 3300–3470)

READY

(Continued)

RUN

Input the following data for plotting
a sinusoidal waveform

Signal amplitude is ? <u>20</u>

Phase angle(degrees) =? <u>30</u>

Data stored for 51 points

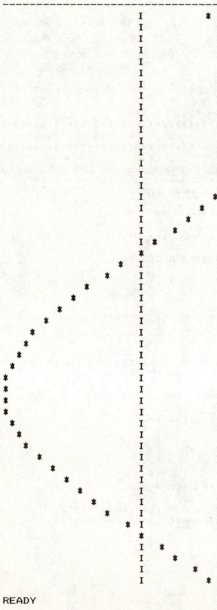

READY

of the sinusoidal function. For the particular case of $D(1) = 10$, $P = INT(20 + (20/20)10 + 0.5) = INT(20 + 10 + 0.5) = INT(30.5) = 30$, or 10 divisions to the right of the center line.

The resulting value of P is compared to CP on lines 3420 through 3440. If P equals the zero or reference axis defined by CP, then only the star is printed. If $P < CP$, then the star is printed at P and the bar (representing the zero axis) is printed at CP. If $P > CP$, the bar is printed at CP and the star at P. The sequence of operations of lines 3430 and 3440 is such because the printer moves from left to right and must print out the indicated symbols in that sequence. The plot provided clearly indicates an initial phase angle of $+30$ degrees and an equal magnitude above and below the center axis. If the waveform had a lagging phase angle of 30 degrees (that is,

```
 10 REM *****   PROGRAM 13-7  *****
 20 REM ****************************************************************
 30 REM Program to plot two sinusoidal waveforms
 40 REM on the same axis with various amplitudes and phase angles.
 50 REM ****************************************************************
 60 REM
100 DIM D1(100),D2(100)
110 I=0
120 INPUT "Amplitude of 1:";A1
130 INPUT "Phase angle (in degrees) of 1:";P1
140 INPUT "Amplitude of 2:";A2
150 INPUT "Phase angle (in degrees) of 2:";P2
160 FOR A=0 TO 360 STEP 10
170 I=I+1
180 R=A/57.296
190 D1(I)=A1*SIN(R+P1/57.296)
200 D2(I)=A2*SIN(R+P2/57.296)
210 NEXT A
220 N=I
230 PRINT "Data now stored for";N;"points"
240 PRINT
250 GOSUB 3500
260 END
```

Plot Routines
```
3500 REM Module to plot data in arrays D1(I) and D2(I)
3510 FS=60 :CP=FS/2
3520 REM Find the largest values in arrays D1(I) and D2(I)
3530 M1=ABS(D1(1)) :M2=ABS(D2(1))
3540 FOR I=2 TO N
3550 IF ABS(D1(I))>M1 THEN M1=ABS(D1(I))
3560 IF ABS(D2(I))>M2 THEN M2=ABS(D2(I))
3570 NEXT I
3580 REM M1 & M2 have largest values of respective arrays
3582 IF M1>M2 THEN F1=FS:F2=INT(M2*FS/M1):GOTO 3590
3584 F2=FS:F1=INT(M1*FS/M2)
3590 FOR I=1 TO FS
3600 PRINT "-";
3610 NEXT I :REM Print axis
3615 PRINT
3620 FOR I=1 TO N
3630 P1=INT(CP+F1*D1(I)/(2*M1)+0.5)
3640 P2=INT(CP+F2*D2(I)/(2*M2)+0.5)
3650 REM Now plot P1 and P2
3660 GOSUB 3700
3670 PRINT
3680 NEXT I
3690 RETURN :REM Exit from plot subroutine
3700 REM Routine to print points in order
3710 IF P1<=CP AND P2<=CP THEN GOTO 3750
3720 IF P1>=CP AND P2>=CP THEN GOTO 3770
3730 IF P1 <= CP THEN GOSUB 3800:GOSUB 3820:GOSUB 3810:RETURN
```

(Continued)

```
3740 IF P1 >= CP THEN GOSUB 3810:GOSUB 3820:GOSUB 3800:RETURN
3750 IF P1 < P2 THEN GOSUB 3800:GOSUB 3810:GOSUB 3820:RETURN
3760 IF P1=P2 THEN GOTO 3800 ELSE GOSUB 3810:GOSUB 3800:GOSUB 3820:RETURN
3770 GOSUB 3820
3780 IF P1 < P2 THEN GOSUB 3800:GOSUB 3810:RETURN
3790 IF P1=P2 THEN GOTO 3800 ELSE GOSUB 3810:GOSUB 3800:RETURN
3800 PRINT TAB(P1);"*";:RETURN
3810 PRINT TAB(P2);".";:RETURN
3820 IF CP=P1 OR CP=P2 THEN RETURN
3830 PRINT TAB(CP);"I";:RETURN
```

READY

<u>RUN</u>

```
Amplitude of 1:? 16
Phase angle (in degrees) of 1:? -48
Amplitude of 2:? 24
Phase angle (in degrees) of 2:? 15
Data now stored for 37 points
```

```
---------------------------------------------------------
                  *               I         .
                    *           I            .
                      *         I
                        *       I
                          *     I              .
                                I*
                                I   *               .
                                I      *             .
                                I        *          .
                                I          *       .
                                I            *
                                I             *  .
                                I              *
                                I        .     *
                                I      .        *
                                I            *
                  .             I          *
                      .         I        *
                         .      I      *
                            .   I    *
                              . I  *
                               *I
                         *      I
                      *         I
                    *           I
                  *             I
                *               I
              . *               I
            *   .               I
            *                   I
            *                   I
              *       .         I
                *               I . 
                  *             I      .
```

READY

```
RUN

Amplitude of 1:? 16
Phase angle (in degrees) of 1:? +48
Amplitude of 2:? 24
Phase angle (in degrees) of 2:? -15
Data now stored for 37 points
```

```
------------------------------------------------------------------
                     .       I           *
                    .        I        *
                             I  .         *
                             I  .        *
                             I        .   *
                             I          . *
                             I           *  .
                             I           * .
                             I         *     .
                             I       *          .
                             I     *
                             I   *             .
                            I*
                        *    I
                      *      I.
                   .       * I            .
                        *    I
                      *      I          .
                    *        I        .
                   *         I       .
                   *         I
                   *         I
                 .  *        I
          .        *         I
       .         *           I
      .        *             I
     .       *               I
             *               I
          .*                 I
              *  I           *
                *I           *
             .   I      *
                 I        *
              .  I          *
          .      I            *
```

READY

passed through the axis 30 degrees to the right of the vertical axis), then *PH* would simply be entered as − 30°. The "flat" appearance at the positive and negative peaks is because the peak values did not result in an integer increase of *P* beyond the values surrounding the maximum amplitude.

 The next program (13-7) will plot two sinusoidal functions on the same graph. Each pass of the printer across the page must therefore indicate two data points and the location of the zero axis. Remember from the previous program that the data points have to be printed out in sequential order. Lines 120 through 150 request the components of the sinusoidal waveforms necessary to plot the curves against an axis measured in radians. There is no need to know the frequency or the magnitude at any particular instant of time during the plotting process. Line 160 specifies 37 data points for each plot. Line 180 will convert the interval (in degrees) defined by line 160 to one measured in radians. For each waveform, the value of ωt (defined by line 160) will be added to the initial phase angle, and the magnitude of each waveform determined by lines 190 and 200. For each pass of the loop A loop, the value of I will be incremented by 1 until $N = I = 37$ and the arrays of $D1(I)$ and $D2(I)$ are determined. Lines 3520 through 3570 simply determine the largest value of each waveform using a general procedure (applicable to any waveform) developed earlier for a single plot. For the sinusoidal waveform, they are clearly the peak values. That waveform with the largest peak value will use the full 60 divisions (line 3510), while the smaller sinusoidal function will use a portion of the 60 divisions determined by the ratio of the peak values. This is demonstrated by line 3582, where the condition $M1 > M2$ would result in $F1 = FS = 60$ divisions and $F2 = INT(M_2/M_1 \times FS)$ with M_2/M_1 being the ratio referenced above. If $M2$ is the larger, line 3584 is appropriate.

The equations of lines 3630 and 3640 are the same as in the previous program. The sequence of commands beginning at line 3700 is quite complex, however, and requires some additional comment. As we pass through the statements, keep in mind that, if the condition of a line is not met, the GO TO statement is ignored. For instance, if *both* $P1$ and $P2$ are not less than or greater than CP, the program continues to line 3730 to test if only $P1$ is less than or equal to CP. If either line 3710 or 3720 is satisfied, lines 3730 and 3740 will be skipped. It would be impossible to examine in detail all the possibilities that exist between $P1$, $P2$, and CP, but let us examine a particular case of $P1 < CP$ and $P2 > CP$. Line 3730 is, therefore, satisfied, and we proceed to the subroutine at 3800 to first plot $P1$ and then to test if $P1$ or $P2$ equals

CP at 3820. Since this is not the case, the program moves on to line 3830 where the zero line is printed, and then returns to line 3730, where it is directed to go to the subroutine at 3810. The plot point of $P2$ is then printed to complete that line of the printout. At the completion of the subroutine at 3810, it will return to line 3730, where the final RETURN would send it back to 3670, where a PRINT command moves it to the next line and the next I (data point) is determined and plotted. The above description should permit a review of all the possibilities that exist between the three quantities. Choosing sets of conditions and following the path of the program as provided above is one of the best ways to develop a feeling for the flow of the program. It is particularly useful when the program initially appears complex and difficult.

The two provided runs of the program demonstrate how the two plot symbols clearly identify the sinusoidal function. Note in the first plot that the conditions $P1 = P2 = CP$ or $P1 = P2$ did not occur. In the second run, the magnitude remains the same for both functions, but the sign of the phase angles was reversed.

13.7 TEST ROUTINE-SINUSOIDAL SIGNAL

The last program (13-8) of the chapter will test the response to a number of questions regarding the mathematical description of the sinusoidal function. The peak value (M), angular velocity (W), and phase angle in degrees (PH) will be randomly generated and the user will be asked to calculate the frequency (F), the effective value (ME), and the phase angle in radians (PR). Lines 440 and 450 will generate a negative phase angle 50% of the time. Based on the number of testing routines described in the manual thus far, we assumed that the content of this program is fairly easy to understand.

```
      10 REM *****  PROGRAM 13-8  *****
      20 REM **********************************************************
      30 REM Test the calculations involving the sinusoidal signal.
      40 REM **********************************************************
      50 REM
     100 PRINT "Answer the following questions about"
     110 PRINT "a given sinusoidal signal."
     120 FOR K=1 TO 3
     130 PRINT
     140 GOSUB 400 :REM Select values for test
     150 PRINT "For the following sinusoidal signal:"
     160 GOSUB 500 :REM print equation
     170 PRINT
     180 GOSUB 600 :REM Do calculations
     190 INPUT "What is the value of the frequency (in Hertz)";FX
     200 IF ABS(FX-F)<ABS(0.01*F) THEN PRINT "Correct" :GOTO 220
     210 PRINT "No, it's";F;"Hertz"
     220 INPUT "What is the effective value of the voltage";MX
     230 IF ABS(MX-ME)<=ABS(0.01*ME) THEN PRINT "Correct" :GOTO 250
     240 PRINT "No, it's";ME;"volts"
     250 INPUT "What is the phase angle of the voltage (in radians)";PX
     260 IF SGN(PX)<>SGN(PR) THEN GOTO 280
     270 IF ABS(PX-PR)<ABS(0.01*PR) THEN PRINT "Correct" : GOTO 290
     280 PRINT "No, it's";PR;"radians"
     290 NEXT K
     300 END
     400 REM Module to select values of test
     410 M=INT(1+100*RND(X)) :REM 1<M<100
     420 W=100*INT(1+100*RND(X)) :REM 100<W<10,000
     430 PH=INT(180*RND(X)) :REM Angle between 0 and 180 degrees
     440 S=INT(1+2*RND(X)) :REM Negative phase angle 50% of the time
     450 IF S=1 THEN PH=-PH
     460 RETURN
     500 REM Module to print sinusoidal equation
     510 PRINT "v(t)=";
     520 PRINT USING "###",M;
     530 PRINT "sin(";W;"t";
     540 IF PH=0 THEN PRINT ")" :RETURN
     550 IF PH>0 THEN PRINT "+";
     560 PRINT PH;")"
     570 RETURN
     600 REM Module to do calculations
     610 P=3.14159 :REM PI predefined in some BASICs
     620 F=W/(2*P)
     630 ME=0.707*M
     640 PR=PH/57.296
     650 RETURN


READY


RUN


Answer the following questions about
a given sinusoidal signal.

For the following sinusoidal signal:
v(t)= 18sin( 7900 t-109 )

What is the value of the frequency (in Hertz)? 1258
Correct
```

```
What is the effective value of the voltage? 12.7
Correct
What is the phase angle of the voltage (in radians)? -1.9
Correct

For the following sinusoidal signal:
v(t)= 55sin( 6700 t+ 154 )

What is the value of the frequency (in Hertz)? 6700
No, it's 1066.339 Hertz
What is the effective value of the voltage? 55
No, it's 38.885 volts
What is the phase angle of the voltage (in radians)? 154
No, it's 2.6878 radians

For the following sinusoidal signal:
v(t)=  5sin( 200 t-124 )

What is the value of the frequency (in Hertz)? 31.85
Correct
What is the effective value of the voltage? 3.54
Correct
What is the phase angle of the voltage (in radians)? -2.18
Correct


READY
```

Note the interesting range of values for the three continuous runs of the program and the common incorrect answers provided in the second run.

EXERCISES

Write a program to perform each of the following tasks.

1. It is sometimes necessary to plot a sinusoidal function on a time scale axis, rather than degrees or radians. Develop a routine to tabulate Degrees, Radians, and Time for a sinusoidal function of period T. Use an increment in θ of 45°. Establish the following printout for a provided value of T.

Degrees	Radians	Time
0°	0	0
45°	$\pi/4$	$\_\_ = \dfrac{T}{8}$
90°	$\pi/2$	$\_\_ = \dfrac{T}{4}$
135°	$3\pi/2$	$\_\_ = \dfrac{3T}{8}$
·	·	·
·	·	·
·	·	·

etc.

2. Tabulate f, w, and T for a range of f extending from 100 to 1000 Hz in steps of 100 Hz.

3. Plot f versus T for a range of f from 10 to 400 Hz in increments of 10 Hz.

4. Develop a scheme for finding the average value of the curve of Figure 13.3 using the provided discrete data points.

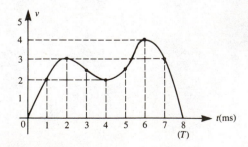

FIGURE 13.3

5. Develop a scheme for determining the effective value of the curve of Figure 13.3.

6. Find the derivative of a function provided in either of the following forms:

$$v = V_m \sin (wt \pm \theta)$$

or

$$v = V_m \cos (wt \pm \theta)$$

7. Given $v(t)$ and $i(t)$ in sinusoidal form, determine whether it is a resistor, inductor, or capacitor, and then calculate the magnitude of the impedance (R, X_L, or X_C), followed by the value of C or L if the impedance is a reactive element. For the first run use:

$$v(t) = 550 \sin (377t + 40°)$$
$$i(t) = 11 \sin (377t - 50°)$$

8. Given $v(t)$ and $i(t)$, determine which is leading or lagging and by how much. The format of the input is limited to $v(t) = V_m \sin (wt \pm \theta)$ and $i(t) = I_m \sin (wt \pm \theta)$. In addition, determine the power delivered to the load, the power factor of the load, and whether the load has a leading or lagging power factor. For the first run use:

$$v(t) = 60 \sin (\omega t + 30°)$$
$$i(t) = 15 \sin (\omega t + 60°)$$

9. Develop a plotting routine that will plot the sum of a dc voltage and a sinusoidal voltage. Plot $v(t) = E_o + E_1 \sin \omega t$, which for the first run should be $v(t) = 5 + 10 \sin 1000t$.

10. Develop a testing routine for determining the average and effective values of a pulsed voltage signal. Randomly generate the magnitude and period of each pulse and include negative values for one out of three pulses (average). Limit the magnitude to the range $0 \rightarrow 10V$ and the period of each pulse to 1 to 5 milliseconds. Test the calculation for a 5 pulse train.

Complex Numbers

14

14.1 INTRODUCTION

Chapter 14 is particularly important because it establishes the foundation for the use of complex numbers in the ac analysis to follow. We also placed the developed modules where they would not be disturbed by the program content particular to the area under investigation. That is, the routines start at locations higher than the last command statement of the programs to be examined. The first few programs will simply perform conversions between forms, while the remaining programs will perform particular operations, such as addition, subtraction, multiplication, and division with complex numbers of either form.

14.2 CONVERSION

The conversion programs to be examined employ the notation and defined quantities of Figure 14.1. In Program 14-1, the polar form $Z < \theta$ will be converted to the rectangular form: $X \pm$

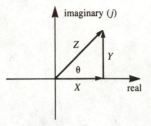

FIGURE 14.1

```
 10 REM *****  PROGRAM 14-1  *****
 20 REM **************************************
 30 REM Program to convert from the polar form
 40 REM to the rectangular form.
 50 REM ************************************************
 60 REM
100 PRINT "Provide input of the polar form and the program"
110 PRINT "will convert to the corresponding rectangular form."
120 PRINT "(Enter 0 to stop)."
130 PRINT
140 INPUT "Polar form: Magnitude=";Z          | Input
145 IF Z=0 THEN END
150 PRINT TAB(12);"and angle=";
160 INPUT TH
170 GOSUB 2100
180 PRINT "Rectangular form=";X;              | Output
190 IF Y>=0 THEN PRINT "+j";Y
200 IF Y<0 THEN PRINT "-j";ABS(Y)
210 GOTO 130
2100 REM Module converts from polar into rectangular form   | Polar
2110 REM Enter with Z, TH(eta) - return with X, Y           | ↓
2120 X=Z*COS(TH*3.14159/180)                                | Rect.
2130 Y=Z*SIN(TH*3.14159/180)
2140 RETURN
```

```
READY

RUN

Provide input of the polar form and the program
will convert to the corresponding rectangular form.
(Enter 0 to stop).

Polar form: Magnitude=? 5
            and angle=? 53.13
Rectangular form= 3 +j 4
```

```
Polar form: Magnitude=? 10
            and angle=? -45
Rectangular form= 7.0711 -j 7.0711

Polar form: Magnitude=? 50
            and angle=? 30
Rectangular form= 43.3013 +j 25

Polar form: Magnitude=? 0


READY
```

jY. The parameters of the rectangular form can be determined from the polar form using the following equations: $X = Z\cos\theta$ and $Y = Z\sin\theta$, as appearing on lines 2120 and 2130 respectively. Note again, however, the need to first convert to radians before using the sine or cosine operators. Lines 190 and 200 determine whether $+j$ or $-j$ should be incorporated with the magnitude of Y as determined by line 2130. Although line 2130 can result in a negative value, the result for $5/-53.13°$ would appear as $3 + j - 4$ rather than $3 - j4$, if it were not for line 200. Remember also that X can be negative, as shown in Figure 14.2, but it will be printed out as $-3 - j4$ for an input such as $5/233.13°$.

Program 14-2 determines the polar form from the rectangular form. The Pythagorean theorem will result in $Z = \sqrt{X^2 + Y^2}$ appear-

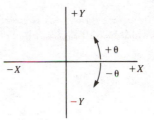

FIGURE 14.2

ing on line 2020. If $X > 0$, as noted on line 2030, then θ (radians) $= \tan^{-1} Y/X$ and θ (degrees) $= (180°/3.14159)\tan^{-1} Y/X$. For each quadrant, the angle θ determined by the program appears in Figure 14.3. If X is less than 0, then θ (degrees) is determined by $(180°)$(sign of Y) $+ (180°/3.14159)\tan^{-1} Y/X$, as on line 2040. If $X = 0$, then θ is either $+$ or $- 90°$, as appearing on line 2050. If $Y = 0$ and X is less than 0, $\theta =$

```
10 REM *****  PROGRAM 14-2  *****
20 REM *********************************
30 REM Program to convert the rectangular
40 REM to the polar form.
50 REM *********************************
60 REM
100 PRINT "Provide the rectangular form and the program"
110 PRINT "will convert to the corresponding polar form."
120 PRINT :INPUT "Rectangular form: X=";X
130 PRINT TAB(14);"and Y=";
140 INPUT Y
150 GOSUB 2000 :REM Conversion into polar form
160 PRINT "   Polar form is: ";
170 PRINT Z;"at an angle of";TH;"degrees"
180 PRINT:PRINT
190 INPUT "More (YES or NO)";A$
200 IF A$="YES" THEN GOTO 100
210 END
```

Input — lines 120, 130, 140
Output — lines 160, 170

(Continued)

```
     ┌2000 REM Module to calculate the polar form from the rectangular form.
      2010 REM Enter with X,Y - Return with Z, TH(eta)
Polar 2020 Z=SQR(X^2+Y^2)
  ↓   2030 IF X>O THEN TH=(180/3.14159)*ATN(Y/X)
Rect. 2040 IF X<O THEN TH=180*SGN(Y)+(180/3.14159)*ATN(Y/X)
      2050 IF X=O THEN TH=90*SGN(Y)
      2060 IF Y=O THEN IF X<O THEN TH=180
     └2070 RETURN


     READY
     *
     RUN


     Provide the rectangular form and the program
     will convert to the corresponding polar form.

     Rectangular form: X=? 3
                   and Y=? 4
        Polar form is:  5 at an angle of 53.1301 degrees


     More (YES or NO)? YES
     Provide the rectangular form and the program
     will convert to the corresponding polar form.

     Rectangular form: X=? -3
                   and Y=? -4
        Polar form is:  5 at an angle of-126.8699 degrees


     More (YES or NO)? YES
     Provide the rectangular form and the program
     will convert to the corresponding polar form.

     Rectangular form: X=? 10
                   and Y=? -10
        Polar form is:  14.1421 at an angle of-45 degrees


     More (YES or NO)? NO


     READY

     *Example 14.1, ICA
```

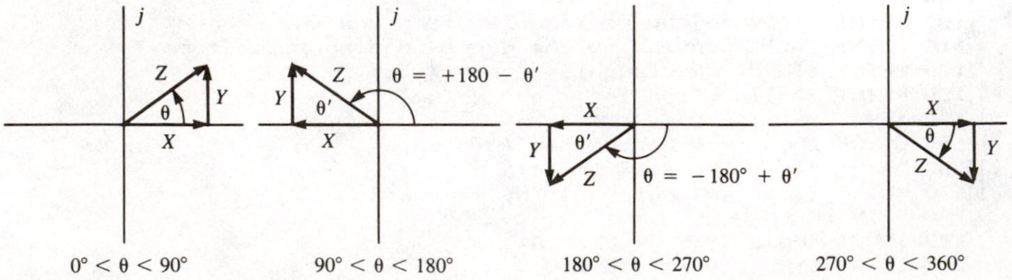

$$0° < \theta < 90° \qquad 90° < \theta < 180° \qquad 180° < \theta < 270° \qquad 270° < \theta < 360°$$

FIGURE 14.3

180°. Each case was described above because this particular subroutine appears frequently in the analysis to follow and should be clearly understood. Be aware, however, that we are now assuming you understand it thoroughly. Note in the provided runs that the resulting angles fall within the guidelines of Figure 14.3.

Program 14-3 includes both conversion formulas and the option of choosing which conversion will be performed. A review of the program will reveal that it is almost totally a "cut and paste" version of the last two programs. Note also that the conversion routines are in the same locations as in Programs 14-1 and 14-2.

```
10 REM *****  PROGRAM 14-3  *****
20 REM *****************************************
30 REM Program to perform selected conversions
40 REM *****************************************
50 REM
100 PRINT
110 PRINT "Enter (1) for rectangular to polar conversion"
120 PRINT "      (2) for polar to rectangular conversion"
130 PRINT TAB(20);
140 INPUT "Choice=";C :REM C is choice 1 or 2
150 IF C<0 OR C>2 THEN GOTO 110
160 ON C GOSUB 200,300
170 PRINT:INPUT "More(YES or NO)";A$
180 IF A$="YES" THEN GOTO 100
190 END
200 REM Use rectangular to polar conversion module
210 PRINT:PRINT:PRINT "Enter rectangular data:"
220 INPUT "X=";X :INPUT "Y=";Y
230 GOSUB 2000
240 PRINT:PRINT "Polar form is";Z;"at an angle of";TH;"degrees"
250 RETURN
300 REM Use polar to rectangular conversion
310 PRINT:PRINT "Enter polar data:":PRINT:INPUT "Z=";Z
320 INPUT "Angle(degrees), TH=";TH
330 GOSUB 2100
340 PRINT:PRINT "Rectangular form is";X;
350 IF Y>=0 THEN PRINT "+j";Y
360 IF Y<0 THEN PRINT "-j";ABS(Y)
370 RETURN
2000 REM Module to convert from rectangular to polar form.
2010 REM Enter with X, Y - Return with Z, TH(eta)
2020 Z=SQR(X^2+Y^2)
2030 IF X<0 THEN TH=(180/3.14159)*ATN(Y/X)
2040 IF X<0 THEN TH=180*SGN(Y)+(180/3.14159)*ATN(Y/X)
2050 IF X=0 THEN TH=90*SGN(Y)
2060 IF Y=0 THEN IF X<0 THEN TH=180
2070 RETURN
2100 REM Module to convert from polar to rectangular form.
2110 REM Enter with Z, TH(eta) - return with X, Y
2120 X=Z*COS(TH*3.14159/180)
2130 Y=Z*SIN(TH*3.14159/180)
2140 RETURN
```

Margin labels:
Input- (Rect.) — lines 220, 230
Output- (Polar) — lines 240, 250
Input (Polar) — lines 310, 320
Output (Rect.) — lines 340, 350, 360
Rect. ↓ Polar — lines 2000–2070
Polar ↓ Rect. — lines 2100–2140

```
READY

RUN


Enter (1) for rectangular to polar conversion
      (2) for polar to rectangular conversion
              Choice=? 2
```

```
Enter polar data:

Z=? 5
Angle(degrees), TH=? -53.13

Rectangular form is 3 -j 4

More(YES or NO)? YES

Enter (1) for rectangular to polar conversion
      (2) for polar to rectangular conversion
                  Choice=? 1


Enter rectangular data:
X=? -10
Y=? 20

Polar form is 22.3607 at an angle of 116.565 degrees

More(YES or NO)? YES

Enter (1) for rectangular to polar conversion
      (2) for polar to rectangular conversion
                  Choice=? 2

Enter polar data:

Z=? 12
Angle(degrees), TH=? 35

Rectangular form is 9.8298 +j 6.8829

More(YES or NO)? NO


READY
```

14.3 ADDITION OF VECTORS

The following program (14-4) will find the sum of vectors input in either the rectangular or polar form. Line 120 will initially set the variables XS and YS to zero. XS is the algebraic sum of the real components and YS is the algebraic sum of the imaginary components to be used in the sum process of lines 540 and 550. Since the addition operation is normally performed in the rectangular form, any input in the polar form is first converted to the rectangular form before the addition takes place. The subroutines at 2000 and 2100 are the conversion routines used earlier. For explanation purposes,

consider the case of adding a vector in the rectangular form to one in the polar form. For the rectangular input, $C = 1$ on line 140, and line 160 will send the program to the subroutine at line 500. At 500 the data is accepted and $XS = XS + X_1 = 0 + X_1 = X_1$, with $YS = Y_1$. It will then RETURN to line 170 to determine if another vector will be added (this program is good for the addition of any number of vectors). In this case, it will return to line 130 to pick up the data on the vector in polar form. It will then move on to the subroutine at 300, where Z and θ are requested, and the conversion to the rectangular form performed using the subroutine starting at 2100. It will then determine the new sum of $XS = XS + X_2 = X_1 + X_2$ and $YS = Y_1$

$+ Y_2$, where any one of the variables can be positive or negative. Since there are only two vectors to be added, the user will respond with a "no" to the question of line 170 and move on to 200, where the answer in rectangular form is printed out. It will then move on to the sub-routine at 2000 before printing out the same answer in polar form.

In the provided run, two vectors in the rectangular form are added to one in the polar form.

```
10 REM *****  PROGRAM 14-4  *****
20 REM *********************************
30 REM Program to sum vectors in either the
40 REM polar or rectangular form.
50 REM *********************************
60 REM
100 PRINT "The program will sum vectors in either"
110 PRINT "the polar or rectangular form."
120 XS=0 :YS=0
130 PRINT
140 INPUT "Enter (1) Rectangular, or (2) Polar input data";C
150 IF C<0 OR C>2 THEN GOTO 130
160 ON C GOSUB 500,300
170 PRINT :INPUT "More(YES or NO)";A$
180 IF A$="YES" THEN GOTO 130
190 PRINT
200 PRINT "Sum in rectangular form is:";XS;
210 IF YS>=0 THEN PRINT "+j";YS ELSE PRINT "-j";ABS(YS)
220 X=XS:Y=YS
230 GOSUB 2000 :REM Convert to polar form
240 PRINT "and in polar form :";
250 PRINT Z;"at an angle of";TH;"degrees"
260 END
300 REM Accept polar input
310 PRINT:PRINT "Enter polar form:"
320 INPUT "Z=";Z
330 INPUT "at an angle (degrees), TH=";TH
340 GOSUB 2100 :REM Convert to rectangular form
350 XS=XS+X :YS=YS+Y :REM Add input to present sum
360 RETURN
500 REM Accept rectangular data
510 PRINT :PRINT "Enter rectangular form:"
520 INPUT "X=";X
530 INPUT "Y=";Y
540 XS=XS+X
550 YS=YS+Y
560 RETURN
2000 REM Module to convert from rectangular to polar form
2010 REM Enter with X, Y - Return with Z, TH(eta)
2020 Z=SQR(X^2+Y^2)
2030 IF X>0 THEN TH=(180/3.14159)*ATN(Y/X)
2040 IF X<0 THEN TH=180*SGN(Y)+(180/3.14159)*ATN(Y/X)
2050 IF X=0 THEN TH=90*SGN(Y)
2060 IF Y=0 THEN IF X<0 THEN TH=180
2070 RETURN
2100 REM Module to convert from polar to rectangular form
2110 REM Enter with Z, TH(eta) - return with X, Y
2120 X=Z*COS(TH*3.14159/180)
2130 Y=Z*SIN(TH*3.14159/180)
2140 RETURN

READY
```

```
RUN

The program will sum vectors in either
the polar or rectangular form.

Enter (1) Rectangular, or (2) Polar input data? 2

Enter polar form:
Z=? 10
at an angle (degrees), TH=? 50

More(YES or NO)? YES

Enter (1) Rectangular, or (2) Polar input data? 1

Enter rectangular form:
X=? 3
Y=? 4

More(YES or NO)? YES

Enter (1) Rectangular, or (2) Polar input data? 1

Enter rectangular form:
X=? -2
Y=? 6

More(YES or NO)? NO

Sum in rectangular form is: 7.4279 +j 17.6604
and in polar form : 19.1589 at an angle of 67.1887 degrees

READY
```

14.4 ARITHMETIC OPERATIONS WITH VECTORS

The addition, subtraction, multiplication, and division of vectors will occur quite frequently in the analysis of ac networks. It would therefore be extremely useful to have a program that will perform all of these functions and only require (as input) the vectors in either form and the operation to be performed. Program 14-5 will perform all of these functions, using that form most applicable to each operation: addition and subtraction are performed in rectangular form and multiplication and division in polar form. If the input data is in the wrong form, it will be converted to the other form using one of the routines developed in the past few programs. Beginning at line 2400, a number of modules will now be introduced for the first time that will prove very useful in the analysis to follow.

The content is sufficiently important to require that we review each of the four arithmetic operations. The addition and subtraction routines are very similar to those discussed in the previous chapters. Addition will result in $OP = 1$ on line 260, which will send it to the subroutine at 2400 as directed by line 280. The rectangular components of each will then be added to obtain the real and imaginary components of the sum vector $X = XA$ and $Y = YA$. The result is then converted to the polar form by line 2430 to be sure that both forms are available when we print out the results. If the sub-

```
          10 REM *****  PROGRAM 14-5  *****
          20 REM ********************************************
          30 REM Program to perform vector arithmetic
          40 REM with the input in either polar or
          50 REM rectangular form.
          60 REM ********************************************
          70 REM
          100 PRINT "This program will add, subtract, multiply or"
          110 PRINT "divide two vectors in the polar or rectangular form."
          120 PRINT:PRINT
        ┌─130 PRINT "Enter data for vector Z1"
        │ 140 GOSUB 450 :REM Provide input values for Z1
   Z₁   │ 150 X1=X :Y1=Y
        └─160 Z1=Z :T1=TH :REM Save rectangular and polar forms of Z1
        ┌─170 PRINT:PRINT "Enter data for vector Z2"
        │ 180 GOSUB 450 :REM Provide input values for Z2
   Z₂   │ 190 X2=X :Y2=Y
        └─200 Z2=Z :T2=TH :REM Save rectangular and polar forms of Z2
        ┌─210 PRINT:PRINT "Select operation"
        │ 220 PRINT TAB(20);"(1) Add Z1+Z2"
 Choice │ 230 PRINT TAB(20);"(2) Subtract Z1-Z2"
        │ 240 PRINT TAB(20);"(3) Multiply Z1*Z2"
        └─250 PRINT TAB(20);"(4) Divide Z1/Z2"
          260 INPUT OP
          270 IF OP<0 OR OP>4 THEN GOTO 220
          280 ON OP GOSUB 2400,2470,2500,2570
        ┌─290 PRINT "Answer: In rectangular form";XA;
 Output │ 300 IF YA>=0 THEN PRINT "+j";YA ELSE PRINT "-j";ABS(YA)
        └─310 PRINT TAB(11);"and in polar form";ZA;"at an angle of";TA;"degrees"
          320 END
        ┌─350 REM Accept polar input
        │ 360 PRINT :PRINT "Enter polar data:"
  Polar │ 370 INPUT "Z=";Z
  Input │ 380 INPUT "at an angle of (degrees), TH=";TH
        │ 390 GOSUB 2100 :REM Convert input to rectangular form
        └─400 RETURN
        ┌─450 REM Module to choose rectangular or polar input
        │ 460 INPUT "Enter (1) Rectangular or (2) Polar input ";C
  Input │ 470 IF C<>1 THEN IF C<>2 THEN GOTO 460
  form  │ 480 ON C GOSUB 500,350
        └─490 RETURN
        ┌─500 REM Accept rectangular input
        │ 510 PRINT :PRINT "Enter rectangular data:"
  Rect. │ 520 INPUT "X=";X
  Input │ 530 INPUT "Y=";Y
        │ 540 GOSUB 2000 :REM Convert input to polar form
        └─550 RETURN
        ┌─2000 REM Module converts from the rectangular into the polar form
        │ 2010 REM Enter with X,Y - return with Z,TH(eta)
  Rect. │ 2020 Z=SQR(X^2+Y^2)
   ↓    │ 2030 IF X>0 THEN TH=(180/3.14159)*ATN(Y/X)
  Polar │ 2040 IF X<0 THEN TH=180*SGN(Y)+(180/3.14159)*ATN(Y/X)
        │ 2050 IF X=0 THEN TH=90*SGN(Y)
        │ 2060 IF Y=0 THEN IF X<0 THEN TH=180
        └─2070 RETURN
        ┌─2100 REM Module converts from the polar form to the rectangular form
  Polar │ 2110 REM Enter with Z,TH - return with X,Y
   ↓    │ 2120 X=Z*COS(TH*3.14159/180)
  Rect. │ 2130 Y=Z*SIN(TH*3.14159/180)
        └─2140 RETURN
          2400 REM Module to Add/Subtract Z1 and Z2
```

(Continued)

```
        ⌈2410 XA=X1+X2 :YA=Y1+Y2 :REM Add vectors
        │2420 X=XA :Y=YA
Z₁ + Z₂ │2430 GOSUB 2000 :REM Convert answer to polar form
        │2440 ZA=Z :TA=TH
        ⌊2450 RETURN
        ⌈2460 REM Enter here for subtraction of vectors
Z₁ - Z₂ │2470 XA=X1-X2 :YA=Y1-Y2 :REM Subtract vectors
        ⌊2480 GOTO 2420 :REM Convert to polar form and return
         2500 REM Module to Multiply/Divide Z1 by Z2
        ⌈2510 ZA=Z1*Z2 :TA=T1+T2 :REM Multiply polar vectors
        │2520 Z=ZA :TH=TA
Z₁ × Z₂ │2530 GOSUB 2100 :REM Convert answer to rectangular form
        │2540 XA=X :YA=Y
        ⌊2550 RETURN
        ⌈2560 REM Enter here for vector division
  Z₁    │2570 TA=T1-T2
  ──    │2580 IF Z2=0 THEN ZA=1E30 ELSE ZA=Z1/Z2
  Z₂    ⌊2590 GOTO 2520
```

```
READY

RUN

This program will add, subtract, multiply or
divide two vectors in the polar or rectangular form.


Enter data for vector Z1
Enter (1) Rectangular or (2) Polar input ? 2

Enter polar data:
Z=? 10
at an angle of (degrees), TH=? 45


Enter data for vector Z2
Enter (1) Rectangular or (2) Polar input ? 2

Enter polar data:
Z=? 2
at an angle of (degrees), TH=? 20

Select operation
                (1) Add Z1+Z2
                (2) Subtract Z1-Z2
                (3) Multiply Z1*Z2
                (4) Divide Z1/Z2
? 4
Answer: In rectangular form 4.5315 +j 2.1131
        and in polar form 5 at an angle of 25 degrees


READY



This program will add, subtract, multiply or
divide two vectors in the polar or rectangular form.


Enter data for vector Z1
Enter (1) Rectangular or (2) Polar input ? 1
```

```
Enter rectangular data:
X=? 10
Y=? 10

Enter data for vector Z2
Enter (1) Rectangular or (2) Polar input ? 2

Enter polar data:
Z=? 2
at an angle of (degrees), TH=? 45

Select operation
                    (1) Add Z1+Z2
                    (2) Subtract Z1-Z2
                    (3) Multiply Z1*Z2
                    (4) Divide Z1/Z2
? 3
Answer: In rectangular form 1.8764E-05 +j 28.2843
        and in polar form 28.2843 at an angle of 90 degrees


           READY
```

traction operation were chosen, then $OP = 2$, and the subroutine at 2470 is applied to first find $XA = X_1 - X_2$ and $YA = Y_1 - Y_2$ and then return to line 2420 to define X and Y and proceed to convert it to the polar form. The use of lines 2420 through 2450 for the subtraction operation eliminates having to rewrite the four lines. Both operations are therefore included in the REM statement of line 2400 that defines the role of the module extending from 2400 to 2480. The module from 2500 to 2590 is very similar in format to the module just described. Both operations of multiplication and division are included, and lines 2520 through 2550 are used to convert to rectangular form. In general, therefore, both modules cover two mathematical operations, but the point of entry will determine which operation will be performed. The subroutines appearing from line 2000 through 2590 will reappear in the chapters to follow. In fact, to save space they will often be omitted, since the content will not change and you can simply refer to the last use of the routine. Looking back to the original program, you can see that lines 10 through 550 are there simply to title the program, request the input data, and display the results in rectangular and polar form. Fi-

nally, note the use of XA, YA, ZA, and TA throughout the module of 2400 through 2590 to permit using these variables for the printout of the results on lines 290 through 310, even though the answers may have been the result of one of four different operations.

In the second run of the program, the magnitude of XA as compared to YA suggests that the answer can read as $+j28.28$ which directly compares with the polar form.

14.5 VECTOR SUM OF SINUSOIDAL FUNCTIONS

The vector algebra described in the previous programs will now be applied to determine the sum of two sinusoidal signals. The peak values of the signals will be used when the vector sum is determined because the effective values will not be displayed. It removes the necessity to calculate the effective value of each waveform and then convert back to the peak value when the solution is printed out. Before analyzing Program 14-6 in detail, note the presence of the

```
  10 REM *****  PROGRAM 14-6  *****
  20 REM ************************************************
  30 REM Program accepts input in the sinusoidal form
  40 REM and provides the sum in the same format.
  50 REM ************************************************
  60 REM
 100 PRINT "Input the following sinusoidal signal information:"
 110 INPUT "Signal frequency (Hertz)=";F
 120 W=2*3.14159*F
 130 PRINT
 140 PRINT "For signal 1:"
 150 GOSUB 700 :REM Obtain input data
 160 GOSUB 900 :REM Print in sinusoidal form
 170 GOSUB 2100 :REM Convert to rectangular form
 180 X1=X :Y1=Y :Z1=Z :T1=TH :REM Save values in polar and rectangular form
 190 PRINT
 200 PRINT "For signal 2:"
 210 GOSUB 700 :REM Obtain input data
 220 GOSUB 900 :REM Print in sinusoidal form
 230 GOSUB 2100 :REM Convert to rectangular form
 240 X2=X :Y2=Y :Z2=Z :T2=TH :REM Save values for signal 2
 250 REM Sum signals
 260 X=X1+X2
 270 Y=Y1+Y2
 280 GOSUB 2000 :REM Convert to polar form
 290 PRINT
 300 PRINT "The sinusoidal expression for the sum is:"
 310 PRINT
 320 GOSUB 900
 330 END
 700 REM Module to accept input data
 710 INPUT "Amplitude=";Z
 720 INPUT "with phase angle (degrees)=";TH
 730 PRINT
 740 RETURN
 900 REM Print sinusoidal equation form
 910 PRINT "v(t)=";Z;"sin(";W;"t";
 920 IF TH>=0 THEN PRINT "+";
 930 PRINT TH;")"
 940 RETURN
2000 REM Module converts from the rectangular form to the polar form
2010 REM Enter with X,Y - return with Z,TH(eta)
2020 Z=SQR(X^2+Y^2)
2030 IF X>0 THEN TH=(180/3.14159)*ATN(Y/X)
2040 IF X<0 THEN TH=180*SGN(Y)+(180/3.14159)*ATN(Y/X)
2050 IF X=0 THEN TH=90*SGN(Y)
2060 IF Y=0 THEN IF X<0 THEN TH=180
2070 RETURN
2100 REM Module converts from the polar form to the rectangular form
2110 REM Enter with Z,TH - return with X,Y
2120 X=Z*COS(TH*3.14159/180)
2130 Y=Z*SIN(TH*3.14159/180)
2140 RETURN


READY

RUN

Input the following sinusoidal signal information:
Signal frequency (Hertz)=? 120
```

```
For signal 1:
Amplitude=? 70
with phase angle (degrees)=? 25

v(t)= 70 sin( 753.9816 t+ 25 )

For signal 2:
Amplitude=? 100
with phase angle (degrees)=? -40

v(t)= 100 sin( 753.9816 t-40 )

The sinusoidal expression for the sum is:

v(t)= 144.2798 sin( 753.9816 t-13.9145 )

READY
```

conversion routines on lines 2000 and 2100. Realize also that lines 150 and 210 are requesting the input data in polar form using the module at 700. The combination of the sequence beginning at 300 and continuing at 900 is limited solely to printing out the result. The heart of the program is lines 260 and 270, for the addition of the waveforms, and 280, for providing the result in a form convenient for printing out the sinusoidal function for the sum of the two input signals.

Note in the printout the requirement that the two signals have the same angular velocity. Otherwise, there are no limitations on the peak values and the associated phase angles. Although the process appears quite simple, mak-

ing the conversions and performing the addition by hand would be very time-consuming.

14.6 TESTING THE CONVERSION PROCESS

The process of converting from one form to another is important in analyzing ac networks. It is a process that will be performed many times when analyzing even the simplest of networks. Program 14-7 will test your ability to convert from one form to the other in any one of the four quadrants. On line 1220, an angle in the

```
10 REM ***** PROGRAM 14-7 *****
20 REM *************************************************
30 REM Program selects polar or rectangular coordinates
40 REM and tests conversion.
50 REM *************************************************
60 REM
100 PRINT "This program tests polar and rectangular conversions"
110 PRINT "(Enter X=0,Y=0 or Z=0,TH=0 to END)"
120 N=5 :PRINT
130 FOR I=1 TO N
140 C=INT(1+2*RND(XV))
150 ON C GOSUB 1200,1400 :REM C=1 Polar-to-Rect., C=2 Rect.-to-Polar
160 PRINT
170 IF I<N THEN PRINT "Thus far you have answered";R;"question(s) correctly"
180 PRINT
190 NEXT I
200 PRINT
210 PRINT "You have answered";R;"questions out of";I;"correctly"
220 END
```

(Continued)

```
1200 REM Select polar coordinates - test conversion into rectangular
1210 Z=INT(1+20*RND(XV)) :REM Select magnitude from 1 to 20
1220 TH=INT(1+180*RND(X)) :REM Select THeta from 1 to 180
1230 SG=+1
1240 S=INT(1+4*RND(XV))
1250 IF S=3 THEN SG=-1
1260 TH=SG*TH :REM Include -sign 1 out of 4 times
1270 PRINT "Polar form of";Z;"at an angle of";TH;"degrees"
1280 PRINT "Rectangular form:"
1290 INPUT "X=";XA
1300 INPUT "Y=";YA
1310 IF XA=0 AND YA=0 THEN I=I-1 :GOTO 200
1320 GOSUB 2100 :REM Convert polar form to the rectangular form
1330 IF SGN(X)<>SGN(XA) THEN GOTO 1380
1340 IF ABS(X-XA)>ABS(0.01*X) THEN GOTO 1380
1350 IF SGN(Y)<>SGN(YA) THEN GOTO 1380
1360 IF ABS(Y-YA)>ABS(0.01*Y) THEN GOTO 1380
1370 PRINT "Correct" :R=R+1 :RETURN
1380 PRINT "No, it's";X; :IF Y>=0 THEN PRINT "+j";Y :RETURN
1390 IF Y<0 THEN PRINT "-j";ABS(Y) :RETURN
1400 REM Select Rectangular coordinates - test polar conversion
1410 PRINT
1420 X=INT(1+25*RND(XV)) :REM Select values from 1 to 25
1430 Y=INT(1+25*RND(XV))
1440 SG=+1
1450 S=INT(1+4*RND(XV))
1460 IF S=3 THEN SG=-1
1470 X=SG*X :REM Include - sign one of four times
1480 SG=+1
1490 S=INT(1+4*RND(XV))
1500 IF S=3 THEN SG=-1
1510 Y=SG*Y :REM Include - sign one of four times
1520 PRINT "Rectangular form of";X;
1530 IF Y>=0 THEN PRINT "+j";Y ELSE PRINT "-j";ABS(Y)
1540 PRINT "Polar form:"
1550 INPUT "Magnitude, Z=";ZA
1560 INPUT "at an angle of, Theta (degrees)=";TA
1570 IF ZA=0 AND TA=0 THEN I=I-1 :GOTO 200
1580 GOSUB 2000 :REM Convert to polar form
1590 IF ABS(Z-ZA)>ABS(0.01*Z) THEN GOTO 1620
1600 IF SGN(TH)<>SGN(TA) THEN GOTO 1620
1605 IF ABS(TH-TA)>ABS(0.01*TH) THEN GOTO 1620
1610 PRINT "Correct" :R=R+1 :RETURN
1620 PRINT "No, it's";Z;"at angle";TH;"degrees"
1630 RETURN
2000 REM Module to convert from the rectangular form to the polar form
2010 REM Enter with X,Y - return with Z,TH(eta)
2020 Z=SQR(X^2+Y^2)
2030 IF X>0 THEN TH=(180/3.14159)*ATN(Y/X)
2040 IF X<0 THEN TH=180*SGN(Y)+(180/3.14159)*ATN(Y/X)
2050 IF X=0 THEN TH=90*SGN(Y)
2060 IF Y=0 THEN IF X<0 THEN TH=180
2070 RETURN
2100 REM Module converts from the polar form to the rectangular form
2110 REM Enter with Z,TH - return with X,Y
2120 X=Z*COS(TH*3.14159/180)
2130 Y=Z*SIN(TH*3.14159/180)
2140 RETURN
```

READY

```
RUN

This program tests polar and rectangular conversions
(Enter X=0,Y=0 or Z=0,TH=0 to END)


Rectangular form of 15 +j 25
Polar form:
Magnitude, Z=? 29
at an angle of, Theta (degrees)=? 59
Correct

Thus far you have answered 1 question(s) correctly


Rectangular form of 24 -j 7
Polar form:
Magnitude, Z=? 24
at an angle of, Theta (degrees)=? 16
No, it's 25 at angle-16.2602 degrees

Thus far you have answered 1 question(s) correctly


Rectangular form of 3 -j 18
Polar form:
Magnitude, Z=? 18.2
at an angle of, Theta (degrees)=? -80.5
Correct

Thus far you have answered 2 question(s) correctly

Polar form of 1 at an angle of 4 degrees
Rectangular form:
X=? 1
Y=? .07
Correct

Thus far you have answered 3 question(s) correctly

Polar form of 20 at an angle of-176 degrees
Rectangular form:
X=? -20
Y=? -1.4
Correct


You have answered 4 questions out of 5 correctly


READY
```

range 1 to 180° is chosen. For the polar form, lines 1230 through 1260 will include a minus sign one out of four times (on an average basis). The value of Z will fall in the range of one to twenty, while the values of X and Y, as defined by lines 1420 through 1510, will include a negative value for X or Y one out of four times. For each solution in either form, two testing routines will have to be performed: in rectangular form for both X and Y and in the polar form for Z, and θ. If either is wrong, the answer will be considered incorrect, and the correct solution printed out by line 1380 or line 1620. Note that once the program is RUN, it will proceed on its own to line 1290 or line 1550 before asking for an input to test against its own solution. That is, it will randomly choose which conversion will be performed on line 140, and then choose components starting at lines 1200 or 1400. In summary, lines 10 through 220 choose the conversion to be performed and provide the number of correct answers. Lines 1200 through 1390 generate a polar form and test the conversion in rectangular form. Lines 1400 through 1630 generate a rectangular form and test the conversion in the polar form. By now lines 2000 through 2140 should be quite familiar.

Note in the run that the rectangular to polar conversion was chosen three out of five times and the angle negative two out of five times.

EXERCISES

Write a program to perform the following operations.

1. Given the input vectors in rectangular form, determine the sum and difference of the vectors and print out the solution in rectangular form. Develop your own routine. Do not refer to Chapter 14.

2. Repeat problem 1 for the multiplication and division of the vectors.

3. Given the input vectors in polar form, determine the product and division of the vectors and print out the solution in polar form.

4. Develop your own routine for converting vectors in the first quadrant from one form to the other. Provide a choice of which operation is to be performed.

5. Develop a general solution for the following operation, given the forms of the vectors appearing in the equation.

$$\frac{(A + jB)(C - jD)}{E(F + jG)}$$

For the first run, use $A = 3$, $B = 4$, $C = 5$, $D = 5$, $E = 10$, $F = 12$, and $G = 24$ and provide the result in rectangular form.

6. Repeat problem 5 for the following:

$$\frac{(A\underline{/\theta_A})(B\underline{/\theta_B})}{C(D\underline{/\theta_D})}$$

For the first run, use $A = 3$, $\theta_A = 30°$, $B = 4$, $\theta_B = 60°$, $C = 6$, $D = 12$, and $\theta_D = -80$ and provide the result in polar form.

7. Given a voltage in phasor or time domain, convert the voltage to the other form. That is, permit a choice of which operation is to be performed and print out the results in the proper format. For the first run, input $\mathbf{V} = 10\underline{/20°}$ and, for the second run, input $v = 30 \sin(377t - 60°)$.

8. Determine the total current entering a junction if the current splits into two components with the following format: $i_1 = I_1\sin(\omega t \pm \theta_1)$, $i_2 = I_2 \sin(\omega t \pm \theta_2)$. Print out the total current in the sinusoidal format. For the first run, use $i_1 = 5 \sin(300t + 30°)$ and $i_2 = 20 \sin(300t - 60°)$.

9. The total voltage across a branch is determined by the sum of two sinusoidal voltages. Determine one of the series voltages if the total voltage is given by $e = E_m \sin(\omega t \pm \theta_1)$ and one of the series voltages is $v = V_m \sin(\omega t \pm \theta_2)$. For the first run, use $e = 120 \sin \omega t$ and $v = 60 \sin(200t + 50°)$.

10. Develop a testing routine for the following:

$$\frac{(A\underline{/60°})(B + j5)}{20\underline{/\theta_D}}$$

That is, determine the general solution in the rectangular form and then randomly generate values of A, B, and θ_D in the following ranges: $5 < A < 25$, $1 < B < 50$, $30° < \theta_D < 60°$. Test the solution in the rectangular form. If the real and imaginary parts are within 1% of the computer-calculated solution, the results can be considered correct.

Ohm's Law - AC

15

15.1   INTRODUCTION

Applying Ohm's law to the single resistive, inductive, or capacitive element permits the development of a variety of programs in the ac domain sufficient in breadth to warrant a complete chapter. The impact of the phase angle on the analysis will be included throughout and a routine will be introduced that will plot both the input and output signals.

15.2   OHM'S LAW - RESISTOR

The analysis of Program 15-1 is limited to the resistor. The sinusoidal expression for the voltage will be determined from the input quantities of the resistor R, frequency f, the *rms* value of the current, and the phase angle.

Note in the program the logic applied to label the various currents, voltages, and phase angles. The letter capital E is applied to the current and voltage as IE and VE to designate effective values. The letter M is reserved for the peak (maximum amplitude) value of each as IM and VM, *while* IA and VA refer to the phase angles of the current and voltage respectively. Since both the voltage and current will be printed out in the phasor and sinusoidal forms, the subroutines at 700 and 900 were developed to eliminate the necessity of reproducing the steps for each quantity. The variables MA and PA are simply set equal to VM or IM and VA or

```
          10 REM *****   PROGRAM 15-1  *****
          20 REM ********************************************
          30 REM Program to use Ohm's law in the form V=I*R
          40 REM for ac sinusoidal signals.
          50 REM ********************************************
          60 REM
         100 PRINT "This program calculates the voltage across"
         110 PRINT "a resistor, R, for a sinusoidal current."
         120 PRINT
        ┌130 INPUT "For the input current, what is the frequency f";F
  Input │140 INPUT "What is the value(rms) of the current";IE
        │150 INPUT "and the phase angle (in degrees)";IA
        └160 INPUT "The value of the resistor, R";R
         170 PRINT
        ┌180 PRINT "The phasor form of the current is:"
        │190 W=2*3.14159*F
        │200 IM=IE*SQR(2)
        │210 MA=IM :PA=IA :REM Set maximum current and phase angle
  iR    │220 GOSUB 700 :REM To print in phasor form
        │230 PRINT "The current in sinusoidal form is:"
        │240 PRINT TAB(10);"i(t)=";
        └250 GOSUB 900 :REM Print in sinusoidal form
        ┌260 REM Now calculate and output V and v(t)
        │270 VM=IM*R
        │280 VE=IE*R
        │290 VA=IA+0 :REM Phase angle of R is 0 degrees
        │300 MA=VM :PA=VA
  vR    │310 PRINT
        │320 PRINT "The phasor form of the voltage across R is:"
        │330 GOSUB 700
        │340 PRINT "and the sinusoidal voltage across R is:"
        │350 PRINT TAB(10);"v(t)=";
        └360 GOSUB 900
         370 PRINT
        ┌380 REM Power calculation
  P     │390 P=IE^2*R
        └400 PRINT "Power dissipated by R is";P;"watts"
         410 END
```

```
Phasor       ┌700 REM Module to print phasor form of signal
form         │710 REM MA=magnitude, PA=phase angle
             │720 PRINT MA/SQR(2);"at an angle of";PA;"degrees"
             └730 RETURN
             ┌900 REM Print sinusoidal form
             │910 REM MA is the peak value, PA is the phase angle
Sinusoidal   │920 PRINT MA;"sin(";W;"t";
format       │930 IF PA>0 THEN PRINT "+";PA;")"
             │940 IF PA=0 THEN PRINT ")"
             │950 IF PA<0 THEN PRINT PA;")"
             └960 RETURN
```

READY

RUN

This program calculates the voltage across
a resistor, R, for a sinusoidal current.

For the input current, what is the frequency f? <u>100</u>
What is the value(rms) of the current? <u>4</u>
and the phase angle (in degrees)? <u>30</u>
The value of the resistor, R? <u>2</u>

The phasor form of the current is:
 4 at an angle of 30 degrees
The current in sinusoidal form is:
        i(t)= 5.6569 sin(628.318 t+ 30)

The phasor form of the voltage across R is:
 8 at an angle of 30 degrees
and the sinusoidal voltage across R is:
        v(t)= 11.3137 sin(628.318 t+ 30)

Power dissipated by R is 32 watts

READY

RUN

This program calculates the voltage across
a resistor, R, for a sinusoidal current.

For the input current, what is the frequency f? <u>60</u>
What is the value(rms) of the current? <u>25E-3</u>
and the phase angle (in degrees)? <u>-40</u>
The value of the resistor, R? <u>2.2E3</u>

The phasor form of the current is:
 .025 at an angle of-40 degrees
The current in sinusoidal form is:
        i(t)= .035 sin(376.9908 t-40)

The phasor form of the voltage across R is:
 55 at an angle of-40 degrees
and the sinusoidal voltage across R is:
        v(t)= 77.7817 sin(376.9908 t-40)

Power dissipated by R is 1.375 watts

READY

IA respectively before entering the printout routines. Of course, $v(t) =$ or $i(t) =$ must appear before the module at 900.

Reviewing the program, we find that lines 100 through 170 title the program and ask for the input data. Lines 180 through 220 and the subroutine from 700 to 730 will calculate the angular velocity and peak value of the current and print the result in phasor form. It will then return to line 230, where lines 230 and 240 and the subroutine at 900 will print out the sinusoidal form.

Lines 270 and 280 calculate the peak (*VM*) and effective (*VE*) values of the voltage across the resistor using Ohm's law. Line 290 will set the angle of the voltage equal to the phase angle of the current, since the phase shift introduced by a resistor is zero degrees. The process described for the current is then repeated to provide the phasor and sinusoidal forms of the resulting voltage.

Lines 380 through 400 will determine the power and print out its magnitude before the program ends at 410. Note in the provided runs that the phase angle and angular velocity are the same for $v(t)$ and $i(t)$ and that the peak values are related by the resistor R.

15.3 *R-L-C*

Program 15-2 is an extension of Program 15-1 that will generate the sinusoidal current of a resistor, inductor, or capacitor due to an applied voltage. Aside from the Ohm's law calculation required for each, the phase angle adjustments for the inductor and capacitor will also be included in the output expression. As with any long program, it is usually wise to review the entire program and break it down into sections that define a particular operation. For instance, lines 100 through 260 will title the program, request the voltage data, and print out the applied voltage in sinusoidal form. Lines 270 through 300 will ask for the element to be examined and send the program to the appropriate subroutine (one for each element). For $X\$ = L$, the subroutine at 450 will ask for the value of L and then apply Ohm's law in the form $I_M = V_M/X_L = V_M / \omega L$ on line 470. Since the current of an inductor lags the voltage across the inductor by 90 degrees, the phase angle of the current must be 90 degrees less than the phase angle of the applied voltage. Therefore, line 470 also includes the fact that $\theta_I = \theta_V - 90°$. The program will then proceed to line 700 to print

```
       10 REM *****  PROGRAM 15-2  *****
       20 REM **************************************
       30 REM Program to apply Ohm's law to an
       40 REM R, L, or C component.
       50 REM **************************************
       60 REM
      100 PRINT "This program will calculate the"
      110 PRINT "current through an R,L, or C component"
      120 PRINT
      130 PRINT "For the applied voltage:"
      140 INPUT "What is the signal frequency, f";F
      150 IF F=0 THEN END
      160 W=INT(2*3.14159*F+0.5)
      170 INPUT "The rms input voltage is";VE
      180 INPUT "at phase angle (degrees)";VA
      190 VM=VE*SQR(2)
      200 PRINT
      210 PRINT "The voltage applied is";
      220 MA=VM :PA=VA
      230 GOSUB 700
      240 PRINT "In sinusoidal form, v(t)=";
      250 GOSUB 900
      260 PRINT
```

Input $v(t)$ — lines 140–180

Print $v(t)$ — lines 190–250

```
R,L   ┌270  INPUT "Which device is the voltage applied to (R,L, or C)";X$
or    │ 280  IF  X$="R"  THEN GOTO 350
C     │ 290  IF  X$="L"  THEN GOTO 450
      └300  IF  X$="C"  THEN GOTO 550
       310  GOTO 270
      ┌350  INPUT "R=";R :REM Component is R
      │ 360  PRINT:PRINT "The current through resistor R is";
R     │ 370  MA=VM/R  :PA=VA+0
      │ 380  GOSUB 700
      │ 390  PRINT "The sinusoidal current through R is";
      └400  GOTO 510
      ┌450  INPUT "L=";L :REM Component is L
      │ 460  PRINT :PRINT "The current through inductor L is";
L     │ 470  MA=VM/(W*L) :PA=VA-90
      │ 480  GOSUB 700
      └490  PRINT "The sinusoidal current through XL is"
i(t)  ┌500  REM Output sinusoidal form of resulting current
      └510  PRINT " i(t)="; :GOSUB 900
       520  PRINT :INPUT "More(YES or NO)";A$
       530  IF A$="YES" THEN GOTO 120
       540  END
      ┌550  INPUT "C=";C :REM COMPONENT IS C
      │ 560  PRINT:PRINT "The current through capacitor C, is";
C     │ 570  MA=VM*(W*C) :PA=VA+90
      │ 580  GOSUB 700
      └590  GOTO 510
Phasor┌700  REM Print phasor form
form  │ 710  REM MA=maximum, PA=phase angle
      │ 720  PRINT MA/SQR(2);"at an angle of";PA;"degrees"
      └730  RETURN
      ┌900  REM Print sinusoidal form
      │ 910  REM MA is the peak value
Sinusoidal│ 920  PRINT MA;"sin(";W;"t";
Format│ 930  IF PA>0 THEN PRINT "+";PA;")"
      │ 940  IF PA=0 THEN PRINT ")"
      │ 950  IF PA<0 THEN PRINT PA;")"
      └960  RETURN
```

```
READY


RUN


This program will calculate the
current through an R,L, or C component

For the applied voltage:
What is the signal frequency, f? 100
The rms input voltage is? 24
at phase angle (degrees)? 10

The voltage applied is 24 at an angle of 10 degrees
In sinusoidal form, v(t)= 33.9411 sin( 628 t+ 10 )

Which device is the voltage applied to (R,L, or C)? L
L=? 5E-3

The current through inductor L is 7.6433 at an angle of-80 degrees
The sinusoidal current through XL is
 i(t)= 10.8093 sin( 628 t-80 )

More(YES or NO)? YES
```

```
For the applied voltage:
What is the signal frequency, f? 60
The rms input voltage is? 100
at phase angle (degrees)? 25

The voltage applied is 100 at an angle of 25 degrees
In sinusoidal form, v(t)= 141.4214 sin( 377 t+ 25 )

Which device is the voltage applied to (R,L, or C)? C
C=? 2E-6

The current through capacitor C, is .075 at an angle of 115 degrees
 i(t)= .1066 sin( 377 t+ 115 )

More(YES or NO)? YES


For the applied voltage:
What is the signal frequency, f? 1000
The rms input voltage is? 50
at phase angle (degrees)? 30

The voltage applied is 50 at an angle of 30 degrees
In sinusoidal form, v(t)= 70.7107 sin( 6283 t+ 30 )

Which device is the voltage applied to (R,L, or C)? R
R=? 100

The current through resistor R is .5 at an angle of 30 degrees
The sinusoidal current through R is i(t)= .7071 sin( 6283 t+ 30 )

More(YES or NO)? NO


READY
```

out the result in the phasor and sinusoidal form (900). A similar description can be applied for the resistor starting at line 350 and the capacitor at line 550. Line 520 will ask if another element is to be considered before ending the program at line 540.

15.4   PLOTTING ROUTINE

The plotting routines of Chapter 13 will now be used in Program 15-3 to plot both the current and voltage of a resistor, inductor, and capacitor. We explained the routine from 3500 to 3830 in some detail in Program 13-7; the module at 900 has also appeared in a number of recent programs. Lines 10 through 250 simply title the program and request the input data for the si-

nusoidal voltage across the element. For a capacitive element, $X\$ = C$ and line 250 directs us to the subroutine at line 750, where the value of C is requested and reactance (X_C) determined at the applied frequency. Note that the angle (XA) associated with the reactance is $-90°$. The program will then return to line 270, where the sinusoidal expression for the voltage is initiated, the peak value determined, and the general variables $MA = V_M$ (the peak or maximum value) and $PA = \theta_V$ (the input phase angle) defined. At 900 the sinusoidal expression for the voltage will be completed before returning to line 300, where the peak value of the current is determined by Ohm's law $I_M = V_M/X_C$, and the effective value from $I_E = I_M/\sqrt{2}$. Line 310 specifies the phase angle of the current as equal to the input phase angle of the voltage less the phase shift ($XA = -90°$)

introduced by the capacitor. In general, I_M/θ_I $= V_M/\theta_V / Z/\theta_Z = V_M/\theta_V / X_C/\theta_C = V_M/X_C/\theta_v - \theta_C = V_M/X_C/\theta_v - Xa = V_M/X_C/\theta_v - (-90°) = V_M/X_C/\theta_v + 90° = I_M/\theta_v + 90° = IM/\underline{IA}$. Once MA and PA are defined, 900 can provide the sinusoidal expression for $i(t)$.

Note the choice of ωt as the variable on line 370, rather than A as in Program 13-7. Once the $D1(I)$ and $D2(I)$ arrays are establishd as in

Program 13-7, the routine beginning at 3500 will plot the results.

The program is run for each element to show the resulting change in phase shift. Note the phase shift of 90° between $v(t)$ and $i(t)$ for the inductor and capacitor and the in-phase relationship for the resistor. In addition, note how a change in frequency and, therefore, ωt has not affected the number of cycles printed in the space of 37 data points.

```
10 REM *****  PROGRAM 15-3  *****
20 REM ********************************************
30 REM Program to apply Ohm's law and plot
40 REM the input and output sinusoidal functions.
50 REM ********************************************
60 REM
100 DIM D1(100),D2(100)
110 PRINT "This program determines the sinusoidal"
120 PRINT "equation for the current through a"
130 PRINT "single R,L, or C component -"
140 PRINT "and then plots the input and output functions"
150 PRINT "on a single axis."
160 PRINT
170 INPUT "The frequency of the applied voltage f=";F
180 W=2*3.14159*F
190 INPUT "with an effective value of, V=";VE
200 INPUT "and a phase angle (degrees)";VA
210 PRINT :INPUT "The voltage is applied to an R,L, or C element";X$
220 PRINT
230 IF X$="R" THEN GOTO 550
240 IF X$="L" THEN GOTO 650
250 IF X$="C" THEN GOTO 750
260 GOTO 210
270 PRINT "The sinusoidal voltage is v(t)=";
280 VM=VE*SQR(2)  :MA=VM :PA=VA
290 GOSUB 900
300 IM=VM/X :IE=IM/SQR(2)
310 IA=VA-XA
320 PRINT "The sinusoidal current is i(t)=";
330 MA=IM :PA=IA
340 GOSUB 900
350 PRINT
360 REM Now determine points to plot graphs
370 FOR WT=0 TO 360 STEP 10
380 I=I+1
390 D1(I)=VE*SQR(2)*SIN((WT+VA)/57.296)
400 D2(I)=IE*SQR(2)*SIN((WT+IA)/57.296)
410 NEXT WT
420 N=I :REM N is the number of stored data points
430 PRINT "Data stored for";N;"points"
440 PRINT
450 GOSUB 3500 :REM Use plot module
460 END
550 REM Component is R
560 INPUT "R=";X
570 XA=0
580 GOTO 270
```

Annotations in left margin:
- R,L or C → lines 230–250
- v(t) → lines 270–290
- i(t) → lines 300–340
- ωt Loop for Plot Points → lines 360–410
- R → lines 550–580

(Continued)

```
L    ┌ 650 REM Component is L
     │ 660 INPUT "L=";L
     │ 670 X=2*3.14159*F*L
     │ 680 XA=90
     └ 690 GOTO 270
     ┌ 750 REM Component is C
     │ 760 INPUT "C=";C
C    │ 770 X=1/(2*3.14159*F*C)
     │ 780 XA=-90
     └ 790 GOTO 270
             ┌ 900 REM Module to print sinusoidal equation
             │ 910 REM MA=peak value, PA=phase angle
Sinusoidal   │ 920 PRINT MA;"sin(";W;"t";
Format       │ 930 IF PA>0 THEN PRINT "+";PA;")"
             │ 940 IF PA=0 THEN PRINT ")";
             │ 950 IF PA<0 THEN PRINT PA;")"
             └ 960 RETURN
        ┌ 3500 REM Module to plot data in arrays D1(I) and
        │ 3510 REM D2(I) with N entries
        │ 3520 FS=60
        │ 3525 CP=FS/2
        │ 3530 M1=ABS(D1(1)) :M2=ABS(D2(1)) :REM Find largest value
        │ 3540 FOR I=2 TO N
        │ 3550 IF ABS(D1(I))>M1 THEN M1=ABS(D1(I))
        │ 3560 IF ABS(D2(I))>M2 THEN M2=ABS(D2(I))
Plot    │ 3570 NEXT I
Routines│ 3580 REM M1 & M2 are largest values of respective arrays
        │ 3582 IF M1>M2 THEN F1=FS:F2=INT(M2*FS/M1):GOTO 3590
        │ 3584 F2=FS:F1=INT(M1*FS/M2)
        ▼ 3590 FOR I=1 TO FS
          3600 PRINT "-";
          3610 NEXT I   :REM Print axis
          3615 PRINT
          3620 FOR I=1 TO N
          3630 P1=INT(CP+F1*D1(I)/(2*M1)+0.5)
          3640 P2=INT(CP+F2*D2(I)/(2*M2)+0.5)
          3650 REM Plot P1 and P2
          3660 GOSUB 3700
          3670 PRINT
          3680 NEXT I
          3690 RETURN :REM Exit from plot subroutine
          3700 REM Routine to print points in order
          3710 IF P1<=CP AND P2<=CP THEN GOTO 3750
          3720 IF P1>=CP AND P2>=CP THEN GOTO 3770
          3730 IF P1 <= CP THEN GOSUB 3800:GOSUB 3820:GOSUB 3810:RETURN
          3740 IF P1 >= CP THEN GOSUB 3810:GOSUB 3820:GOSUB 3800:RETURN
          3750 IF P1 < P2 THEN GOSUB 3800:GOSUB 3810:GOSUB 3820:RETURN
          3760 IF P1=P2 THEN GOSUB 3800:GOTO 3820
          3765 GOSUB 3810:GOSUB 3800:GOTO 3820
          3770 GOSUB 3820
          3780 IF P1 < P2 THEN GOSUB 3800:GOSUB 3810:RETURN
          3790 IF P1=P2 THEN GOTO 3800 ELSE GOSUB 3810:GOSUB 3800:RETURN
          3800 PRINT TAB(P1);"*";:RETURN
          3810 PRINT TAB(P2);".";:RETURN
          3820 IF CP=P1 OR CP=P2 THEN RETURN
          3830 PRINT TAB(CP);"I";:RETURN


READY
```

RUN

This program determines the sinusoidal
equation for the current through a
single R,L, or C component -
and then plots the input and output functions
on a single axis.

The frequency of the applied voltage f=? <u>60</u>
with an effective value of, V=? <u>10</u>
and a phase angle (degrees)? <u>30</u>

The voltage is applied to an R,L, or C element? <u>R</u>

R=? <u>2</u>
The sinusoidal voltage is v(t)= 14.1421 sin(376.9908 t+ 30)
The sinusoidal current is i(t)= 7.0711 sin(376.9908 t+ 30)

Data stored for 37 points

```
------------------------------------------------
                           I       .        *
                           I         .        *
                           I           .        *
                           I             .        *
                           I              .        *
                           I               .        *
                           I                .        *
                           I                .       *
                           I               .        *
                           I              .       *
                           I             .       *
                           I           .       *
                           I         .      *
                           *        .      *
               *  .        I     .    *
            *        .     I    .
          *            .   I
        *                . I
       *                 . I
     *                   . I
    *                    . I
    *                    . I
    *                    . I
     *                   . I
        *                . I
          *              . I
            *           .  I
              *        .   I
                *     .    I
                 *  .      *
                           I  .  *
                           I    .    *
                           I      .        *
```

READY

RUN

This program determines the sinusoidal
equation for the current through a
single R,L, or C component -
and then plots the input and output functions
on a single axis.

The frequency of the applied voltage f=? 1000
with an effective value of, V=? 20
and a phase angle (degrees)? 60

The voltage is applied to an R,L, or C element? L

L=? 0.796E-3
The sinusoidal voltage is v(t)= 28.2843 sin(6283.18 t+ 60)
The sinusoidal current is i(t)= 5.6553 sin(6283.18 t-30)

Data stored for 37 points

```
----------------------------------------------------------------
                              .   I                        *
                              .   I                          *
                              . I                             *
                              .                               *
                              I.                              *
                              I .                           *
                              I   .                       *
                              I     .                   *
                              I       .              *
                              I        .          *
                              I         .      *
                              I          .  *
                           *  I          .
                              I         .
                       *      I        .
                    *         I       .
                  *           I      .
                *             I    .
               *              I  .
              *               I.
             *                .
             *               .I
             *               .I
              *               .   I
                *             .    I
                   *        .      I
                      *   .        I
                         *         I
                       .  *        I
                      .     I
                     .      I      *
                    .       I         *
                   .        I           *
                  .         I             *
                 .          I               *
                .           I                 *
```

READY

<u>RUN</u>

This program determines the sinusoidal
equation for the current through a
single R,L, or C component -
and then plots the input and output functions
on a single axis.

The frequency of the applied voltage f=? <u>1000</u>
with an effective value of, V=? <u>20</u>
and a phase angle (degrees)? <u>60</u>

The voltage is applied to an R,L, or C element? <u>C</u>

C=? <u>0.318E-3</u>
The sinusoidal voltage is v(t)= 28.2843 sin(6283.18 t+ 60)
The sinusoidal current is i(t)= 56.5134 sin(6283.18 t+ 150)

Data stored for 37 points

```
------------------------------------------------------------
                        I           *  .
                        I              .     *
                        I        .           *
                        .                     *
                 .      I                     *
                        I                     *
            .           I                    *
                        I                   *
        .               I                  *
      .                 I                 *
   .                    I               *
                        I              *
.                       *
.                    *  I
.                 *     I
     .          *       I
            *           I
          *             I
        *               I
        *         .     I
        *               I
        *               I
         *              I
          *             .
           *            I        .
            *           I           .
             *          I
              *         I                  .
               *        I                        .
                *       I                           .
                 *      I                              .
                  *     I                                .
                   *    I                                  .
                    *   I                                    .
                     *  I           *                          .
                        I         *                              .
                        I       *                                  .
                        I     *                                      .
                        I   *                              .
                        I *
                        *
                        I *
                        I   *
                        I     *                    .
                        I       *
                        I         *          .
                        I            *  .
```

READY

15.5   TEST ROUTINE - OHM'S LAW

The test routine of Program 15-4 is essentially the inverse of Program 15-2, in that the sinusoidal voltage across an element is given and the frequency, effective value, and phase angle of the resulting current determined. The program from line 500 to 640 is limited to randomly generating the required values of the sinusoidal expression for the voltage and determining which element will be employed. Rather than generate a random value for X_L, X_C, and R as required, the impedance variable Z_M represents the magnitude of the chosen resistance or reactance. Depending on which element was randomly chosen by line 590, the phase angle of the impedance is determined by lines 610 through 630.

The program will use line 150 to print out the sinusoidal voltage and the magnitude of the chosen element. Lines 260 through 310 will then test your calculation of the frequency as determined from ω and compare it to the computer-calculated value. The routines starting at 320 and 370 test your response for the effective value of the current and its phase angle. Note that the sign check was not applied to the frequency calculation, but it is appropriate for the other two. The program will END when zero is input as the solution to the frequency calculation. Note in the provided run that a resistor was chosen four out of six times, with the capacitor and inductor both once. A negative phase angle also appeared 50% of the time. As with any randomly generated set of values, a continuance of the program may result in a totally different set of elements.

```
10 REM *****  PROGRAM 15-4  *****
20 REM*****************************************
30 REM Program to test the use of Ohm's law
40 REM for a single ac component
50 REM *****************************************
60 REM
100 PRINT "This program tests the use of Ohm's law"
110 PRINT "as applied to a single component - R,L, or C."
120 PRINT
130 GOSUB 500 :REM Select circuit values
140 PRINT
150 PRINT "A sinusoidal voltage,";
160 PRINT "v(t)=";VE*SQR(2);"sin(";W;"t";
170 IF VA<0 THEN PRINT VA;")"
180 IF VA=0 THEN PRINT ")"
190 IF VA>0 THEN PRINT "+";VA;")"
200 PRINT "is applied to ";
210 IF C=1 THEN PRINT "a resistor, R=";
220 IF C=2 THEN PRINT "an inductor of impedance, XL=";
230 IF C=3 THEN PRINT "a capacitive impedance, XC=";
240 PRINT ZM;"ohms"
250 PRINT
260 PRINT "What is the applied signal frequency (Enter 0 to stop)";
270 INPUT ",f=";FA
280 IF FA=0 THEN END
290 F=W/(2*3.14159)
300 IF ABS(F-FA)<ABS(0.01*F) THEN PRINT "Correct":GOTO 320
310 PRINT "No, it's";F;"Hertz"
320 INPUT "What is the effective value of the resulting current";IA
330 IE=VE/ZM
340 IF SGN(IE)<>SGN(IA) THEN GOTO 360
350 IF ABS(IE-IA)<ABS(0.01*IE) THEN PRINT "Correct":GOTO 370
360 PRINT "No, it's";IE;"amps"
370 PRINT "What is the phase angle (in degrees)"
380 INPUT "of the current";IX
```

```
390 PA=VA-ZA
400 IF SGN(PA)<>SGN(IX) THEN GOTO 420
410 IF ABS(PA-IX)<ABS(0.01*PA) THEN PRINT "Correct":GOTO 430
420 PRINT "No, it's ";PA;"degrees"
430 PRINT
440 GOTO 120
500 REM Module to select circuit values
510 W=10*INT(1+1000*RND(XV)) :REM 10<W<10,000
520 VE=2*INT(1+25*RND(XV)) :REM 2<Veff<50
530 VA=5*INT(1+18*RND(XV)) :REM 5<Angle<90
540 SG=+1
550 S=INT(1+4*RND(XV)) :REM Allow minus sign 1 in 4 times
560 IF S=3 THEN SG=-1
570 VA=SG*VA
580 REM Select (1) R, (2) XL, (3) XC
590 C=INT(1+3*RND(XV))
600 ZM=10*INT(1+99*RND(XV)) :REM 10<ZM<1000
610 IF C=1 THEN ZA=0
620 IF C=2 THEN ZA=90
630 IF C=3 THEN ZA=-90
640 RETURN
```

```
READY

RUN


This program tests the use of Ohm's law
as applied to a single component - R,L, or C.


A sinusoidal voltage,v(t)= 67.8823 sin( 3360 t+ 90 )
is applied to an inductor of impedance, XL= 830 ohms

What is the applied signal frequency (Enter 0 to stop),f=? 535
Correct
What is the effective value of the resulting current? 67.88
No, it's .058 amps
What is the phase angle (in degrees)
of the current? 90
No, it's  0 degrees



A sinusoidal voltage,v(t)= 56.5685 sin( 7380 t+ 65 )
is applied to a resistor, R= 50 ohms

What is the applied signal frequency (Enter 0 to stop),f=? 128.8
No, it's 1174.5645 Hertz
What is the effective value of the resulting current? 1.13
No, it's .8 amps
What is the phase angle (in degrees)
of the current? 65
Correct



A sinusoidal voltage,v(t)= 62.2254 sin( 5000 t+ 5 )
is applied to a resistor, R= 270 ohms

What is the applied signal frequency (Enter 0 to stop),f=? 796
Correct
```

```
What is the effective value of the resulting current? .163
Correct
What is the phase angle (in degrees)
of the current? 5
Correct


A sinusoidal voltage,v(t)= 53.7401 sin( 9370 t-15 )
is applied to a resistor, R= 160 ohms

What is the applied signal frequency (Enter 0 to stop),f=? 1492
Correct
What is the effective value of the resulting current? .237
Correct
What is the phase angle (in degrees)
of the current? -15
Correct



A sinusoidal voltage,v(t)= 59.397 sin( 9820 t+ 25 )
is applied to a capacitive impedance, XC= 30 ohms

What is the applied signal frequency (Enter 0 to stop),f=? 0


READY
```

EXERCISES

Write a program to perform the following tasks.

1. Using the results of Program 15-1, tabulate the magnitude of R and the full sinusoidal expression for $i(t)$ for a specified range of R. For a first run, choose $f = 60$ Hz, $I_{rms} = 3A$, $\theta = 60°$, with R extending from 2 to 10 ohms in increments of 2 ohms. Note the impact of R on the resulting sinusoidal expression.

2. Develop a routine for duplicating the process of problem 15-1 for an inductive element. For the first run, choose $f = 1000$ Hz, $I_{rms} = 2A$, $\theta = 0°$, with L extending from 2 to 10 mH in increments of 2 mH.

3. Given an inductive or capacitive element, permit a choice of whether $v(t)$ or $i(t)$ is to be determined. Once the choice is made, request the appropriate data and perform the required analysis. For the first run, request $v(t)$ and pro-vide the following data for $i(t)$: $f = 10^4$ Hz, $I_{peak} = 10A$, $\theta = 20°$, and $C = 1\mu F$. For the second run, request $i(t)$ and provide the following data for $v(t)$: $f = 5 \times 10^3$ Hz, $V_{peak} = 18 \times 10^{-3}$ V, $\theta = 60°$, and $L = 5$mH.

4. Repeat Program 15-3 for $\omega t = 0°$ to $360°$ in steps of $5°$ and note whether there is an improvement in the resulting plots.

5. Repeat Program 15-4 for $\omega t = 0°$ to $360°$ in steps of $20°$ and comment on the quality of the results.

6. Develop a plotting routine that will plot only the positive portion of a sinusoidal curve across the full width of the printout (use $FS = 60$). Comment on whether the increased number of data points has created a more accurate plot. For the first run, use $V_{rms} = 10$V, $f = 60$Hz, and $\theta = 30°$ to permit a comparison with $v(t)$ of the second run of Program 15-3.

7. Modify Program 15-3 to plot only the positive portion of the resulting curves using the same *FS* deflection.

8. Develop a routine for plotting the power to a resistive load given the voltage across the load in sinusoidal form. That is, for each value of ωt in the range 0° to 360° in steps of 10°, determine the magnitude of the product $v(t) \times i(t)$ and plot the resulting data points. Use the full width of the printout for the resulting positive values. For the first run, use $R = 4$ ohms and $v_R(t) = 31 \sin(1000t + 30°)$.

Series and Parallel AC Networks

16

16.1　INTRODUCTION

The previous three chapters have established the foundation necessary to examine the series and parallel combination of elements with an applied sinusoidal signal. We will first examine the series *R-L* and parallel *R-C* network, followed by methods that will determine the total impedance and admittance of more complex series and parallel systems.

16.2　SERIES *R-L* CIRCUITS

Program 16-1 will perform a complete analysis of the series *R-L* circuit of Figure 16.1, using the subroutines at 2000 and 2100 to convert

between vector forms. Lines 100 through 190 will title the program and request the input data regarding the component values and applied sinusoidal signal. The routine beginning at 200 (which uses the subroutine at 2000) will calculate the polar form of the total impedance and then calculate the effective value of the current and its phase angle on lines 250 and 260. The subroutine at 2100 will determine the rectangular form of the current, so that both forms can be printed out by lines 290 and 300. The effective value and phase angle of the voltage across *R* and *L* will be determined by lines 320 and 330. Note the letter *E* to indicate effective value and the *A* for the phase angle. The first letter identifies the elements being examined. In each case, the polar form is printed out first, followed by a conversion at 2100 and a printout of the rectangular form. As an added measure, the power is calculated on line 450 and the power factor on line 480.

The true value of such a program is apparent from the second run, where the values chosen require a lengthy set of longhand calculations. If you change the input data, the entire result is printed out in seconds.

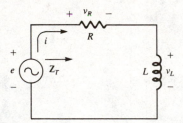

FIGURE 16.1

```
 10 REM *****  PROGRAM 16-1  *****
 20 REM ********************************************
 30 REM Program to analyze the series R-L circuit.
 40 REM ********************************************
 50 REM
100 PRINT "This program analyzes a series R-L circuit."
110 PRINT
120 INPUT "The input signal frequency f=";F
130 W=2*3.14159*F
140 INPUT "Resistor, R=";R
150 INPUT "and inductor, L=";L
160 XL=W*L
170 INPUT "Supply voltage (rms) .";VE
180 INPUT "At an angle of";VA
190 PRINT
200 REM Now calculate ZT
210 X=R  :Y=XL
220 GOSUB 2000 :REM Convert to polar form
230 ZT=Z  :ZA=TH
240 REM Calculate I
250 IE=VE/ZT
260 IA=VA-ZA
270 Z=IE  :TH=IA
280 GOSUB 2100 :REM Convert to rectangular form
290 PRINT :PRINT "The current is";IE;"at an angle of";IA;"degrees"
300 PRINT TAB(10); :GOSUB 980
```

The labels at the left of the program listing read: "Input" beside lines 120–180, and "Z_T" beside lines 200–230.

```
        ┌310 REM Calculate VR and VL
Calc.   │320 RE=IE*R  :RA=IA
        └330 LE=IE*XL :LA=IA+90
         340 PRINT
        ┌350 PRINT "The voltage across R is";RE;"at an angle of";RA;"degrees"
v_R     │360 Z=RE  :TH=RA
        │370 GOSUB 2100
        └380 PRINT TAB(19); :GOSUB 980
         390 PRINT
        ┌400 PRINT "And voltage across L is";LE;"at an angle of ";LA;"degrees"
v_L     │410 Z=LE  :TH=LA
        │420 GOSUB 2100
        └430 PRINT TAB(10); :GOSUB 980
        ┌440 REM Calculate power
P       │450 P=IE^2*R
        │460 PRINT
        └470 PRINT "The power dissipated is";P;"watts"
F_p     ┌480 PF=R/ZT
        └490 PRINT "and the circuit has a power factor of=";PF
         500 END
Rect.   ┌980 PRINT "=";X; :IF Y>0 THEN PRINT "+j";Y :RETURN
Output  └990 IF Y<0 THEN PRINT "-j";ABS(Y) :RETURN
         2000 REM Module converts from rectangular to polar form.
         2010 REM Enter with X, Y - return with Z, TH (eta)
         2020 Z=SQR(X^2+Y^2)
         2030 IF X>0 THEN TH=(180/3.14159)*ATN(Y/X)
         2040 IF X<0 THEN TH=180*SGN(Y)+(180/3.14159)*ATN(Y/X)
         2050 IF X=0 THEN TH=90*SGN(Y)
         2060 IF Y=0 THEN IF X<0 THEN TH=180
         2070 RETURN
         2100 REM Module converts from polar to rectangular form
         2110 REM Enter with Z, TH(eta) - return with X, Y
         2120 X=Z*COS(TH*3.14159/180)
         2130 Y=Z*SIN(TH*3.14159/180)
         2140 RETURN


READY

RUN
  .
This program analyzes a series R-L circuit.

The input signal frequency f=? 31.831
Resistor, R=? 3
and inductor, L=? 20E-3
Supply voltage (rms) ? 100
At an angle of? 0


The current is 20 at an angle of-53.1301 degrees
        = 12 -j 16

The voltage across R is 60 at an angle of-53.1301 degrees
                = 36 -j 48

And voltage across L is 80 at an angle of  36.8699 degrees
        = 64 +j 47.9999

The power dissipated is 1200.0007 watts
and the circuit has a power factor of= .6
```

*R–L network, ICA pg. 426 (4th Edition)

```
READY


This program analyzes a series R-L circuit.

The input signal frequency f=? 60E3
Resistor, R=? 2000
and inductor, L=? 15.916E-3
Supply voltage (rms) ? 120
At an angle of? 0


The current is .019 at an angle of-71.5656 degrees
       = 5.9997E-03 -j .018

The voltage across R is 37.9463 at an angle of-71.5656 degrees
                = 11.9993 -j 35.9991

And voltage across L is 113.8423 at an angle of  18.4344 degrees
           = 108.0007 +j 35.999

The power dissipated is .72 watts
and the circuit has a power factor of= .3162


READY
```

16.3 PARALLEL *R-C* NETWORK

Program 16-2 is the dual of 16-1, except it must first calculate the conductance on line 200 before the total admittance of Figure 16.2 can be determined. Note that the basic equations for E, I_C, and the power factor were written in the admittance format, $V = I/Y_T$, $I_C = V_C Y_C$ and $F_p = G/Y_T$, to remove the necessity to calculate Z_T and X_C.

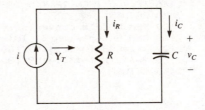

FIGURE 16.2

The phasor diagram of the results of the second run appear in Figure 16.3. Note that the phase angle between I_C and I_R is 25.75° +

```
10 REM *****  PROGRAM 16-2  *****
20 REM ****************************************
30 REM Program to analyze a parallel RC circuit.
40 REM ****************************************
50 REM
100 PRINT "This program analyzes a parallel R-C network."
110 PRINT
120 INPUT "The input angular velocity, w=";W
130 INPUT "Resistor, R=";R
140 INPUT "and capacitor, C=";C
150 YC=W*C
160 INPUT "Supply current(rms)";IE
170 INPUT "at an angle of";IA
180 PRINT
```

Input { (lines 120–180)

```
       ┌190 REM Now calculate YT
       │200 G=1/R
       │210 X=G :Y=YC
 Y_T   │220 GOSUB 2000 :REM Convert to polar form
       │230 YT=Z :YA=TH
       └240 PRINT "Equivalent admittance is";YT;"at an angle of";YA;"degrees"
        250 PRINT TAB(24); :GOSUB 980
       ┌260 REM Calculate voltage across parallel network
       │270 VE=IE/YT
       │280 VA=IA-YA
  v    │290 Z=VE :TH=VA
       │300 GOSUB 2100 :REM Convert to rectangular form
       │310 PRINT "Voltage is";VE;"at an angle of";VA;"degrees"
       └320 PRINT TAB(10); :GOSUB 980
       ┌330 REM Calculate IR and IC
       │340 RE=VE/R
 Calc. │350 RA=VA
       │360 CE=VE*YC
       └370 CA=VA+90
       ┌380 PRINT "Current through R is";RE;"at an angle of";RA;"degrees"
 i_R   │390 Z=RE :TH=RA
       │400 GOSUB 2100
       └410 PRINT TAB(10); :GOSUB 980
       ┌420 PRINT "and current through C is";CE;"at an angle of";CA;"degrees"
 i_C   │430 Z=CE :TH=CA
       │440 GOSUB 2100
       └450 PRINT TAB(10); :GOSUB 980
       ┌460 REM Calculate power
  P    │470 P=RE^2*R
       │480 PRINT
       └490 PRINT "The power dissipated is";P;"watts"
 F_p   ┌500 PF=G/YT
       └510 PRINT "and the network has a power factor of";PF
        520 END
 Rect. ┌980 PRINT "=";X;:IF Y>0 THEN PRINT "+j";Y :RETURN
Format └990 IF Y<0 THEN PRINT "-j";ABS(Y) :RETURN
        2000 REM Module converts rectangular form to polar form.
        2010 REM Enter with X, Y - return with Z, TH(eta)
        2020 Z=SQR(X^2+Y^2)
        2030 IF X>0 THEN TH=(180/3.14159)*ATN(Y/X)
        2040 IF X<0 THEN TH=180*SGN(Y)+(180/3.14159)*ATN(Y/X)
        2050 IF X=0 THEN TH=90*SGN(Y)
        2060 IF Y=0 THEN IF X<0 THEN TH=180
        2070 RETURN
        2100 REM Module converts from polar to rectangular form.
        2110 REM Enter with Z, TH(eta) - return with X, Y
        2120 X=Z*COS(TH*3.14159/180)
        2130 Y=Z*SIN(TH*3.14159/180)
        2140 RETURN


READY

RUN
.
This program analyzes a parallel R-C network.

The input angular velocity, w=? 100000
Resistor, R=? 1.67
and capacitor, C=? 8E-6
```

*R–C Network, pg. 442, ICA

```
Supply current(rms)? 10
at an angle of? 0

Equivalent admittance is .9993 at an angle of 53.1851 degrees
                        = .5988 +j .8
Voltage is 10.0072 at an angle of-53.1851 degrees
             = 5.9966 -j 8.0115
Current through R is 5.9923 at an angle of-53.1851 degrees
            = 3.5908 -j 4.7973
and current through C is 8.0057 at an angle of 36.8149 degrees
            = 6.4092 +j 4.7973

The power dissipated is 59.9663 watts
and the network has a power factor of .5992

READY

RUN

This program analyzes a parallel R-C network.

The input angular velocity, w=? 377
Resistor, R=? 2.2E3
and capacitor, C=? 2.5E-6
Supply current(rms)? 2E-3
at an angle of? 0

Equivalent admittance is 1.0464E-03 at an angle of 64.2531 degrees
                        = 4.5455E-04 +j 9.425E-04
Voltage is 1.9113 at an angle of-64.2531 degrees
           = .8303 -j 1.7216
Current through R is 8.6879E-04 at an angle of-64.2531 degrees
          = 3.774E-04 -j 7.8254E-04
and current through C is 1.8014E-03 at an angle of 25.7469 degrees
          = 1.6226E-03 +j 7.8254E-04

The power dissipated is 1.6606E-03 watts
and the network has a power factor of .4344

READY
```

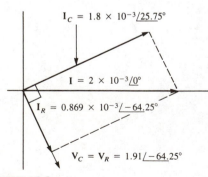

FIGURE 16.3

$64.25° = 90°$ (as it should be) and the current through the resistor is in phase with the volt-

age across it, while the current I_C leads the voltage across it by 90°. A quick calculation will also show that the magnitude of the current $I = \sqrt{(1.8 \times 10^{-3})^2 + (0.869 \times 10^{-3})^2} = 2 \times 10^{-3} = 2mA$.

16.4   EQUIVALENT FORMS

The series configuration of Figure 16.4a has the equivalent parallel form appearing in Figure 16.4b. The resistor and reactance values are different, but the reactance in each case will always be capacitive or inductive. Keep in mind that the networks are only terminally equiv-

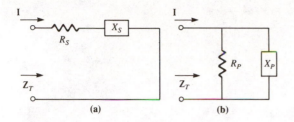

FIGURE 16.4

alent. For an applied voltage, the resulting current I will be the same in each case—but internally the networks are quite different.

Program 16-3 will determine the component values for one form given the other. The choice of routines is made on lines 130 through 150 and the input of the series or parallel ele-

ments occurs at 500 and 700 respectively. The actual conversion takes place at 900 and 950, and the printout of the results at 720 and 760.

Note in the provided runs that the X_S or X_P are not identified as inductive or capacitive, since they must be the same in each form. If we calculate the total impedance for each form for the first run, we find that Z_T (parallel) = $5 \times 10^3 \underline{/0°} \parallel 8 \times 10^3 \underline{/90°} = 3595.51 + j2247.19 = Z_T$(series).

16.5 PARALLEL *R-L-C* NETWORK

The input impedance and admittance will now be determined for the complex parallel network

```
10 REM *****  PROGRAM 16-3  *****
20 REM *****************************************
30 REM Program converts from series R-X form
40 REM to parallel R-X form, and vice-versa.
50 REM *****************************************
60 REM
100 PRINT "This program converts from the series R-X form"
110 PRINT "to the parallel R-X form, or vice-versa."
120 PRINT
130 PRINT "Choose (1) Series to Parallel"
140 PRINT "       (2) Parallel to Series"
150 PRINT TAB(7);"Choice (0 to stop)=";
160 INPUT C
170 IF C=0 THEN END
180 IF C>2 THEN GOTO 150
190 ON C GOSUB 500,700
200 GOTO 120
```

Series ↓ Parallels
```
500 REM Accept input of series R-X data.
510 PRINT
520 PRINT "For series Rs,Xs elements:"
530 INPUT "Rs=";RS
540 INPUT "Xs=";XS
550 GOSUB 900
560 PRINT "Parallel Rp,Xp elements are then:"
570 PRINT "Rp=";RP;"and Xp=";XP
580 RETURN
```

Parallel ↓ Series
```
700 REM Accept input of parallel R-X data
710 PRINT
720 PRINT "For Parallel Rp,Xp elements:"
730 INPUT "Rp=";RP
740 INPUT "Xp=";XP
750 GOSUB 950
760 PRINT "Series Rs,Xs elements are then:"
770 PRINT "Rs=";RS;"and Xs=";XS
780 RETURN
```

Calc. $S \to P$
```
900 REM Convert Series to Parallel form
910 RP=(RS^2+XS^2)/RS
920 XP=(RS^2+XS^2)/XS
930 RETURN
```

(Continued)

Calc.
$P \to S$

```
950 REM Convert Parallel to Series form
960 D=RP^2+XP^2
970 RS=RP*XP^2/D
980 XS=RP^2*XP/D
990 RETURN
```

```
READY

RUN
.
This program converts from the series R-X form
to the parallel R-X form, or vice-versa.

Choose (1) Series to Parallel
       (2) Parallel to Series
       Choice (0 to stop)=? 2

For Parallel Rp,Xp elements:
Rp=? 5E3
Xp=? 8E3
Series Rs,Xs elements are then:
Rs= 3595.5056 and Xs= 2247.191

Choose (1) Series to Parallel
       (2) Parallel to Series
       Choice (0 to stop)=? 1

For series Rs,Xs elements:
Rs=? 3.5
Xs=? 4.5
Parallel Rp,Xp elements are then:
Rp= 9.2857 and Xp= 7.2222

Choose (1) Series to Parallel
       (2) Parallel to Series
       Choice (0 to stop)=? 0


READY
```

*Example 15.17, ICA

of Figure 16.5 using Program 16-4. The network can be modified (reduced) by simply substituting $1E30$ (10^{30}) for the value of R_1, R_2, R_3, or L, and $1E30$(10^{-30}) for C. Note on lines 220 through 240 that the total conductance of the parallel resistors is determined first, since the associated impedance angle is the same and the total conductance is simply the sum of the individual values. The susceptance of each reactive element is then determined, followed by

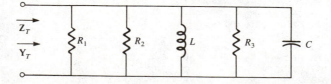

FIGURE 16.5

a calculation of the total susceptance on line 300 and a printout of the results on lines 310 through 340.

The total admittance is printed out in rectangular form at 380 and in the polar form on line 400, after the conversion to polar form is completed using the subroutine at 2000. The total impedance is determined and printed out by lines 420 through 460. Note, in the provided runs, the difference in sign for the angle of the total admittance and impedance and that B_T is the difference of the two susceptance values. In the second run, the resistor R_3 and the inductor L are non-existent. Note that, even though a value of B_L does appear, it is essentially zero compared to the other susceptances.

```
          10 REM *****  PROGRAM 16-4  *****
          20 REM *************************************
          30 REM Program calculates YT and ZT for the
          40 REM parallel network of Figure 16-5.
          50 REM *************************************
          60 REM
         100 PRINT "This program calculates YT and ZT"
         110 PRINT "for the parallel network of"
         120 PRINT "Figure 16-5."
         130 PRINT
        ┌140 INPUT "Signal frequency, f=";F
         150 W=2*3.14159*F
         160 PRINT "Resistor elements are:"
  Input  170 REM Use 1E30 for all non-existant resistors
         180 INPUT "R1=";R1
         190 INPUT "R2=";R2
        └200 INPUT "R3=";R3
         210 PRINT
        ┌220 G1=1/R1 :G2=1/R2 :G3=1/R3
    G    230 GT=G1+G2+G3
        └240 PRINT "Total conductance is";GT;"siemens"
         250 PRINT
        ┌260 INPUT "Inductance, L=";L :REM Substitute L=1E30 if non-existant
         270 BL=1/(W*L)
         280 INPUT "Capacitance, C=";C :REM Substitute C=1E-30 if non-existant
         290 BC=W*C
    B    300 BT=BC-BL
         310 PRINT
         320 PRINT "Resulting in BC=";BC;"siemens,"
         330 PRINT "              BL=";BL;"siemens"
        └340 PRINT "and,          BT=";BT;"siemens"
         350 PRINT
        ┌360 REM Now calculate YT and ZT
         370 X=GT :Y=BT
   Y_T   380 PRINT "YT=";  :GOSUB 980
         390 GOSUB 2000 :REM Convert to polar form
        └400 PRINT "which is";Z;"at an angle of";TH;"degrees"
         410 PRINT
        ┌420 ZT=1/Z :ZA=-TH
         430 PRINT "And ZT=";ZT;"at an angle of";ZA;"degrees"
   Z_T   440 Z=ZT :TH=ZA
         450 GOSUB 2100 :REM Convert to rectangular form
        └460 PRINT "which is"; :GOSUB 980
         470 END
  Rect. ┌980 PRINT X; :IF Y>=0 THEN PRINT "+j";Y :RETURN
 Format └990 IF Y<0 THEN PRINT "-j";ABS(Y) :RETURN
        2000 REM Module converts from rectangular to polar form.
        2010 REM Enter with X, Y - return with Z, TH(eta)
        2020 Z=SQR(X^2+Y^2)
        2030 IF X>0 THEN TH=(180/3.14159)*ATN(Y/X)
        2040 IF X<0 THEN TH=180*SGN(Y)+(180/3.14159)*ATN(Y/X)
```

(Continued)

```
2050 IF X=0 THEN TH=90*SGN(Y)
2060 IF Y=0 THEN IF X<0 THEN TH=180
2070 RETURN
2100 REM Module converts from polar to rectangular form.
2110 REM Enter with Z, TH(eta) - return with X, Y
2120 X=Z*COS(TH*3.14159/180)
2130 Y=Z*SIN(TH*3.14159/180)
2140 RETURN
```

READY

RUN

This program calculates YT and ZT
for the parallel network of
Figure 16-5.

Signal frequency, f=? 100
Resistor elements are:
R1=? 2
R2=? 4
R3=? 5

Total conductance is .95 siemens

Inductance, L=? 8E-3
Capacitance, C=? 0.05

Resulting in BC= 31.4159 siemens,
            BL= .1989 siemens
and,   BT= 31.217 siemens

YT= .95 +j 31.217
which is 31.2314 at an angle of 88.257 degrees

And ZT= .032 at an angle of-88.257 degrees
which is 9.7396E-04 -j .032

READY

RUN

This program calculates YT and ZT
for the parallel network of
Figure 16-5.

Signal frequency, f=? 1000
Resistor elements are:
R1=? 2.2E3
R2=? 5.6E3
R3=? 1E30

Total conductance is 6.3312E-04 siemens

Inductance, L=? 1E30
Capacitance, C=? 0.04E-6

Resulting in BC= 2.5133E-04 siemens,
            BL= 1.5916E-34 siemens
and,   BT= 2.5133E-04 siemens
```
```

```
YT= 6.3312E-04 +j 2.5133E-04
which is 6.8118E-04 at an angle of 21.6515 degrees

And ZT= 1468.0468 at an angle of-21.6515 degrees
which is 1364.4691 -j 541.6507


READY
```

16.6 EQUIVALENT SERIES IMPEDANCE

The next program (16-5) will determine the total impedance of any number of series impedances in either the rectangular or polar form. The total impedance will be determined and printed out when the response is zero to the request of line 190. The program begins with a total impedance in rectangular and polar form set equal to zero by lines 160 and 170. It will then react to

the form of the input data and move to the subroutine at 800 (if rectangular) and 900 (if polar). If rectangular, the value of X and Y is requested on lines 810 and 820 and the polar form determined by the subroutine at 2000.

On lines 240 and 250 the rectangular forms of the first rectangular input are set equal to $X1$, $Y1$ (rectangular) and $Z1$ and $T1$ (polar form from subroutine at 2000). Since it is the first pass, XT and YT equal zero, and X_2 and Y_2 for the subroutine at 2400 are correspondingly

```
         10 REM *****  PROGRAM 16-5  *****
         20 REM *****************************************
         30 REM Program to calculate ZT, the equivalent
         40 REM impedance of series components.
         50 REM *****************************************
         60 REM
        100 PRINT "This program calculates the equivalent"
        110 PRINT "impedance of a number of components in series."
        120 PRINT "You may enter each impedance in either"
        130 PRINT "rectangular or polar form."
        150 PRINT
Initialize ⎡160 XT=0 :YT=0 :REM Setting each quantity initially to zero
        ⎣170 ZT=0 :TT=0
        180 PRINT
        ⎡190 PRINT "Enter impedance as (0 to obtain ZT) ";
        │200 INPUT "(1) Rectangular, (2) Polar";C
   Input │210 IF C=0 THEN GOTO 310 :REM Print equivalent resistance
        │220 IF C>2 THEN GOTO 190
        ⎣230 ON C GOSUB 800,900
        ⎡240 X1=X :Y1=Y
        │250 Z1=Z :T1=TH
     ZT │260 X2=XT :Y2=YT
        │270 GOSUB 2400 :REM Add to total
        │280 XT=XA :YT=YA
        ⎣290 ZT=ZA :TT=TA :REM Save as total
        300 GOTO 190 :REM Get next input
        310 PRINT
        ⎡320 PRINT "The total equivalent impedance is: ";
        │330 PRINT TAB(30);
        │340 PRINT XT; :IF YT>=0 THEN PRINT "+j";YT
  Output│350 IF YT<0 THEN PRINT "-j";ABS(YT)
        │360 PRINT TAB(22);
        ⎣370 PRINT "which is";ZT;"at an angle of";TT;"degrees"
        380 END
```

(Continued)

```
         ┌ 800 REM Accept rectangular input
           810 INPUT "X=";X
Modules    820 INPUT "Y=";Y
           830 IF X=0 AND Y=0 THEN C=0
         │ 840 GOSUB 2000 :REM Convert to polar form
           850 RETURN
           900 REM Accept Polar input
         │ 910 INPUT "Z=";Z
           920 INPUT "At angle ";TH
         ↓ 930 IF Z=0 AND TH=0 THEN C=0
           940 GOSUB 2100 :REM Convert to rectangular form
           950 RETURN
          2000 REM Module to convert from rectangular to polar form.
          2010 REM Enter with X, Y - return with Z, TH(eta)
          2020 Z=SQR(X^2+Y^2)
          2030 IF X>0 THEN TH=(180/3.14159)*ATN(Y/X)
          2040 IF X<0 THEN TH=180*SGN(Y)+(180/3.14159)*ATN(Y/X)
          2050 IF X=0 THEN TH=90*SGN(Y)
          2060 IF Y=0 THEN IF X<0 THEN TH=180
          2070 RETURN
          2100 REM Module converts from polar to rectangular form.
          2110 REM Enter with Z, TH(eta) - return with X, Y
          2120 X=Z*COS(TH*3.14159/180)
          2130 Y=Z*SIN(TH*3.14159/180)
          2140 RETURN
          2400 REM Module to Add Z1 and Z2
          2410 REM Enter with Z1=X1+jY1, Z2=X2+jY2
          2420 REM Return with ZA in both polar and rectangular form
          2430 XA=X1+X2 :YA=Y1+Y2 :REM Add vectors
          2440 X=XA :Y=YA
          2450 GOSUB 2000 :REM Convert answer to polar form
          2460 ZA=Z :TA=TH
          2470 RETURN
```

READY

RUN


```
This program calculates the equivalent
impedance of a number of components in series.
You may enter each impedance in either
rectangular or polar form.


Enter impedance as (0 to obtain ZT) (1) Rectangular, (2) Polar? 2
Z=? 6
At angle ? 0
Enter impedance as (0 to obtain ZT) (1) Rectangular, (2) Polar? 2
Z=? 10
At angle ? 90
Enter impedance as (0 to obtain ZT) (1) Rectangular, (2) Polar? 2
Z=? 12
At angle ? -45
Enter impedance as (0 to obtain ZT) (1) Rectangular, (2) Polar? 0

The total equivalent impedance is:   14.4853 +j 1.5147
                        which is 14.5643 at an angle of 5.9697 degrees
```


READY

zero. The module at 2400 was used for the first time in Program 14-5, where it was explained in some detail. The output is the rectangular and polar form of the sum using the notation $XA + jYA = ZA \underline{/TH}$. The current values of XT, YT, (rectangular) and ZT, TT (polar) are then stored for the next pass. The total impedance will not be printed out until the response to line 190 is zero, and $C = 0$ on line 210 sends the program to 310 to print out the results.

If the next input were in polar form, it would first be converted to rectangular form before the program repeats the same process as encountered for the rectangular form, starting at line 240. During this pass, however, X_2 and Y_2 will represent the previous sum of the impedances. In summary, the program will take

each impedance in sequence and convert it (if necessary) to the rectangular form and add it to the present total using the subroutine at 2400. The provided run clearly indicates why this program is valuable.

16.7 THE EQUIVALENT Z_T AND Y_T

Program 16-6 is essentially the dual of Program 16-5. It will calculate the total impedance of any number of parallel impedances in rectangular or polar form. In this case, the general equation

$$Y_T = \frac{1}{Z_T} = \frac{1}{Z_1} + \frac{1}{Z_2} + \frac{1}{Z_3} + \cdots.$$

```
         10 REM ***** PROGRAM 16-6 *****
         20 REM ******************************************
         30 REM Program accepts input data on parallel
         40 REM components and calculates equivalent
         50 REM admittance, YT, and impedance, ZT.
         60 REM ******************************************
         70 REM
        100 PRINT "Enter data on components in parallel - "
        110 PRINT "program will calculate YT and ZT."
        120 PRINT
        130 XT=0 :YT=0 :REM Start with ZT=0+j0
        140 PRINT
       ┌150 PRINT "Enter choice of input impedance form (0 to get YT,ZT):"
        160 PRINT "Choice is (1) Rectangular input"
 Input │170 PRINT "          (2) Polar input"
        180 PRINT TAB(10);"Choice=";
       └190 INPUT C
        200 IF C=0 THEN GOTO 400
        210 IF C>2 THEN GOTO 180
        220 ON C GOSUB 600,800
  Sum   230 XT=XT+X :YT=YT+Y
        240 GOTO 140
        400 REM Output results
        410 PRINT
       ┌420 PRINT "Total admittance, YT is ";
        430 PRINT XT; :IF YT>=0 THEN PRINT "+j";YT ELSE PRINT "-j";ABS(YT)
  Y_T  │440 X=XT :Y=YT
        450 GOSUB 2000 :REM Convert to polar form
        460 YM=Z :YA=TH
       └470 PRINT "which is";YM;"at an angle of";YA;"degrees"
       ┌480 REM Now calculate ZT
        490 ZM=1/YM :ZA=-YA
        500 PRINT :PRINT "Total impedance, ZT is";
  Z_T  │510 PRINT ZM;"at an angle of";ZA;"degrees"
        520 Z=ZM :TH=ZA
        530 GOSUB 2100 :REM Convert to rectangular form
       └540 PRINT "which is";X; :IF Y>=0 THEN PRINT "+j";Y ELSE PRINT "-j";ABS(Y)
        550 END
```

(Continued)

```
Rect.    ┌600 REM Module to accept input in rectangular form
↓        │610 INPUT "X=";X
Polar    │620 INPUT "Y=";Y
&        │630 GOSUB 2000 :REM Convert to polar form
1/Z      │640 GOSUB 900 :REM Calculate 1/Z
         └650 RETURN
Polar    ┌800 REM Module to accept input in polar form
↓        │810 INPUT "Magnitude=";Z
Rect.    │820 INPUT "At angle";TH
&        │830 GOSUB 2100 :REM Convert to rectangular form
1/Z      │840 GOSUB 900 :REM Calculate 1/Z
         └850 RETURN
Calc.    ┌900 REM Module to calculate 1/Z
1/Z      └910 Z=1/Z :TH=-TH
          920 GOSUB 2100 :REM Convert to rectangular form
          930 RETURN
         2000 REM Module converts from rectangular to polar form.
         2010 REM Enter with X, Y - return with Z, TH(eta)
         2020 Z=SQR(X^2+Y^2)
         2030 IF X>0 THEN TH=(180/3.14159)*ATN(Y/X)
         2040 IF X<0 THEN TH=180*SGN(Y)+(180/3.14159)*ATN(Y/X)
         2050 IF X=0 THEN TH=90*SGN(Y)
         2060 IF Y=0 THEN IF X<0 THEN TH=180
         2070 RETURN
         2100 REM Module converts from polar to rectangular form.
         2110 REM Enter with Z, TH(eta) - return with X< Y
         2120 X=Z*COS(TH*3.14159/180)
         2130 Y=Z*SIN(TH*3.14159/180)
         2140 RETURN


READY

RUN

Enter data on components in parallel -
program will calculate YT and ZT.


Enter choice of input impedance form (0 to get YT,ZT):
Choice is (1) Rectangular input
          (2) Polar input
          Choice=? 2
Magnitude=? 10
At angle? 0

Enter choice of input impedance form (0 to get YT,ZT):
Choice is (1) Rectangular input
          (2) Polar input
          Choice=? 1
X=? 3
Y=? 4

Enter choice of input impedance form (0 to get YT,ZT):
Choice is (1) Rectangular input
          (2) Polar input
          Choice=? 1
X=? 8
Y=? -6

Enter choice of input impedance form (0 to get YT,ZT):
Choice is (1) Rectangular input
          (2) Polar input
          Choice=? 0
```

```
Total admittance, YT is  .3 -j .1
which is .3162 at an angle of-18.435 degrees

Total impedance, ZT is 3.1623 at an angle of 18.435 degrees
which is 3 +j 1
```

READY

will be employed, rather than combining each impedance with the current total using the general equation for two parallel elements. For each impedance, the polar form will first be determined (if necessary) and $1/Z$ determined by the subroutine at 900. The result will then be added to the rectangular form of the current total of Y_T to determine the new Y_T. The total Y_T will eventually be printed out in rectangular and polar form, followed by a calculation of $1/Z_T$ on line 490, so that the total impedance can be printed out in both forms. In the program a number of steps were saved (including the subroutine at 2400 in Program 16-5) by ignoring the polar form of the output until requested and fully using the power of line 230 with all susceptances in the rectangular form for the addition process.

An admittance diagram of the results was provided in Figure 16.6 to demonstrate the accuracy of the printout for the provided run.

16.8   TESTING ROUTINE

Programs 16-7 and 16-8 will test your skills calculating the total impedance of up to five series or parallel impedances. In each case, the num-

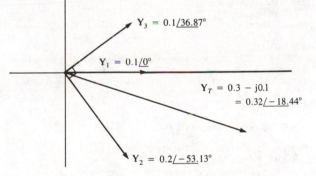

FIGURE 16.6

```
 10 REM *****  PROGRAM 16-7  *****
 20 REM *********************************************
 30 REM Program selects up to 5 series impedances
 40 REM and tests calculation of ZT.
 60 REM *********************************************
 70 REM
100 PRINT "Program will select up to 5 series impedances"
110 PRINT "and ask for the calculation of the total"
120 PRINT "impedance, ZT."
130 XT=0 :YT=0 :REM Start with ZT=0+j0
140 PRINT
150 N=INT(2+4*RND(XV)) :REM For 2 to 5 series impedances
160 PRINT "For";N;"impedances in series, calculate ZT."
```

(Continued)

```
170 PRINT "The impedances are:"
180 PRINT
190 FOR I=1 TO N
200 PRINT "Z";I;"=";
210 GOSUB 500
220 PRINT
230 XT=XT+X :YT=YT+Y
240 NEXT I
250 X=XT :Y=YT
260 GOSUB 2000 :REM Convert to polar form
270 ZT=Z :TT=TH
280 INPUT "Magnitude, Z=";ZM
290 INPUT "At angle ";ZA
300 IF ABS(ZT-ZM)>ABS(0.01*ZT) THEN GOTO 340
310 IF SGN(TT)<>SGN(ZA) THEN GOTO 340
320 IF ABS(TT-ZA)>ABS(0.01*TT) THEN GOTO 340
330 PRINT "Correct!!"  :END
340 PRINT :PRINT "No, it's";Z;"at an angle of";TH;"degrees"
350 GOSUB 2100 :REM Convert to rectangular form
360 PRINT "which is"; :GOSUB 980
370 END
500 REM Select impedance
510 C=INT(1+2*RND(XV)) :REM C=1 or 2
520 ON C GOSUB 600,800
530 IF C=1 THEN GOSUB 980
540 IF C=2 THEN PRINT Z;"at an angle of";TH
550 RETURN
600 REM Module to select input in rectangular form
610 X=5*INT(1+10*RND(XV)) :REM X between 5 and 50
620 Y=5*INT(1+10*RND(XV))
630 GOSUB 700 :Y=SG*Y :REM Value of Y plus or minus between 5 and 50
640 GOSUB 2000 :REM Convert to polar form
650 RETURN
700 REM Select sign SG=+1 or SG=-1
710 SG=+1
720 S=INT(1+4*RND(XV))
730 IF S=3 THEN SG=-1 :REM SG is negative once in four times
740 RETURN
800 REM Module to accept input in polar form
810 Z=5*INT(1+10*RND(XV)) :REM 5<Z<50
820 TH=5*INT(18*RND(XV)) :REM 0<TH<90
830 GOSUB 700 :TH=SG*TH
840 GOSUB 2100 :REM Convert to rectangular form
850 RETURN
980 PRINT X; :IF Y>=0 THEN PRINT "+j";Y :RETURN
990 IF Y<0 THEN PRINT "-j";ABS(Y) :RETURN
2000 REM Module converts from rectangular to polar form.
2010 REM Enter with X, Y - return with Z, TH(eta)
2020 Z=SQR(X^2+Y^2)
2030 IF X>0 THEN TH=(180/3.14159)*ATN(Y/X)
2040 IF X<0 THEN TH=180*SGN(Y)+(180/3.14159)*ATN(Y/X)
2050 IF X=0 THEN TH=90*SGN(Y)
2060 IF Y=0 THEN IF X<0 THEN TH=180
2070 RETURN
2100 REM Module converts from polar to rectangular form.
2110 REM Enter with Z, TH(eta) - return with X, Y
2120 X=Z*COS(TH*3.14159/180)
2130 Y=Z*SIN(TH*3.14159/180)
2140 RETURN
```

READY

<u>RUN</u>

Program will select up to 5 series impedances
and ask for the calculation of the total
impedance, ZT.

For 3 impedances in series, calculate ZT.
The impedances are:

Z 1 = 15 -j 50

Z 2 = 15 at an angle of 40

Z 3 = 45 +j 45

Magnitude, Z=? <u>71.6</u>
At angle ? <u>3.7</u>
Correct!!

READY

<u>RUN</u>

Program will select up to 5 series impedances
and ask for the calculation of the total
impedance, ZT.

For 3 impedances in series, calculate ZT.
The impedances are:

Z 1 = 15 at an angle of-20

Z 2 = 10 at an angle of 20

Z 3 = 25 at an angle of 55

Magnitude, Z=? <u>42</u>
At angle ? <u>36</u>

No, it's 42.2316 at an angle of 26.3865 degrees
which is 37.8317 +j 18.7687

READY

```
10 REM *****  PROGRAM 16-8  *****
20 REM ********************************************
30 REM Program selects up to 5 parallel impedances
40 REM and tests calculation of ZT.
60 REM ********************************************
70 REM
100 PRINT "Program will select up to 5 parallel impedances"
110 PRINT "and ask for the total impedance, ZT."
120 XT=0 :YT=0 :REM Start with ZT=0+j0
130 PRINT
140 N=INT(2+4*RND(XV)) :REM From 2 to 5 parallel impedances
150 PRINT "For";N;"impedances in parallel, calculate ZT."
160 PRINT "The impedances are:"
170 PRINT
180 FOR I=1 TO N
190 PRINT "Z";I;"=";
```

(Continued)

```
200 GOSUB 500
210 PRINT
220 XT=XT+X :YT=YT+Y
230 NEXT I
240 PRINT
250 INPUT "Magnitude, ZT=";ZM
260 INPUT "At angle ";ZA
270 REM Calculate ZT
280 X=XT :Y=YT
290 GOSUB 2000 :REM Convert to polar form
300 YM=Z :YA=TH
310 ZT=1/YM :TT=-YA
320 IF ABS(ZT-ZM)>ABS(0.01*ZT) THEN GOTO 360
330 IF SGN(ZA)<>SGN(TT) THEN GOTO 360
340 IF ABS(TT-ZA)>ABS(0.01*TT) THEN GOTO 360
350 PRINT "Correct!!" :END
360 PRINT
370 PRINT "No, it's";ZT;"at an angle of";TT;"degrees"
380 Z=ZT :TH=TT :GOSUB 2100 :REM Convert to rectangular form
390 PRINT "which is"; :GOSUB 980
400 END
500 REM Select impedance
510 C=INT(1+2*RND(XV)) :REM C=1 or 2
520 ON C GOSUB 600,800
530 IF C=1 THEN GOSUB 980
540 IF C=2 THEN PRINT Z;"at an angle of ";TH;"degrees"
550 GOSUB 900 :REM Calculate 1/Z
560 RETURN
600 REM Module to select input in rectangular form
610 X=5*INT(1+10*RND(XV)) :REM X between 5 and 50
620 Y=5*INT(1+10*RND(XV))
630 GOSUB 700 :Y=SG*Y :REM Value of Y plus or minus between 5 and 50
640 GOSUB 2000 :REM Convert to polar form
650 RETURN
700 REM Select sign SG=+1 or SG=-1
710 SG=+1
720 S=INT(1+4*RND(XV))
730 IF S=3 THEN SG=-1 :REM SG negative once in four times
740 RETURN
800 REM Module to accept input in polar form
810 Z=5*INT(1+10*RND(XV))
820 TH=5*INT(1+10*RND(XV))
830 GOSUB 700 :TH=SG*TH
840 GOSUB 2100 :REM Convert to rectangular form
850 RETURN
900 REM Module to calculate 1/Z
910 Z=1/Z :TH=-TH
920 GOSUB 2100 :REM Convert to rectangular form
930 RETURN
980 PRINT X; :IF Y>=0 THEN PRINT "+j";Y
990 IF Y<0 THEN PRINT "-j";ABS(Y) :RETURN
2000 REM Module converts from rectangular to polar form.
2010 REM Enter with X, Y - return with Z, TH(eta)
2020 Z=SQR(X^2+Y^2)
2030 IF X>0 THEN TH=(180/3.14159)*ATN(Y/X)
2040 IF X<0 THEN TH=180*SGN(Y)+(180/3.14159)*ATN(Y/X)
2050 IF X=0 THEN TH=90*SGN(Y)
2060 IF Y=0 THEN IF X>0 THEN TH=180
2070 RETURN
2100 REM Module converts from polar to rectangular form.
2110 REM Enter with Z, TH(eta) - return with X, Y
2120 X=Z*COS(TH*3.14159/180)
```

```
2130 Y=Z*SIN(TH*3.14159/180)
2140 RETURN
```

READY

RUN

Program will select up to 5 parallel impedances
and ask for the total impedance, ZT.

For 4 impedances in parallel, calculate ZT.
The impedances are:

Z 1 = 50 at an angle of 5 degrees

Z 2 = 30 +j 10

Z 3 = 50 at an angle of -10 degrees

Z 4 = 20 at an angle of -40 degrees

Magnitude, ZT=? 5.5
At angle ? 42.5

No, it's 9.0473 at an angle of-12.4714 degrees
which is 8.8338 -j 1.9538

READY

RUN

Program will select up to 5 parallel impedances
and ask for the total impedance, ZT.

For 3 impedances in parallel, calculate ZT.
The impedances are:

Z 1 = 5 -j 45

Z 2 = 40 at an angle of 15 degrees

Z 3 = 25 +j 35

Magnitude, ZT=? 24.8
At angle ? 4.9
Correct!!

READY

ber of impedances is randomly generated as between two and five, and each impedance is either provided in rectangular form by the subroutine at 600 or in polar form by the subroutine at 800. The program for the parallel elements is only slightly longer because it must calculate $1/Z$ for each element to calculate Z_T from Y_T. For each program, the provided polar form for the answer is tested against the computer solution. The accuracy of the result for the second

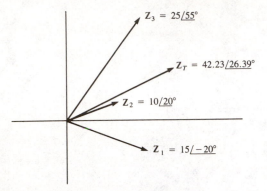

FIGURE 16.7

run of the Program 16-7 is demonstrated by Figure 16.7.

EXERCISES

Write a program to perform the following tasks.

1. Given a series R-C circuit and the magnitude and angle of the applied voltage, determine the total impedance, the series current, and the voltage across each element. For the first run, use $R = 6$ Ω, $X_C = 8\Omega$, and $\mathbf{E} = 50\underline{/0°}$.

2. Given a series R-C circuit with an applied series current source, determine the total impedance, the source voltage, and the voltage across each element. Print all voltages in the sinusoidal format. For the first run, use $R = 6\Omega$, $X_C = 8\Omega$, and $\mathbf{I} = 5\underline{/53.13°}$ and compare the results with those of problem 16-1.

3. Given a parallel R-L network with an applied current source, determine the total impedance and admittance, the source voltage, and the current through each branch. For the first run, use $R = 3.33\Omega$, $X_L = 2.5\Omega$, and $\mathbf{I} = 10\underline{/0°}$.

4. Given a parallel R-L network with an applied voltage source, determine the total impedance and admittance, the source current, and the current through each branch. For the first run, use $R = 3.33\Omega$, $X_L = 2.5\Omega$, and $\mathbf{E} = 20\underline{/53.13°}$ and compare the results with those of problem 16-3.

5. Develop a general solution for the series R-L-C network with an applied voltage source. That is, print out the total impedance, series current, voltage across each element, the total power delivered, and the power factor of the network. For the first run, use $R = 3\Omega$, $X_L = 7\Omega$, $X_C = 3\Omega$, and $\mathbf{E} = 50\underline{/0°}$.

6. Develop a general solution for the parallel R-L-C network with an applied voltage source. That is, print out the total impedance and admittance, source current, current through each element, the total power delivered, and the power factor of the network. For the first run, use $R = 3.33\Omega$, $X_L = 1.43\Omega$, $X_C = 3.33\Omega$, and $\mathbf{E} = 100\underline{/53.13°}$.

7. Repeat problem 16-6 with a current source input, modifying it to determine the source voltage rather than the current. For the first run, use $R = 3.33\Omega$, $X_L = 1.43\Omega$, $X_C = 3.33\Omega$, and $\mathbf{I} = 50\underline{/0°}$ and compare the results with those of problem 16-6.

8. For an L-C series network with a voltage source input, tabulate the magnitude of Z_T, I_T, V_L, and V_C for a specified frequency range. For the first run, use $L = 10$mH, $C = 1\mu F$, $\mathbf{E} = 10\underline{/0°}$ and a frequency range from 100 Hz to 3000 Hz in increments of 100 Hz.

9. Use the results of problem 16-8 to plot Z_T versus frequency and I_T versus frequency on two separate graphs. Label both axes to the degree possible.

10. Develop a testing routine for the series R-L-C circuit that has randomly generated values of R, L, and C, the applied frequency, and the magnitude and angle of the applied voltage. Request the total impedance, series current, and the voltage across each element. For the first run, use the following ranges: 1kΩ < R < 10kΩ, 1mH < L < 10mH, 1μF < C < 10μF, 10^3Hz < f < 10^4 Hz, 2 × 10^{-3}V < E < 20 × 10^{-3}V, 30° < θ_E < 60°.

Methods of AC Analysis

17

17.1 INTRODUCTION

A number of modules for performing various vector operations will be developed in this chapter in preparation for analyzing complex ac systems. Although the programs necessarily become longer and more complex, they employ a number of modules that, if understood, will reduce the overall complexity of the analysis. To insure you understand the modules, the step-by-step process for particular operations will be examined in detail.

17.2 ADDITION AND SUBTRACTION OF VECTORS

The first program (17-1) will perform the addition and subtraction of vectors provided in either the rectangular or polar form. Note that the modules starting at line 4000 are the major part of the program. In fact, an excellent starting point in the review of the program would be to isolate the various modules in that region and identify their role in the program.

The routines from 4000 through 4070 and 4100 through 4140 should now be quite familiar from previous chapters. However, note that the routines use subscripted values of V, rather than the X, Y, and TH (θ) notation of earlier programs. The reason for the change will become obvious as we progress through the chapter. For now, simply realize that the four subscripted values of V defined by $V(I)$ represent $V(1) = X$, $V(2) = Y$, $V(3) = Z$, $V(4) = TH(\theta)$. A comparison between the subroutine at 4000 and the one used earlier will then be quite direct.

As indicated by the REM statements of lines 4400 and 4410, this subroutine will add or subtract two vectors, $\mathbf{V}_A$ and $\mathbf{V}_B$, defined by $VA(1) = X_A$, $VA(2) = Y_A$, $VA(3) = Z_A$, $VA(4) = TH_A$, and $VB(1) = X_B$, $VB(2) = Y_B$, $VB(3)$

```
 10 REM *****  PROGRAM 17-1  *****
 20 REM ********************************************
 30 REM Program to demonstrate the use of a module
 40 REM to add and subtract vectors.
 50 REM ********************************************
 60 REM
100 DIM V(4),VA(4),VB(4),VC(4)
110 PRINT "This program accepts vector input in"
120 PRINT "either the rectangular or the polar form "
130 PRINT "and provides as output the sum and the"
140 PRINT "difference of these vectors."
150 PRINT
160 PRINT "Input vector 1:"
170 GOSUB 5500
180 FOR I=1 TO 4
190 VA(I)=V(I)
200 NEXT I
210 PRINT "Input vector 2:"
220 GOSUB 5500
230 FOR I=1 TO 4
240 VB(I)=V(I)
250 NEXT I
260 GOSUB 4400 :REM Add vectors
270 PRINT "Sum of vectors is"; :GOSUB 5980
280 PRINT TAB(10);"which is";VC(3);"at an angle of";VC(4);"degrees"
290 GOSUB 4500 :REM Subtract vectors
300 PRINT "Difference of vectors is"; :GOSUB 5980
310 PRINT TAB(10);"which is";VC(3);"at an angle of";VC(4);"degrees"
320 END
```

V_1 — lines 160–200

V_2 — lines 210–250

Sum — lines 260–280

Difference — lines 290–310

```
        ┌4000 REM Module to convert from rectangular to polar form
         4010 REM Enter with V(1),V(2) - return with V(3),V(4)
Rect.    4020 V(3)=SQR(V(1)^2+V(2)^2)
  ↓      4030 IF V(1)>0 THEN V(4)=(180/3.14159)*ATN(V(2)/V(1))
Polar    4040 IF V(1)<0 THEN V(4)=180*SGN(V(2))+(180/3.14159)*ATN(V(2)/V(1))
         4050 IF V(1)=0 THEN V(4)=90*SGN(V(2))
         4060 IF V(2)=0 THEN IF V(1)<0 THEN V(4)=180
        └4070 RETURN
        ┌4100 REM Conversion from polar to rectangular form
Polar    4110 REM Enter with V(3),V(4) - return with V(1),V(2)
  ↓      4120 V(1)=V(3)*COS(V(4)*3.14159/180)
Rect.    4130 V(2)=V(3)*SIN(V(4)*3.14159/180)
        └4140 RETURN
        ┌4400 REM Module to Add or Subtract VC=VA + or - VB
         4410 REM Enter with VA,VB - return with VC
         4420 VC(1)=VA(1)+VB(1)
  +      4430 VC(2)=VA(2)+VB(2)
         4440 V(1)=VC(1)  :V(2)=VC(2)
         4450 GOSUB 4000 :REM Convert to polar form
         4460 VC(3)=V(3) :VC(4)=V(4)
        └4470 RETURN
        ┌4500 REM Enter here for subtraction operation
         4510 REM VC=VA-VB
         4520 VC(1)=VA(1)-VB(1)
  -      4530 VC(2)=VA(2)-VB(2)
        └4540 GOTO 4440 :REM Convert to polar form and return
        ┌4600 REM Module to multiply and divide VA by VB
  ×      4610 REM VC=VA*VB
         4620 VC(3)=VA(3)*VB(3)
        └4630 VC(4)=VA(4)+VB(4)
Polar   ┌4640 V(3)=VC(3)  :V(4)=VC(4)
  ↓      4650 GOSUB 4100 :REM Convert to rectangular form
Rect.    4660 VC(1)=V(1)  :VC(2)=V(2)
        └4670 RETURN :REM Return from both multiply and divide
        ┌4700 REM Enter here for divide
         4710 REM VC=VA/VB
  ÷      4720 VC(4)=VA(4)-VB(4)
         4730 IF VB(3)=0 THEN VC(3)=1E10 :REM VC(3) very large
         4740 IF VB(3)<>0 THEN VC(3)=VA(3)/VB(3)
        └4750 GOTO 4640 :REM Convert to rectangular form and return
        ┌5500 REM Module to accept input of vector
         5510 REM (1) in rectangular form, (2) in polar form
         5520 PRINT "(1) enter rectangular form, or (2) to enter polar form:";
Input    5530 INPUT C
Control  5540 IF C=0 THEN END
         5550 IF C>2 THEN GOTO 5520
         5560 ON C GOSUB 5800,5900
         5570 PRINT
        └5580 RETURN
        ┌5800 REM Module to accept input in rectangular form
         5810 REM Accept input of V(1),V(2), and convert to V(3),V(4)
Rect.    5820 INPUT "X=";V(1)
  ↓      5830 INPUT "Y=";V(2)
Polar    5840 GOSUB 4000 :REM Convert to polar form
        └5850 RETURN
        ┌5900 REM Module to accept input in polar form
         5910 REM Accept input of V(3),V(4), and convert to V(1),V(2)
Polar    5920 INPUT "Magnitude=";V(3)
  ↓      5930 INPUT "at angle";V(4)
Rect.    5940 GOSUB 4100 :REM Convert to rectangular form
        └5950 RETURN
```

(Continued)

Rect. ┌5980 PRINT VC(1); :IF VC(2)>=0 THEN PRINT "+j";VC(2) :RETURN
Format └5990 IF VC(2)<0 THEN PRINT "-j";ABS(VC(2)) :RETURN

READY

RUN

This program accepts vector input in
either the rectangular or the polar form
and provides as output the sum and the
difference of these vectors.

Input vector 1:
(1) enter rectangular form, or (2) to enter polar form:? 1
X=? 3
Y=? -4

Input vector 2:
(1) enter rectangular form, or (2) to enter polar form:? 2
Magnitude=? 10
at angle? -90

Sum of vectors is 3 -j 14
          which is 14.3178 at an angle of-77.9053 degrees
Difference of vectors is 3 +j 6
          which is 6.7082 at an angle of 63.4351 degrees

READY

This program accepts vector input in
either the rectangular or the polar form
and provides as output the sum and the
difference of these vectors.

Input vector 1:
(1) enter rectangular form, or (2) to enter polar form:? 2
Magnitude=? 3.08
at angle? -79.6

Input vector 2:
(1) enter rectangular form, or (2) to enter polar form:? 1
X=? 4.06
Y=? -0.34

Sum of vectors is 4.616 -j 3.3694
          which is 5.7149 at an angle of-36.1272 degrees
Difference of vectors is-3.504 -j 2.6894
          which is 4.4171 at an angle of-142.4929 degrees

READY

$= Z_B$, and $VB(4) = TH_B$. The subscripted variables $VC(1) = X_C$, $VC(2) = Y_C$, $VC(3) = Z_C$, $VC(4) = TH_C$ represent the result of the addition or the subtraction process. Looking ahead,

we find that the multiplication and division subroutines beginning at 4600 use the same notation. The multiplication and division routines are included in this program because they are part of the stored modules that extend from 4000 through 5990. Actually, the routine starting at 4600 could be left out, but, since the entire section can be easily listed, we included it. In the next program, the addition and subtraction routines will appear when we examine the multiplication and division operations.

In general, the subscripted variables $V(I)$ define the values of X, Y, Z, and TH for the input quantities. The subscripted variables $VA(I)$ and $VB(I)$ are then defined in terms of $V(I)$ for each vector, and the mathematical operation performed. The resulting subscriptive variables of $VC(I)$ represent the result of the operation and are used to print out the results.

Note, in the addition/subtraction subroutines, the requirement to redefine $V(1)$ and $V(2)$ before the conversion process can be performed using the subroutine of line 4000. That is, the conversion processes are limited to working with the subscripted variables defined by $V(I)$. Be aware that, although the subroutine at 4400 performed both operations, the addition process begins at line 4400 and the subtraction process at 4500. If a subtraction is to be performed, it will enter the program at 4500. The values of $VC(1)$ and $VC(2)$ must then be determined before moving back to 4440 in the subroutine, where the polar form and $VC(3)$ and $VC(4)$ are defined by line 4450. For each operation, therefore, the results in the rectangular form are determined first, followed by a conversion to the polar form, before returning to the controlling part of the program at line 4470. Lines 4440 through 4470 are included together in the same module because they can be used for both the addition and subtraction operations. In the subroutine of 4600, both operations are normally performed in the polar form, resulting in the sharing of lines 4650 through 4670 to convert from the polar to the rectangular form to be sure $VC(1)$ and $VC(2)$ are available. In equation form, the multiplication and division processes are defined by

$$VC \underline{/\theta_C} = VA \underline{/\theta_A} \times VB \underline{/\theta_B}$$
$$= VA \times VB \underline{/\theta_A + \theta_B}$$
$$= VA(3) \times VB(3) \underline{/VA(4) + VB(4)}$$
$$= VC(3) \underline{/VC(4)}$$

and

$$VC \underline{/\theta_C} = VA \underline{/\theta_A} / VB \underline{/\theta_B} = VA/VB \underline{/\theta_A - \theta_B}$$
$$= \frac{VA(3)}{VB(3)} \underline{/VA(4) - VB(4)}$$
$$= VC(3) \underline{/VC(4)}$$

The modules beginning at 5500 will request the input data and convert to the other form to insure that all four subscripted values of $V(I)$ are available.

As an example, consider the case of one vector in rectangular form and another in polar form. When you input the data of $\mathbf{V}_A$ and $\mathbf{V}_B$, always keep in mind that $\mathbf{V}_B$ is being subtracted from $\mathbf{V}_A$. For this example, $\mathbf{V}_A$ will be in polar form, while $\mathbf{V}_B$ will be in rectangular form. $\mathbf{V}_A$ is input first, starting on line 160. Line 170 will then send the program to line 5500 to enter $\mathbf{V}_A$, and convert it to the rectangular form so that $V(1)$ and $V(2)$ are also available. Line 5950 will then direct the process back to line 180, where all the subscripted values of $\mathbf{V}_A$ are defined. Line 210 will initiate the request for $\mathbf{V}_B$ and send the program back to 5500, where it will be directed to 5800, since the input is in rectangular form. $V(3)$ and $V(4)$ will then be determined before returning to line 230 to define the subscripted values of $\mathbf{V}_B$.

Line 260 will automatically send the program to line 4400, where the addition will be performed and printed out by lines 270 and 280. Line 290 will direct the program to line 4500, where it performs the subtraction operation and redefines the subscripted values of $\mathbf{V}_C$, before it prints out the results on line 300 and 310. The program will then END at 320. The results of the

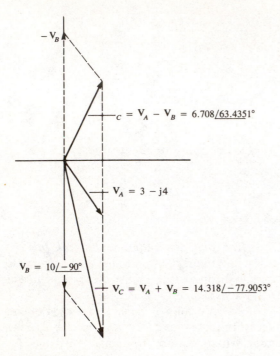

$$C = V_A - V_B = 6.708\underline{/63.4351°}$$

$$V_A = 3 - j4$$

$$V_B = 10\underline{/-90°}$$

$$V_C = V_A + V_B = 14.318\underline{/-77.9053°}$$

FIGURE 17.1

first run are provided in Figure 17.1 for both operations.

17.3   MULTIPLICATION AND DIVISION OF VECTORS

The subroutines introduced in Program 17-1 will now be applied in Program 17-2 to determine the product and quotient of two vectors input in the rectangular or polar form. In fact, the entire sequence from 4000 to 5990 was included, even though the addition and subtraction subroutines will not be employed in this program. A comparison with Program 17-1 will reveal that the only difference is the heading on lines 110 through 140 and the content of lines 260 through 310. The modules at 4600 and 4700 will now perform the product and division operations, rather than the addition (at 4400) and subtraction (at 4500) operations of Program 17-1.

In the second run of the program, the magnitude of V_B, denoted by $VB(3)$, was entered as a negative quantity. Throughout the calculations, $VB(3)$ will remain negative, resulting in negative values for the polar form in the results. The fact that V_A was entered in rec-

```
 10 REM *****  PROGRAM 17-2  *****
 20 REM **********************************************
 30 REM Program to demonstrate the use of a module
 40 REM to multiply and divide vectors.
 50 REM **********************************************
 60 REM
100 DIM V(4),VA(4),VB(4),VC(4)
110 PRINT "This program accepts vector input in"
120 PRINT "either the rectangular or polar form and "
130 PRINT "provides as output the product and the"
140 PRINT "quotient of these vectors."
150 PRINT
160 PRINT "Input vector 1:"
170 GOSUB 5500
180 FOR I=1 TO 4
190 VA(I)=V(I)
200 NEXT I
210 PRINT "Input vector 2:"
220 GOSUB 5500
230 FOR I=1 TO 4
240 VB(I)=V(I)
250 NEXT I
260 GOSUB 4600 :REM Multiply vectors
270 PRINT "Product of vectors is"; :GOSUB 5980
280 PRINT TAB(10);"which is";VC(3);"at an angle of";VC(4);"degrees"
```

The code lines are grouped with brackets labeled: V_1 (lines 160–200), V_2 (lines 210–250), and × (lines 260–280).

```
 ┌290 GOSUB 4700 :REM Divide vectors
÷│300 PRINT "Quotient of vectors is"; :GOSUB 5980
 └310 PRINT TAB(10);"which is";VC(3);"at an angle of";VC(4);"degrees"
  320 END
 4000 REM Module to convert from rectangular to polar form
 4010 REM Enter with V(1),V(2) - return with V(3),V(4)
 4020 V(3)=SQR(V(1)^2+V(2)^2)
 4030 IF V(1)>0 THEN V(4)=(180/3.14159)*ATN(V(2)/V(1))
 4040 IF V(1)<0 THEN V(4)=180*SGN(V(2))+(180/3.14159)*ATN(V(2)/V(1))
 4050 IF V(1)=0 THEN V(4)=90*SGN(V(2))
 4060 IF V(2)=0 THEN IF V(1)<0 THEN V(4)=180
 4070 RETURN
 4100 REM Conversion from polar to rectangular form
 4110 REM Enter with V(3),V(4) - return with V(1),V(2)
 4120 V(1)=V(3)*COS(V(4)*3.14159/180)
 4130 V(2)=V(3)*SIN(V(4)*3.14159/180)
 4140 RETURN
 4400 REM Module to Add or Subtract VC=VA + or - VB
 4410 REM Enter with VA,VB - return with VC
 4420 VC(1)=VA(1)+VB(1)
 4430 VC(2)=VA(2)+VB(2)
 4440 V(1)=VC(1) :V(2)=VC(2)
 4450 GOSUB 4000 :REM Convert to polar form
 4460 VC(3)=V(3) :VC(4)=V(4)
 4470 RETURN
 4500 REM Enter here for subtraction operation
 4510 REM VC=VA-VB
 4520 VC(1)=VA(1)-VB(1)
 4530 VC(2)=VA(2)-VB(2)
 4540 GOTO 4440 :REM Convert to polar form and return
 4600 REM Module to multiply and divide VA by VB
 4610 REM VC=VA*VB
 4620 VC(3)=VA(3)*VB(3)
 4630 VC(4)=VA(4)+VB(4)
 4640 V(3)=VC(3) :V(4)=VC(4)
 4650 GOSUB 4100 :REM Convert to rectangular form
 4660 VC(1)=V(1) :VC(2)=V(2)
 4670 RETURN :REM Return from both multiply and divide
 4700 REM Enter here for divide
 4710 REM VC=VA/VB
 4720 VC(4)=VA(4)-VB(4)
 4730 IF VB(3)=0 THEN VC(3)=1E10 :REM VC(3) very large
 4740 IF VB(3)<>0 THEN VC(3)=VA(3)/VB(3)
 4750 GOTO 4640 :REM Convert to rectangular form and return
 5500 REM Module to accept vector input
 5510 REM (1) in Rectangular form, (2) in Polar form
 5520 PRINT "(1) Enter rectangular form, or (2) Enter polar form:";
 5530 INPUT C
 5540 IF C= 0 THEN STOP
 5550 IF C>2 THEN GOTO 5520
 5560 ON C GOSUB 5800,5900
 5570 PRINT
 5580 RETURN
 5800 REM Module to accept input in rectangular form
 5810 REM Accept input of V(1),V(2), and convert to V(3),V(4)
 5820 INPUT "X=";V(1)
 5830 INPUT "Y=";V(2)
 5840 GOSUB 4000 :REM Convert to the polar form
 5850 RETURN
 5900 REM Module to accept input in polar form
 5910 REM Accept input of V(3),V(4), and convert to V(1),V(2)
 5920 INPUT "Magnitude=";V(3)
 5930 INPUT "at angle";V(4)
```

(Continued)

```
5940 GOSUB 4100 :REM Convert to the rectangular form
5950 RETURN
5980 PRINT V(1); :IF V(2)>=0 THEN PRINT "+j";V(2) :RETURN
5990 IF V(2)<0 THEN PRINT "-j";ABS(V(2)) :RETURN
```

READY

RUN

This program accepts vector input in
either the rectangular or polar form and
provides as output the product and the
quotient of these vectors.

Input vector 1:
(1) Enter rectangular form, or (2) Enter polar form:? 2
Magnitude=? 5
at angle? 30

Input vector 2:
(1) Enter rectangular form, or (2) Enter polar form:? 1
X=? 3
Y=? 4

Product of vectors is 2.9904 +j 24.8205
        which is 25 at an angle of 83.1301 degrees
Quotient of vectors is .9196 -j .3928
        which is 1 at an angle of-23.1301 degrees

READY

This program accepts vector input in
either the rectangular or polar form and
provides as output the product and the
quotient of these vectors.

Input vector 1:
(1) Enter rectangular form, or (2) Enter polar form:? 1
X=? -2
Y=? -5

Input vector 2:
(1) Enter rectangular form, or (2) Enter polar form:? 2
Magnitude=? -4
at angle? 206

Product of vectors is 1.5771 -j 21.4829
        which is-21.5407 at an angle of 94.1986 degrees
Quotient of vectors is-.9974 -j .9043
        which is-1.3463 at an angle of-317.8014 degrees

READY

tangular form will always result in a positive value for $VA(3)$ due to the squaring of $V(1)$ and $V(2)$ on line 4020. The net result is that $VC(3) = VA(3) \times VB(3)$ and $VC(3) = VA(3)/VB(3)$ will generate negative results for the solution. The rectangular and polar forms for the product of the second run appear in Figure 17.2 to show their equivalence.

17.4 PARALLEL IMPEDANCES

The general equation for determining the total impedance of two parallel impedances has the following format: $\mathbf{Z}_T = \mathbf{Z}_1\mathbf{Z}_2/(\mathbf{Z}_1 + \mathbf{Z}_2)$. Note that three distinct mathematical operations are required. First, the denominator is determined by finding the sum of the two vectors, using a routine such as defined in Program 17-1. Program 17-2 would then be applied to determine the product of the vectors and the total impedance by performing the final division operation.

Comparing Program 17-3 to the previous two, we find that the major change is the sequence of commands from line 4200 to 4390,

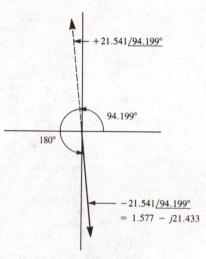

FIGURE 17.2

designed specifically to perform the operations required to find the total impedance of two parallel impedances. The module from 5500 to 5950 will simply request the input data and insure that each vector is available in polar and rectangular form, while the routine from 4000 to 4140 is limited to performing the conversion processes.

```
10 REM *****  PROGRAM 17-3  *****
20 REM *********************************************************
30 REM Program to demonstrate the use of a module to
40 REM calculate the parallel equivalent of two impedances.
50 REM *********************************************************
60 REM
100 DIM V(4),Z1(4),Z2(4),ZP(4)
110 PRINT "This program accepts vector input in"
120 PRINT "either the rectangular or polar form "
130 PRINT "and provides the parallel equivalent"
140 PRINT
150 PRINT
160 PRINT "Input vector 1:"
170 GOSUB 5500
180 FOR I=1 TO 4
190 Z1(I)=V(I)
200 NEXT I
210 PRINT "Input vector 2:"
220 GOSUB 5500
230 FOR I=1 TO 4
240 Z2(I)=V(I)
250 NEXT I
260 GOSUB 4200 :REM Calculate parallel equivalent
```

V_1 — lines 160–200
V_2 — lines 210–250
Calc. — line 260

(Continued)

```
Output          ⎡270  PRINT "Parallel equivalent is";ZP(1);
                │280  IF ZP(2)>=0 THEN PRINT "+j";ZP(2)
                │290  IF ZP(2)<0 THEN PRINT "-j";ABS(ZP(2))
                ⎣300  PRINT TAB(10);"which is";ZP(3);"at an angle of";ZP(4);"degrees"
                 310  END
                ⎡4000 REM Module to convert from rectangular to polar form
                │4010 REM Enter with V(1),V(2) - return with V(3),V(4)
                │4020 V(3)=SQR(V(1)^2+V(2)^2)
Rect.           │4030 IF V(1)>0 THEN V(4)=(180/3.14159)*ATN(V(2)/V(1))
↓               │4040 IF V(1)<0 THEN V(4)=180*SGN(V(2))+(180/3.14159)*ATN(V(2)/V(1))
Polar           │4050 IF V(1)=0 THEN V(4)=90*SGN(V(2))
                │4060 IF V(2)=0 THEN IF V(1)<0 THEN V(4)=180
                ⎣4070 RETURN
                ⎡4100 REM Conversion from polar to rectangular form
Polar           │4110 REM Enter with V(3),V(4) - return with V(1),V(2)
↓               │4120 V(1)=V(3)*COS(V(4)*3.14159/180)
Rect.           │4130 V(2)=V(3)*SIN(V(4)*3.14159/180)
                ⎣4140 RETURN
                ⎡4200 REM Module to combine impedances in parallel
                │4210 REM ZP=Z1 in parallel with Z2
                │4220 V(1)=Z1(1) :V(2)=Z1(2)
                │4230 GOSUB 4000 :REM Convert Z1 to polar form
                │4240 Z1(3)=V(3) :Z1(4)=V(4)
                │4250 V(1)=Z2(1) :V(2)=Z2(2)
                │4260 GOSUB 4000 :REM Convert Z2 to polar form
                │4270 Z2(3)=V(3) :Z2(4)=V(4)
Calc.           │4280 REM Now do parallel combination
Parallel        │4290 ZP(3)=Z1(3)*Z2(3) :REM Numerator magnitude
Imp.            │4300 ZP(4)=Z1(4)+Z2(4) :REM Numerator angle
                │4310 V(1)=Z1(1)+Z2(1) :V(2)=Z1(2)+Z2(2) :REM Sum in rectangular form
                │4320 GOSUB 4000 :REM Convert to polar form
                │4330 ZP(3)=ZP(3)/V(3) :REM Division magnitude
                │4340 ZP(4)=ZP(4)-V(4) :REM Division angle
                │4350 REM Convert solution to rectangular form
                │4360 V(3)=ZP(3) :V(4)=ZP(4)
                │4370 GOSUB 4100
                │4380 ZP(1)=V(1) :ZP(2)=V(2)
                ⎣4390 RETURN
                ⎡5500 REM Module to accept vector input
                │5510 REM (1) in Rectangular form, (2) in Polar form
                │5520 PRINT "(1) Enter rectangular form, or (2) Enter polar form:";
                │5530 INPUT C
Input           │5540 IF C=0 THEN END
Control         │5550 IF C>2 THEN GOTO 5520
                │5560 ON C GOSUB 5800,5900
                │5570 PRINT
                ⎣5580 RETURN
                ⎡5800 REM Module to accept input in rectangular form
                │5810 REM Accept input of V(1),V(2), and convert to V(3),V(4)
Rect.           │5820 INPUT "X=";V(1)
↓               │5830 INPUT "Y=";V(2)
Polar           │5840 GOSUB 4000 :REM Convert to the polar form
                ⎣5850 RETURN
                ⎡5900 REM Module to accept input in polar form
Polar           │5910 REM Accept input of V(3),V(4), and convert to V(1),V(2)
↓               │5920 INPUT "Magnitude=";V(3)
Rect.           │5930 INPUT "at angle";V(4)
                │5940 GOSUB 4100 :REM Convert to the rectangular form
                ⎣5950 RETURN
```

READY

```
RUN

This program accepts vector input in
either the rectangular or polar form
and provides the parallel equivalent


Input vector 1:
(1) Enter rectangular form, or (2) Enter polar form:? 1
X=? 3
Y=? -4

Input vector 2:
(1) Enter rectangular form, or (2) Enter polar form:? 2
Magnitude=? 5
at angle? 30

Parallel equivalent is 3.2735 -j .6699
          which is 3.3413 at an angle of-11.5651 degrees


READY



This program accepts vector input in
either the rectangular or polar form
and provides the parallel equivalent


Input vector 1:
(1) Enter rectangular form, or (2) Enter polar form:? 2
Magnitude=? 9.56
at angle? 37.89

Input vector 2:
(1) Enter rectangular form, or (2) Enter polar form:? 1
X=? -5.08
Y=? 8.46

Parallel equivalent is 1.2788 +j 6.3602
          which is 6.4875 at an angle of 78.6318 degrees


READY
```

Keeping in mind that variables with (1) and (2) refer to the magnitude of X and Y respectively in the rectangular form, and (3) and (4) to the magnitude and angle respectively in the polar form, should remove any difficulty you have understanding the routine from 4200 to 4390. The combination of the REM statements and the subscripted notation for $\mathbf{Z}_1$ and $\mathbf{Z}_2$ to identify each vector should also be helpful. The unique aspects of the program are the twenty lines in the subroutine at 4200. The remainder is subroutines or statements used more than once in earlier programs.

Determining the total impedance of two purely resistive or reactive impedances is not particularly difficult. However, if each impedance has a real and imaginary part with one in the polar form and one in the rectangular form, the process can be lengthy and time-consuming. In particular, note the second run, with numbers accurate to the hundredth place. The program has performed an important task in a minimum of time with a high degree of accuracy; the result is also provided in both forms to prepare for any operation to follow.

17.5 MESH ANALYSIS

The network of Figure 17.3 will now be analyzed using mesh analysis and the routines defined in Programs 17-1 and 17-2. Applying the method will result in the following equations for the network:

$$(Z_1 + Z_2)I_1 \qquad - Z_2 I_2 = \quad E_1$$
$$- Z_2 I_1 + (Z_2 + Z_3)I_2 = - E_2$$

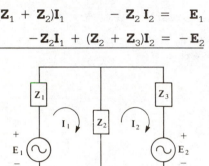

FIGURE 17.3

Applying determinants will result in the following equations for $\tilde{I}_1$:

$$I_1 = \frac{\begin{vmatrix} E_1 & -Z_2 \\ -E_2 & (Z_2 + Z_3) \end{vmatrix}}{\begin{vmatrix} (Z_1 + Z_2) & -Z_2 \\ -Z_2 & (Z_2 + Z_3) \end{vmatrix}}$$

$$= \frac{(E_1 - E_2)Z_2 + E_1 Z_3}{Z_1 Z_2 + Z_1 Z_3 + Z_2 Z_3}$$

Program 17-4 is limited to finding the current I_1. The current I_2 will be left as an exercise for the end of the chapter. The first unique aspect of the program is the long series of dimension statements on lines 100 and 110. Recall that $V(4)$ is set aside for the rectangular and polar form of any vector. $VA(4)$, $VB(4)$ and $VC(4)$

appear in the arithmetic operations of addition, subtraction, multiplication, and division. $I1(4)$ will include the rectangular and polar forms of the solution, while $E1(4)$ and $E2(4)$ will have both forms of the applied voltage. $Z1(4)$, $Z2(4)$, and $Z3(4)$ are the three impedances, while $N(4)$ is set aside for both forms of the numerator, with $D(4)$ for the denominator.

The subroutines on lines 4000 through 5990 have been described in detail in earlier programs. Lines 160 through 360 simply determine the parameters for the rectangular and polar form for each quantity. The routine at line 370 will determine the product $Z_1 \times Z_2$, which appear in the denominator of the expression for I_1. First, the values of $VA(I)$ and $VB(I)$ are defined, and then the subroutine at 4600 is used to determine the product and the subscripted values of $VC(I)$. The next loop of I, starting at 420, will store the results of the routine of 370 and change the parameters of $VB(I)$ to those of $Z3(I)$. Since $VA(I)$ remains the same, the routine at 4600 can now form the $Z_1 \times Z_3$ product which also appears in the denominator of the equation for I_1. The subroutine starting at line 470 will then add the rectangular components of the previous result, stored as $D(I)$, to the rectangular components of $Z_1 \times Z_3$ to determine the real and imaginary part of the sum $Z_1 \times Z_2 + Z_1 \times Z_3$. The subroutine at 4000 determines the polar form of the results thus far. Lines 500 through 590 determine $Z_2 \times Z_3$ and add the result to the above sum before defining the denominator parameters as $D(I)$ on line 600.

The numerator $N(I)$ is then determined in similar fashion from lines 620 through 830 before I_1 is determined starting on line 840. For each operation, the proper value of VA and VB has to be defined, followed by the proper storage of $N(I)$, until the total $N(I)$ is defined. On line 860, $N(I)$ will be determined by the values of $VC(I)$, but then $VA(I)$ and $VB(I)$ will be set equal to $N(I)$ and $D(I)$ respectively. The division operation of line 4700 can then be performed.

```
            10 REM *****  PROGRAM 17-4  *****
            20 REM *********************************************
            30 REM Program to use vector algebra to calculate the
            40 REM mesh current I1 for the network of Figure 17-3.
            50 REM *********************************************
            60 REM
            80REM
           100 DIM V(4),VA(4),VB(4),VC(4),I1(4)
           110 DIM E1(4),E2(4),Z1(4),Z2(4),Z3(4),N(4),D(4)
           120 PRINT "This program calculates the mesh current I1"
           130 PRINT "for the network of Figure 17-3."
           140 PRINT
           150 REM Input network data
         ┌─160 PRINT "E1=:" :GOSUB 5500
         │ 170 FOR I=1 TO 4
  E₁     │ 180 E1(I)=V(I)
         └─190 NEXT I
         ┌─200 PRINT "E2=:" :GOSUB 5500
         │ 210 FOR I=1 TO 4
  E₂     │ 220 E2(I)=V(I)
         └─230 NEXT I
           240 PRINT
         ┌─250 PRINT "Z1=:" :GOSUB 5500
         │ 260 FOR I=1 TO 4
  Z₁     │ 270 Z1(I)=V(I)
         └─280 NEXT I
         ┌─290 PRINT "Z2=:" :GOSUB 5500
         │ 300 FOR I=1 TO 4
  Z₂     │ 310 Z2(I)=V(I)
         └─320 NEXT I
         ┌─330 PRINT "Z3=:" :GOSUB 5500
         │ 340 FOR I=1 TO 4
  Z₃     │ 350 Z3(I)=V(I)
         └─360 NEXT I
         ┌─370 REM Determine Z1*Z2
         │ 380 FOR I=1 TO 4
Z₁·Z₂    │ 390 VA(I)=Z1(I) :VB(I)=Z2(I)
         │ 400 NEXT I
         └─410 GOSUB 4600
         ┌─420 REM Determine Z1*Z3 and store Z1*Z2 as D(I)
         │ 430 FOR I=1 TO 4
Z₁·Z₃    │ 440 D(I)=VC(I)  :VB(I)=Z3(I)
         │ 450 NEXT I
         └─460 GOSUB 4600
Z₁·Z₂  ┌─470 REM Convert sum Z1*Z2+Z1*Z3 to polar form
 +     │ 480 V(1)=D(1)+VC(1)  :V(2)=D(2)+VC(2)
Z₁·Z₃  └─490 GOSUB 4000
         ┌─500 REM Determine Z2*Z3 and store Z1*Z2+Z1*Z3 as D(I)
         │ 510 FOR I=1 TO 4
Z₂·Z₃    │ 520 D(I)=V(I)  :VA(I)=Z2(I)
         │ 530 NEXT I
         └─540 GOSUB 4600
         ┌─550 REM Establish D(I)=Z1*Z2+Z1*Z3+Z2*Z3
         │ 560 V(1)=D(1)+VC(1)
Z₁·Z₂    │ 570 V(2)=D(2)+VC(2)
 +       │ 580 GOSUB 4000
Z₁·Z₃    │ 590 FOR I=1 TO 4
 +       │ 600 D(I)=V(I)
Z₂·Z₃    └─610 NEXT I
           620 REM Now calculate numerator vector
```

(Continued)

$E_1 - E_2$
```
630 REM Determine E1-E2
640 FOR I=1 TO 4
650 VA(I)=E1(I) :VB(I)=E2(I)
660 NEXT I
670 GOSUB 4500
```

$(E_1 - E_2)Z_2$
```
680 REM Determine (E1-E2)*Z2
690 FOR I=1 TO 4
700 VA(I)=VC(I) :VB(I)=Z2(I)
710 NEXT I
720 GOSUB 4600
```

$E_1 \cdot Z_3$
```
730 REM Determine E1*Z3 and store (E1-E2)*Z2 as N(I)
740 FOR I=1 TO 4
750 N(I)=VC(I)
760 VA(I)=E1(I) :VB(I)=Z3(I)
770 NEXT I
780 GOSUB 4600
```

$(E_1 - E_2)Z_2$
$+ E_1 \cdot Z_3$
```
790 REM Establish N(I)=(E1-E2)*Z2+E1*Z3
800 FOR I=1 TO 4
810 VA(I)=VC(I) :VB(I)=N(I)
820 NEXT I
830 GOSUB 4400
```

Calc.
I_1
```
840 REM Calculate I1=N(I)/D(I)
850 FOR I=1 TO 4
860 N(I)=VC(I) :VA(I)=N(I) :VB(I)=D(I)
870 NEXT I
880 GOSUB 4700
890 FOR I=1 TO 4
900 I1(I)=VC(I)
910 NEXT I
```

Print
I_1
```
920 PRINT "Current I1 is"; :GOSUB 5980
930 PRINT "which is";V(3);"at an angle of";V(4);"degrees"
940 END
```

4000-5990
as in Program 17-1

RUN

*
This program calculates the mesh current I1
for the network of Figure 17-3.

E1=:
(1) Enter rectangular form, or (2) Enter polar form:? 2
Magnitude=? 2
at angle? 0

E2=:
(1) Enter rectangular form, or (2) Enter polar form:? 2
Magnitude=? 6
at angle? 0

Z1=:
(1) Enter rectangular form, or (2) Enter polar form:? 1
X=? 0
Y=? 2

*Example 17.5, ICA

```
Z2=:
(1) Enter rectangular form, or (2) Enter polar form:? 1
X=? 4
Y=? 0

Z3=:
(1) Enter rectangular form, or (2) Enter polar form:? 1
X=? 0
Y=? -1

Current I1 is-2 +j 3
which is 3.6055 at an angle of-236.3099 degrees


READY
```

The longhand solution of such a problem is quite time consuming and often incorrect due to the loss of a sign or an incorrect conversion.

17.6   NODAL ANALYSIS

The previous program was an excellent example of the use of modules to solve complex systems. Unfortunately, though, a program written for a particular system, such as the network of Figure 17.4, will be shorter than one using the extended module approach. Writing a program with modules will usually take less time than writing one for a particular system, and you will not need to be as concerned about the details of the module if input as a package.

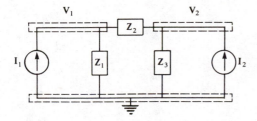

V_1 V_2

I_1 Z_1 Z_2 Z_3 I_2

FIGURE 17.4

Program 17-5 is essentially the dual of program 17-4. It will determine the nodal voltage V_1 of Figure 17.4 using the nodal analysis

technique. Since the program is limited to the network of Figure 17.4, it is significantly shorter than Program 17-4. Applying the nodal analysis technique to the network of Figure 17.4 will result in the following equations:

$$(Y_1 + Y_2)V_1 \quad - Y_2V_2 \qquad = I_1$$
$$- Y_2V_1 \quad (Y_2 + Y_3)V_2 = I_2$$

which results in the following for V_1 when determinants are applied:

$$V_1 = \frac{I_1[Y_2 + Y_3] + I_2Y_2}{Y_1Y_2 + Y_1Y_3 + Y_2Y_3}$$

The following analysis makes it unnecessary to use any of the addition, subtraction, multiplication, or division subroutines. All the required calculations are performed using the defined real and imaginery parts for each quantity. For instance, the numerator is formed in the following manner:

$$
\begin{aligned}
N &= I_1(Y_2 + Y_3) + Y_2I_2 \\
&= (I_{1x} + jI_{1y})(G_2 + jB_2 + G_3 + jB_3) \\
&\quad + (I_{2x} + jI_{2y})(G_2 + jB_2) \\
&= I_{1x}G_2 + jI_{1x}B_2 + jI_{1y}G_2 - I_{1y}B_2 \\
&\quad + I_{1x}G_3 + jI_{1x}B_3 + jI_{1y}G_3 - I_{1y}B_3 \\
&\quad + I_{2x}G_2 + jI_{2x}B_2 + jI_{2y}G_2 - I_{2y}B_2
\end{aligned}
$$

which will result in a real and imaginary part defined by:

$$N_r = I_{1x}[G_2 + G_3] - I_{1y}[B_2 + B_3] + I_{2x}[G_2]$$
$$- I_{2y}[B_2]$$

```
        10 REM *****   PROGRAM 17-5   *****
        20 REM *********************************************
        30 REM Program to calculate the nodal voltage V1
        40 REM for the network of Figure 17-4.
        50 REM *********************************************
        60 REM
       100 DIM V(4),VA(4),VB(4),VC(4)
       110 DIM I1(4),I2(4),Z1(4),Z2(4),Z3(4),N(4),D(4)
       120 DIM V1(4),Y1(4),Y2(4),Y3(4)
       130 PRINT "This program calculates the nodal voltage V1"
       140 PRINT "for the network of Figure 17-4."
       150 PRINT
       160 REM First get circuit data
      ⎡170 PRINT "I1=:" :GOSUB 5500
  I₁  ⎢180 FOR I=1 TO 4
      ⎢190 I1(I)=V(I)
      ⎣200 NEXT I
      ⎡210 PRINT "I2=:" :GOSUB 5500
  I₂  ⎢220 FOR I=1 TO 4
      ⎢230 I2(I)=V(I)
      ⎣240 NEXT I
       250 PRINT
      ⎡260 PRINT "Z1=:" :GOSUB 5500
      ⎢270 FOR I=1 TO 4
      ⎢280 Z1(I)=V(I)
      ⎢290 NEXT I
  Z₁  ⎢300 V(3)=1/V(3)
 (Y₁) ⎢310 V(4)=-V(4)
      ⎢320 GOSUB 4100 :REM Convert Z1 to rectangular form
      ⎢330 FOR I=1 TO 4
      ⎢340 Y1(I)=V(I)
      ⎣350 NEXT I
      ⎡360 PRINT "Z2=:" :GOSUB 5500
      ⎢370 FOR I=1 TO 4
      ⎢380 Z2(I)=V(I)
      ⎢390 NEXT I
  Z₂  ⎢400 V(3)=1/V(3)
 (Y₂) ⎢410 V(4)=-V(4)
      ⎢420 GOSUB 4100 :REM Convert Z2 to rectangular form
      ⎢430 FOR I=1 TO 4
      ⎢440 Y2(I)=V(I)
      ⎣450 NEXT I
      ⎡460 PRINT "Z3=:" :GOSUB 5500
      ⎢470 FOR I=1 TO 4
      ⎢480 Z3(I)=V(I)
      ⎢490 NEXT I
  Z₃  ⎢500 V(3)=1/V(3)
 (Y₃) ⎢510 V(4)=-V(4)
      ⎢520 GOSUB 4100 :REM Covert Z3 to rectangular form
      ⎢530 FOR I=1 TO 4
      ⎢540 Y3(I)=V(I)
      ⎣550 NEXT I
      ⎡560 REM Data now entered
      ⎢570 REM Now begin calculations using
      ⎢580 REM NR,NI and DR,DI approach
      ⎢590 A=I1(1)*(Y2(1)+Y3(1))
      ⎢600 B=I1(2)*(Y2(2)+Y3(2))
      ⎢610 C=I2(1)*Y2(1)-I2(2)*Y2(2)
      ⎢620 N(1)=A-B+C :REM Calculation broken into smaller parts
 Calc.⎢630 A=I1(1)*(Y2(2)+Y3(2))
      ⎢640 B=I2(1)*Y2(2)+I1(2)*(Y2(1)+Y3(1))
      ⎢650 C=I2(2)*Y2(1)
      ⎣660 N(2)=A+B+C
```

```
 670 A=Y1(1)*Y2(1)+Y1(1)*Y3(1)+Y2(1)*Y3(1)
 680 B=Y1(2)*Y2(2)+Y1(2)*Y3(2)+Y2(2)*Y3(2)
 690 D(1)=A-B
 700 A=Y1(1)*Y2(2)+Y1(1)*Y3(2)+Y2(1)*Y1(2)
 710 B=Y2(1)*Y3(2)+Y3(1)*Y1(2)+Y3(1)*Y2(2)
 720 D(2)=A+B
 730 V(1)=(N(1)*D(1)+N(2)*D(2))/(D(1)^2+D(2)^2)
 740 V(2)=(N(2)*D(1)-N(1)*D(2))/(D(1)^2+D(2)^2)
 750 GOSUB 4000 :REM Convert nodal voltage into polar form
 760 PRINT "Nodal voltage V1 is "; :GOSUB 5980
 770 PRINT "which is";V(3);"at an angle of";V(4);"degrees"
 780 END
```

Print V_1 brackets lines 750–780.

```
 4000-5990
 as in Program 17-1
```

RUN

*
This program calculates the nodal voltage V1
for the network of Figure 17-4.

I1=:
(1) Enter rectangular form, or (2) Enter polar form:? 2
Magnitude=? -6
at angle? 0

I2=:
(1) Enter rectangular form, or (2) Enter polar form:? 2
Magnitude=? 4
at angle? 0

Z1=:
(1) Enter rectangular form, or (2) Enter polar form:? 1
X=? 4
Y=? 0

Z2=:
(1) Enter rectangular form, or (2) Enter polar form:? 2
Magnitude=? 5
at angle? 90

Z3=:
(1) Enter rectangular form, or (2) Enter polar form:? 2
Magnitude=? 2
at angle? -90

Nodal voltage V1 is -12.48 -j 16.64
which is 20.8 at an angle of-126.8699 degrees

READY

*Example 17.11, ICA

```
RUN
*
This program calculates the nodal voltage V1
for the network of Figure 17-4.

I1=:
(1) Enter rectangular form, or (2) Enter polar form:? 2
Magnitude=? 1.88
at angle? -48.81

I2=:
(1) Enter rectangular form, or (2) Enter polar form:? 2
Magnitude=? 10
at angle? 20


Z1=:
(1) Enter rectangular form, or (2) Enter polar form:? 2
Magnitude=? 68.03
at angle? 100.15

Z2=:
(1) Enter rectangular form, or (2) Enter polar form:? 2
Magnitude=? 10
at angle? -90

Z3=:
(1) Enter rectangular form, or (2) Enter polar form:? 2
Magnitude=? 8
at angle? 0

Nodal voltage V1 is   83.5508 +j 13.3836
which is 84.6159 at an angle of 9.1007 degrees


READY
```

*Example 17.12, ICA

and

$$N_i = I_{1x}[B_2 + B_3] + I_{2x}[B_2] + I_{1y}[G_2 + G_3] + I_{2y}[G_2]$$

The denominator will provide the following real and imaginary parts:

$$D_r = G_1G_2 + G_1G_3 + G_2G_3 - (B_1B_2 + B_1B_3 + B_2B_3)$$

and

$$D_i = G_1B_2 + G_1B_3 + G_2B_1 + G_2B_3 + G_3B_1 + G_3B_2$$

and the desired voltage $\mathbf{V}_1$ can then be written in the following form:

$$\mathbf{V}_1 = \frac{N_r + jN_i}{D_r + jD_i}$$

$$= \frac{N_rD_r + N_iD_i}{D_r^2 + D_i^2} + j\frac{N_iD_r - N_rD_i}{D_r^2 + D_i^2}$$

$$= V_r + jV_i$$

which in polar form is:

$$\mathbf{V}_1 = V\underline{/\theta} = \sqrt{V_r^2 + V_i^2}\ \underline{/\tan^{-1}V_i/V_r}$$

In the actual program, the real parts are defined by (1) and the imaginary parts by (2). The magnitude and angles are defined by (3) and (4) respectively. For instance, if $\mathbf{Z}_1 = 3 + j4$, line 280 would result in $Z1(1) = 3$, $Z1(2)$

= 4, $Z1(3)$ = 5, and $Z1(4)$ = 53.13°. On lines 300 and 310, the polar form of the admittance parameters are defined by $V(3)$ = Y = $1/Z$ = $1/5$ = 0.2 and $V(4)$ = −53.13°. The program will then determine the rectangular form from the polar form defined by $V(3)$ and $V(4)$. The $Z1(1)$ and $Z1(2)$ above are ignored in determining the **Y** parameters. On line 340 the $Y1(I)$ parameters for $\mathbf{Z}_1$ = 3 + $j4$ are defined by $Y1(1)$ = 0.12, $Y1(2)$ = −0.16, $Y1(3)$ = 0.2, and $Y1(4)$ = −53.13°. The same process will be applied to each value of **Z**.

The equation for N_r will then be written as:

$$N_r = I1(1) \times (Y2(1) + Y3(1)) + I1(2)(Y2(2)$$
$$+ Y3(2)) + I2(1) \times Y2(1)$$
$$- I2(2) \times Y2(2)$$

which is broken up into three lines on lines 590, 600, and 610 in the program. Once the values of N_r, N_i, D_r, and D_i are determined, the value of $\mathbf{V}_1$ is defined by:

$$\mathbf{V}_1 = (N_r + jN_i)/(D_r + jD_i) = (N(1) + jN(2))/$$
$$(D(1) + jD(2))$$

The division process will result in a real part for **V** defined by $V(1)$ and an imaginary part defined by $V(2)$ on lines 730 and 740 respectively. The polar form is then defined by $V(3)$ and $V(4)$ on line 750.

Note the effort required to generate the equations for N_r, N_i, D_r, and D_i above. The program is certainly shorter than Program 17-4, but the use of the addition, subtraction, multiplication, and division modules would have saved some time and effort.

Two runs of the program are provided to show the ease of running the program once it is placed in computer memory through the keyboard or input disk or tape.

The second run is for the more complex network of Figure 17.5.

The network is first redrawn as shown in Figure 17.6, where

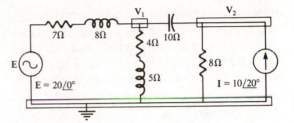

FIGURE 17.5

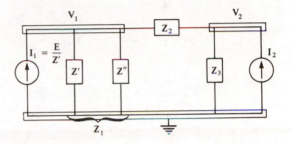

FIGURE 17.6

$\mathbf{I}_1$ = 1.88$\underline{/-48.81°}$, $\mathbf{I}_2$ = 10$\underline{/20°}$,
$\mathbf{Z}_1$ = 68.03$\underline{/100.15°}$, $\mathbf{Z}_2$ = 10$\underline{/-90°}$
and $\mathbf{Z}_3$ = 8$\underline{/0°}$.

The results of the second run clearly indicate that the program can be applied to more complex systems once they are redrawn in the format used to write the program. The nodal voltage of $\mathbf{V}_1$ determined in Figure 17.6 is the same as the $\mathbf{V}_1$ in Figure 17.5, which permits the calculation of a number of quantities for the network of Figure 17.5. The availability of $\mathbf{V}_2$ (left as an exercise at the end of the chapter) would provide a complete solution for the network of Figure 17.5.

17-7 DELTA-WYE CONFIGURATION

The next program is an excellent application of the addition, multiplication, and division subroutines. In fact, since the routines from 4000 through 5990 have appeared so frequently they will not be included in the printout of Program 17-7. Program 17-6 will convert the Δ configuration of Figure 17.7a to the Y configuration of Figure 17.7b. Only $\mathbf{Z}_1$ will be determined; $\mathbf{Z}_2$ and $\mathbf{Z}_3$ are requested as an exercise at the end of the chapter.

The general equation for $\mathbf{Z}_1$ is:

$$\mathbf{Z}_1 = \frac{\mathbf{Z}_A \mathbf{Z}_C}{\mathbf{Z}_A + \mathbf{Z}_B + \mathbf{Z}_C}$$

The denominator will be determined first using the addition routine, followed by a product cal-

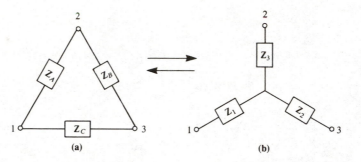

(a) **(b)**

FIGURE 17.7

```
              10 REM *****  PROGRAM 17-6  *****
              20 REM ************************************************
              30 REM Program to perform delta-wye conversion
              40 REM using R-L-C components.
              50 REM ************************************************
              60 REM
             100 DIM V(4),VA(4),VB(4),VC(4)
             110 DIM D(4),N(4)
             120 DIM Z1(4),ZA(4),ZB(4),ZC(4)
             130 PRINT "Enter the following impedance data:"
             140 PRINT
            ┌150 PRINT "ZA=" :GOSUB 5500
            │160 FOR I=1 TO 4
            │170 ZA(I)=V(I)
            │180 NEXT I
            │190 PRINT "ZB=" :GOSUB 5500
     Input  │200 FOR I=1 TO 4
       Δ    │210 ZB(I)=V(I)
            │220 NEXT I
            │230 PRINT "ZC=" :GOSUB 5500
            │240 FOR I=1 TO 4
            │250 ZC(I)=V(I)
            └260 NEXT I
            ┌270 V(1)=ZA(1)+ZB(1)+ZC(1)
 ZA + ZB + ZC│280 V(2)=ZA(2)+ZB(2)+ZC(2)
            └290 GOSUB 4000 :REM Convert sum ZA+ZB+ZC to polar form
            ┌300 REM Determine ZA*ZC and store sum of impedances, D(I)
            │310 FOR I=1 TO 4
            │320 D(I)=V(I)
   ZA · ZC  │330 VA(I)=ZA(I)
            │340 VB(I)=ZC(I)
            │350 NEXT I
            └360 GOSUB 4600
```

$$\frac{Z_A \cdot Z_C}{Z_A + Z_B + Z_C}$$

```
370 REM Determine N/D
380 FOR I=1 TO 4
390 VA(I)=VC(I)
400 VB(I)=D(I)
410 NEXT I
420 GOSUB 4700
```

Print
Z_1

```
430 REM Print value of Z1
440 FOR I=1 TO 4
450 Z1(I)=VC(I)
460 NEXT I
470 PRINT "Impedance Z1="; :GOSUB 5980
480 PRINT "which is";Z1(3);"at an angle of";Z1(4);"degrees"
490 END
```

4000-5990
as in Program 17-1

```
RUN
*
Enter the following impedance data:

ZA=
(1) Enter rectangular form, or (2) Enter polar form:? 2
Magnitude=? 4
at angle? -90

ZB=
(1) Enter rectangular form, or (2) Enter polar form:? 2
Magnitude=? 4
at angle? -90

ZC=
(1) Enter rectangular form, or (2) Enter polar form:? 1
X=? 3
Y=? 4

Impedance Z1= 3.84 +j 1.12
which is 4 at an angle of 16.2602 degrees


READY

RUN

Enter the following impedance data:

ZA=
(1) Enter rectangular form, or (2) Enter polar form:? 2
Magnitude=? 27
at angle? 60

ZB=
(1) Enter rectangular form, or (2) Enter polar form:? 2
Magnitude=? 27
at angle? 60
```

*Example 17.16, ICA

```
ZC=
(1) Enter rectangular form, or (2) Enter polar form:? 2
Magnitude=? 27
at angle? 60

Impedance Z1= 4.5 +j 7.7942
which is 9 at an angle of 60 degrees


READY
```

culation to determine the numerator, and finally by a division operation to determine $\mathbf{Z_1}$.

For each impedance, the real and imaginary parts are determined by lines 150 through 260. Line 270 will then find the sum of the real parts of the three impedances, and line 280 will find the algebraic sum of the imaginary parts. Line 290 will then determine the polar form before line 320 stores the sum of the impedances as $D(1) = V(1)$, $D(2) = V(2)$, $D(3) = V(3)$, and $D(4) = V(4)$. The storage is necessary because $V(1)$ through $V(4)$ will be used later in the program in the conversion processes starting at lines 4000 and 4100.

The program next determines the numerator at lines 300 through 360 and the ratio N/D at lines 370 through 420. Note the reappearance of $D(I)$ as the denominator on line 400. For both operations, remember that VA and VB are the subscripted variables employed in the subroutines and therefore require steps such as 330, 340, 390 and 400. Note in the second run of the program that if $\mathbf{Z_A} = \mathbf{Z_B} = \mathbf{Z_C}$, then $\mathbf{Z_1} = \mathbf{Z_2} = \mathbf{Z_3} = \mathbf{Z_\Delta}/3$, and, for the provided values, $\mathbf{Z_1} = 27\underline{/60°}/3 = 9\underline{/60°}$.

17.8 HAY BRIDGE

The next program (17-7) will test whether balance conditions exist for the bridge of Figure 17.8 and, if not, provide the values of R_4 and L that will result in a balance condition ($V_G = 0$ volts).

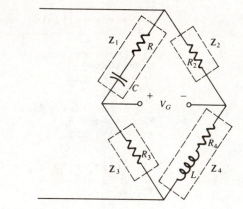

FIGURE 17.8

For a bridge network the balance conditions are defined by:

```
10 REM *****   PROGRAM 17-7  *****
20 REM **************************************
30 REM Program to test a Hay bridge network.
40 REM **************************************
50 REM
100 PRINT "This program will test calculations"
110 PRINT "for a Hay bridge circuit as shown"
120 PRINT "in Figure 17-8."
130 PRINT
140 PRINT "Enter the following network data:"
```

```
      ┌150  INPUT "R1=";R1
      │160  INPUT "R2=";R2
      │170  INPUT "R3=";R3
Input │180  INPUT "R4=";R4
Data  │190  PRINT
      │200  INPUT "C1=";C1
      │210  INPUT "L4=";L4
      └220  INPUT "Desired frequency f=";F
      ┌230  REM Perform network calculations
      │240  W=2*3.14159*F
Calc. │250  X1=1/(W*C1)
      │260  X4=W*L4
      │270  Z1=SQR(R1^2+X1^2)
      └280  Z4=SQR(R4^2+X4^2)
      ┌290  REM Perform test for bridge balance
Balance│300  IF ABS(Z1/R3 - R2/Z4)>ABS(0.01*Z1/R3) THEN GOTO 340
Test  │310  IF ABS(X1/R1-X4/R4)>ABS(0.01*X1/R1) THEN GOTO 340
      └320  PRINT "The bridge is balanced."
       330  END
      ┌340  PRINT:PRINT "The bridge is not balanced."
      │350  PRINT:PRINT "In order to obtain balance use elements:"
Calc. │360  REM Calculate R4 and L4 for balance
R4,L4 │370  R4=W^2*C1^2*R1*R2*R3/(1+W^2*C1^2*R1^2)
for   │380  L4=C1*R2*R3/(1+W^2*C1^2*R1^2)
Balance│390  PRINT "R4=";R4;"ohms"
      │400  PRINT "L4=";L4;"henry"
      └410  PRINT
       420  REM Repeat test of balance to exit program
       430  PRINT "And then,"
       440  GOTO 230
```

READY

<u>RUN</u>

```
*
This program will test calculations
for a Hay bridge circuit as shown
in Figure 17-8.

Enter the following network data:
R1=? 1E3
R2=? 100
R3=? 100
R4=? 10

C1=? 1E-6
L4=? 10E-3
Desired frequency f=? 159

The bridge is not balanced.

In order to obtain balance use elements:
R4= 4.9951 ohms
L4= 5.0049E-03 henry

And then,
The bridge is balanced.
```

READY

*Problem 17.23, ICA

$$\frac{\mathbf{Z}_1}{\mathbf{Z}_3} = \frac{\mathbf{Z}_2}{\mathbf{Z}_4}$$

It is not sufficient for the ratio of the magnitudes to be equal; the total angle of one side must also equal the angle of the other side.

For the Hay Bridge, $\mathbf{Z}_1 = \sqrt{R_1{}^2 + X_C{}^2}$ $\underline{/\theta_1}$, $\mathbf{Z}_2 = R_2 \underline{/0°}$, $\mathbf{Z}_3 = R_3 \underline{/0°}$ and $\mathbf{Z}_4 = \sqrt{R_4{}^2 + X_L{}^2} \underline{/\theta_4}$ and, for balance,

$$\frac{\sqrt{R_1{}^2 + X_C{}^2}}{R_3} = \frac{R_2}{\sqrt{R_4^2 + X_L^2}} \text{ and } \theta_1 = -\theta_4$$

Since $\theta_1 = -\tan^{-1}X_C/R_1$ and $\theta_4 = \tan^{-1}X_L/R_4$, the balance condition is also defined by:

$$\frac{X_C}{R_1} = \frac{X_L}{R_4}$$

The impedances $\mathbf{Z}_1$ and $\mathbf{Z}_4$ are defined by lines 270 and 280, and the magnitude test performed on line 300. The angle test is then made on line 310. If either test fails, the program will go to lines 370 and 380 to determine the values of L and R_4 that will establish a balanced condition. We chose an unbalanced run to demonstrate the capabilities of the program.

17.9 TESTING ROUTINES

Programs 17-8 and 17-9 test the addition, subtraction, multiplication, and division of vectors. The addition and subtraction operations will be tested in the first program, while the second program will test the multiplication and division operations.

In both programs, the form of the vector, the magnitude, and the sign will be randomly generated. All the testing occurs on lines 470 through 590 after the user solutions are provided on lines 450 and 460.

```
10 REM *****  PROGRAM 17-8  *****
20 REM ***********************************
30 REM Program to test the addition and
40 REM subtraction of vectors.
50 REM ***********************************
60 REM
100 DIM V(4),VA(4),VB(4),VC(4)
110 DIM V1(4),V2(4)
120 PRINT "This program tests your addition (V1+V2)"
130 PRINT "or subtraction (V1-V2) of vectors."
140 PRINT
150 PRINT "For vector 1 of:"; :GOSUB 650
160 FOR I=1 TO 4
170 V1(I)=V(I)
180 NEXT I
190 PRINT "and vector 2 of:"; :GOSUB 650
200 FOR I=1 TO 4
210 V2(I)=V(I)
220 NEXT I
230 PRINT
240 PRINT "Sum of vectors is:"
250 FOR I=1 TO 4
260 VA(I)=V1(I) :VB(I)=V2(I)
270 NEXT I
280 GOSUB 4400 :REM Add vectors
290 GOSUB 400 :REM Test answer
300 PRINT "Difference of vectors V1-V2 is:"
310 GOSUB 4500 :REM Subtract vectors
320 GOSUB 400 :REM Test answer
330 END
400 REM Module to test answers.
410 ON C GOSUB 450,550
420 PRINT
```

```
430 RETURN
450 INPUT "X=";X
460 INPUT "Y=";Y
470 IF SGN(X)<>SGN(V(1)) THEN GOTO 530
480 IF ABS(X-V(1))>ABS(0.01*V(1)) THEN GOTO 530
490 IF SGN(Y)<>SGN(V(2)) THEN GOTO 530
500 IF ABS(Y-V(2))>ABS(0.01*V(2)) THEN GOTO 530
510 PRINT "Correct"
520 GOTO 540
530 PRINT "No, it's"; :GOSUB 950
540 RETURN
550 INPUT "Magnitude=";Z
560 INPUT "at angle";TH
570 IF ABS(Z-V(3))>ABS(0.01*V(3)) THEN GOTO 620
580 IF SGN(TH)<>SGN(V(4)) THEN GOTO 620
590 IF ABS(TH-V(4))>ABS(0.01*V(4)) THEN GOTO 620
600 PRINT "Correct"
610 GOTO 630
620 PRINT "No, it's";V(3);"at an angle of";V(4);"degrees"
630 RETURN
650 REM module to select vector
660 REM (1) in rectangular form, (2) in polar form
670 C=INT(1+2*RND(XV))
680 ON C GOSUB 750,850
690 PRINT :RETURN
700 REM Module to select sign
710 S=+1
720 SG=INT(1+4*RND(XV))
730 IF SG=3 THEN S=-1
740 RETURN
750 REM Module to select input in rectangular form
760 REM Select input of V(1),V(2), and convert to V(3),V(4)
770 GOSUB 700
780 V(1)=S*5*INT(1+10*RND(XV))
790 GOSUB 700
800 V(2)=S*5*INT(1+10*RND(XV))
810 GOSUB 4000 :REM Convert to polar form
820 GOSUB 950
830 RETURN
850 REM Module to select input in polar form
860 REM Select input of V(3),V(4), and convert to V(1),V(2)
870 V(3)=5*INT(1+10*RND(XV))
880 GOSUB 700
890 V(4)=S*5*INT(1+36*RND(XV))
900 GOSUB 4100 :REM Convert to rectangular form
910 PRINT V(3);"at an angle of";V(4);"degrees"
920 RETURN
950 PRINT V(1);
960 IF V(2)>=0 THEN PRINT "+j";V(2) :RETURN
970 IF V(2)<0 THEN PRINT "-j";ABS(V(2)) :RETURN
```

```
┌─ 4000-4750
│  as in Program 17-1
│
│
↓
```

(Continued)

READY

<u>RUN</u>

This program tests your addition (V1+V2)
or subtraction (V1-V2) of vectors.

For vector 1 of: 35 at an angle of-170 degrees

and vector 2 of: 40 +j 5

Sum of vectors is:
X=? <u>5.5</u>
Y=? <u>11</u>
No, it's 5.5317 -j 1.0778

Difference of vectors V1-V2 is:
X=? <u>-74.47</u>
Y=? <u>-11.08</u>
Correct

READY

This program tests your addition (V1+V2)
or subtraction (V1-V2) of vectors.

For vector 1 of: 50 +j 50

and vector 2 of: 30 -j 30

Sum of vectors is:
X=? <u>80</u>
Y=? <u>20</u>
Correct

Difference of vectors V1-V2 is:
X=? <u>20</u>
Y=? <u>20</u>
No, it's 20 +j 80

READY

```
 10 REM *****   PROGRAM 17-9   *****
 20 REM ********************************
 30 REM Program to test the multiplication
 40 REM and division of vectors.
 50 REM ********************************
 60 REM
100 DIM V(4),VA(4),VB(4),VC(4)
110 DIM V1(4),V2(4)
120 PRINT "This program tests your ability to multiply (V1*V2)"
130 PRINT "or divide (V1/V2), vectors."
140 PRINT
150 PRINT "For vector 1 of:"; :GOSUB 650
160 FOR I=1 TO 4
170 V1(I)=V(I)
180 NEXT I
190 PRINT "and vector 2 of:"; :GOSUB 650
200 FOR I=1 TO 4
210 V2(I)=V(I)
220 NEXT I
230 PRINT
240 PRINT "Product of vectors is:"
250 FOR I=1 TO 4
260 VA(I)=V1(I) :VB(I)=V2(I)
270 NEXT I
280 GOSUB 4600 :REM Multiply vectors
290 GOSUB 400 :REM Test answer
300 PRINT "Quotient of vectors V1/V2 is:"
310 GOSUB 4700 :REM Divide vectors
320 GOSUB 400 :REM Test answer
330 END
400 REM Module to test answers.
410 ON C GOSUB 450,550
420 PRINT
430 RETURN
450 INPUT "X=";X
460 INPUT "Y=";Y
470 IF SGN(X)<>SGN(V(1)) THEN GOTO 530
480 IF ABS(X-V(1))>ABS(0.01*V(1)) THEN GOTO 530
490 IF SGN(Y)<>SGN(V(2)) THEN GOTO 530
500 IF ABS(Y-V(2))>ABS(0.01*V(2)) THEN GOTO 530
510 PRINT "Correct"
520 GOTO 540
530 PRINT "No, it's"; :GOSUB 950
540 RETURN
550 INPUT "Magnitude=";Z
560 INPUT "at angle";TH
570 IF ABS(Z-V(3))>ABS(0.01*V(3)) THEN GOTO 620
580 IF SGN(TH)<>SGN(V(4)) THEN GOTO 620
590 IF ABS(TH-V(4))>ABS(0.01*V(4)) THEN GOTO 620
600 PRINT "Correct"
610 GOTO 630
620 PRINT "No, it's";V(3);"at an angle of";V(4);"degrees"
630 RETURN
650 REM module to select vector
660 REM (1) in rectangular form, (2) in polar form
670 C=INT(1+2*RND(XV))
680 ON C GOSUB 750,850
690 PRINT :RETURN
700 REM Module to select sign
710 S=+1
720 SG=INT(1+4*RND(XV))
730 IF SG=3 THEN S=-1
740 RETURN
```

(Continued)

```
750 REM Module to select input in rectangular form
760 REM Select input of V(1),V(2), and convert to V(3),V(4)
770 GOSUB 700
780 V(1)=S*5*INT(1+10*RND(XV))
790 GOSUB 700
800 V(2)=S*5*INT(1+10*RND(XV))
810 GOSUB 4000 :REM Convert to polar form
820 GOSUB 950
830 RETURN
850 REM Module to select input in polar form
860 REM Select input of V(3),V(4), and convert to V(1),V(2)
870 V(3)=5*INT(1+10*RND(XV))
880 GOSUB 700
890 V(4)=S*5*INT(1+36*RND(XV))
900 GOSUB 4100 :REM Convert to rectangular form
910 PRINT V(3);"at an angle of";V(4);"degrees"
920 RETURN
950 PRINT V(1);
960 IF V(2)>=0 THEN PRINT "+j";V(2) :RETURN
970 IF V(2)<0 THEN PRINT "-j";ABS(V(2)) :RETURN
```

```
 ┌─4000-4750
 │  as in Program 17-1
 │
 │
 │
 ▼
```

RUN

This program tests your ability to multiply (V1*V2)
or divide (V1/V2), vectors.

For vector 1 of: 15 at an angle of 65 degrees

and vector 2 of: 10 +j 45

Product of vectors is:
X=? -548.6
Y=? 420.96
Correct

Quotient of vectors V1/V2 is:
X=? 1.46
Y=? -.32
No, it's .3177 -j .07

READY

This program tests your ability to multiply (V1*V2)
or divide (V1/V2), vectors.

For vector 1 of: 40 at an angle of 175 degrees

and vector 2 of:-10 -j 20

```
Product of vectors is:
X=? 330
Y=? -830
No, it's 468.2024 +j 762.0935

Quotient of vectors V1/V2 is:
X=? 0.66
Y=? -1.66
Correct


READY
```

EXERCISES

Write a program to perform the following tasks.

1. Find the sum of the three impedances ($\mathbf{Z}_T = \mathbf{Z}_1 + \mathbf{Z}_2 + \mathbf{Z}_3$) provided in either the rectangular or polar form. Provide the results in both forms.

2. The following combination of vectors will appear frequently in the analysis of networks:

$$\frac{\mathbf{Z}_1\mathbf{Z}_2}{\mathbf{Z}_1 + \mathbf{Z}_2 + \mathbf{Z}_3}$$

In particular, if $\mathbf{Z}_3 = 0$, it will provide the total impedance of two parallel impedances, and, if $\mathbf{Z}_2$ is replaced by a voltage, it has the format of the voltage divider rule. If all elements are present, it has the format of the $\Delta - Y$ conversion. Find the general solution for the operation in both the polar and rectangular form.

3. Repeat problem 17–2 for the following:

$$\frac{\mathbf{Z}_1\mathbf{Z}_2 + \mathbf{Z}_1\mathbf{Z}_3 + \mathbf{Z}_2\mathbf{Z}_3}{\mathbf{Z}_2}$$

which forms the general solution for a $Y - \Delta$ conversion and provides the familiar form of Δ_z in the numerator.

4. Determine the current $\mathbf{I}_2$ and the voltages $\mathbf{V}_{Z_1}$, $\mathbf{V}_{Z_2}$, and $\mathbf{V}_{Z_3}$ for the network of Figure 17.3. For the first run, use the same data that appears in the provided run of Program 17-4.

5. Determine the voltage $\mathbf{V}_2$ and the currents $\mathbf{I}_{Z_1}$, $\mathbf{I}_{Z_2}$ and $\mathbf{I}_{Z_3}$ for the network of Figure 17.4. For the first run, use the same data that appears in the provided run of Program 17-5.

6. Determine the impedance $\mathbf{Z}_A$ for the delta configuration of Figure 17.7, given the impedances of the Wye configuration ($\mathbf{Z}_1$, $\mathbf{Z}_2$ and $\mathbf{Z}_3$). For the first run, use $\mathbf{Z}_1 = 3 + j4$, $\mathbf{Z}_2 = 4 + j3$, and $\mathbf{Z}_3 = 5 - j5$.

7. Calculate the current $\mathbf{I}_1$ for the network of Figure 17.3 if a voltage source $\mathbf{E}_3$ is added in series with the impedance $\mathbf{Z}_2$. The negative polarity is closest to ground potential.

8. Calculate the voltage $\mathbf{V}_1$ for the network of Figure 17.4 if a current source $\mathbf{I}_3$ is added in parallel with $\mathbf{Z}_2$. The current source is supplying current to node $\mathbf{V}_2$.

9. Determine the total impedance of Figure 17.8 for the balance condition of $V_g = 0V$. For the first run, use $\mathbf{Z}_1 = 3 + j4$, $\mathbf{Z}_2 = 10\underline{/30°}$, $\mathbf{Z}_3 = 6 + j8$, and $\mathbf{Z}_4 = 20\underline{/30°}$.

10. Determine the total impedance of Figure 17.8 if the voltage V_G is replaced by an impedance $\mathbf{Z}_5$ and the conditions $\mathbf{Z}_3 = \mathbf{Z}_4 = \mathbf{Z}_5$, and $\mathbf{Z}_1 = \mathbf{Z}_2/3$ exist. For the first run, use $\mathbf{Z}_3 = 9\underline{/30°}$ and $\mathbf{Z}_1 = 6\underline{/60°}$.

Network Theorems–AC

18

18.1 INTRODUCTION

The subroutines from 4000 through 5990 developed in the previous chapters will now be used to apply the Superposition and Thevenin theorems. A knowledge of the content of Chapters 16 and 17 is required because the supporting subroutines and some input and output operations will not be examined in detail here.

The networks analyzed are standard and thus provide a general awareness of how the theorems are applied. These also frequently result when a more complex system is redrawn in a simpler form.

18.2 SUPERPOSITION

The superposition theorem will now be applied to the network of Figure 18.1 to determine the current **I**.

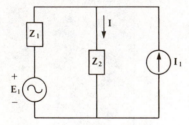

FIGURE 18.1

Examining the effects of the voltage source will result in the network of Figure 18.2

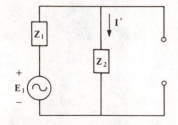

FIGURE 18.2

and $I' = \dfrac{E}{Z_1 + Z_2}$

The network of Figure 18.3 will result when we examine the effect of the current source.

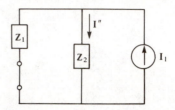

FIGURE 18.3

with $I'' = \dfrac{Z_1 I_1}{Z_1 + Z_2}$

The current **I** is then determined from **I** = **I'** + **I''**.

Lines 130 through 250 simply title Program 18-1 and define the subscripted values of the impedances and sources. REM statements throughout the program define the operations being performed. The subroutine from line 270 to 320 determines the rectangular and polar forms of $Z_1 + Z_2$ for use throughout the program. The result is stored as VC(I) until **I'** is determined by the subroutine starting at line 330; at this point, D(I) is set equal to VC(I). Line 350 is needed because VC(I) will be the location of the result of the division operation. The next line will define the division operators before performing the operation at line 4700. The rectangular and polar forms of the current **I'** are then stored as VC(I) until line 410 makes the transfer to IE(I)—the notation IE indicating the contribution due to the current **I** from the voltage source **E**. In the same routine, at line 4700, VA(I) and VB(I) are again defined and the product $Z_1 \cdot I_1$ formed for the current **I**. The process of determining **I** continues until II(I) is defined on line 530 as the current contribution due to the current I_1. The total current **I** is then determined by the subroutine at line 510 and defined on line 580. The current contribution due to each source and the total current are then printed out by lines 610 through 730.

```
        10 REM *****  PROGRAM 18-1  *****
        20 REM ************************************************
        30 REM Program solves standard network of Figure 18-1
        40 REM using Superposition.
        50 REM ************************************************
        60 REM
        100 DIM V(4),VA(4),VB(4),VC(4)
        110 DIM Z1(4),Z2(4),E1(4),I1(4),N(4),D(4)
        120 DIM IE(4),II(4),IT(4)
        130 PRINT "This program determines the current I"
        140 PRINT "for the network of Figure 18-1 using"
        150 PRINT "the Superposition theorem."
        160 PRINT
        170 PRINT "Input the following network information:"
  Z₁ ⌈180 PRINT "Z1=" :GOSUB 5500
     ⌊190 FOR I=1 TO 4 :Z1(I)=V(I) :NEXT I
  Z₂ ⌈200 PRINT "Z2=" :GOSUB 5500
     ⌊210 FOR I=1 TO 4 :Z2(I)=V(I) :NEXT I
  E₁ ⌈220 PRINT "E1=" :GOSUB 5500
     ⌊230 FOR I=1 TO 4 :E1(I)=V(I) :NEXT I
  I₁ ⌈240 PRINT "I1=" :GOSUB 5500
     ⌊250 FOR I=1 TO 4 :I1(I)=V(I) :NEXT I
     ⌈260 REM
      |270 REM Determine Z1+Z2
      |280 FOR I=1 TO 4
Z₁ + Z₂|290 VA(I)=Z1(I)
      |300 VB(I)=Z2(I)
      |310 NEXT I
     ⌊320 GOSUB 4400
     ⌈330 REM Determine E1/(Z1+Z2)
      |340 FOR I=1 TO 4
   I' |350 D(I)=VC(I)
      |360 VA(I)=E1(I) :VB(I)=D(I)
      |370 NEXT I
     ⌊380 GOSUB 4700
     ⌈390 REM Determine Z1*I1
      |400 FOR I=1 TO 4
      |410 IE(I)=VC(I)
 Z₁I₁ |420 VA(I)=Z1(I) :VB(I)=I1(I)
      |430 NEXT I
     ⌊440 GOSUB 4600
     ⌈450 REM Determine Z1*I1/(Z1+Z2)
      |460 FOR I=1 TO 4
   I''|470 N(I)=VC(I)
      |480 VA(I)=N(I) :VB(I)=D(I)
      |490 NEXT I
     ⌊500 GOSUB 4700
     ⌈510 REM Determine IT=IE+II
      |520 FOR I=1 TO 4
      |530 II(I)=VC(I)
      |540 VA(I)=IE(I) :VB(I)=II(I)
      |550 NEXT I
   I  |560 GOSUB 4400
      |570 FOR I=1 TO 4
      |580 IT(I)=VC(I)
      |590 NEXT I
     ⌊600 PRINT
```

(Continued)

```
      ┌610 REM Prepare to print IT
      │620 PRINT "The current through Z2 is";
      │630 FOR I=1 TO 4
      │640 V(I)=IT(I)
      │650 NEXT I
Print │660 GOSUB 5980 :PRINT "which is";V(3);"at an angle of";V(4);"degrees"
Results│670 FOR I=1 TO 4 :V(I)=IE(I) :NEXT I
      │680 PRINT :PRINT "This total current resulting from the following:"
      │690 PRINT :PRINT "Due only to E: IE="; :GOSUB 5980
      │700 PRINT "which is";V(3);"at an angle of";V(4);"degrees"
      │710 FOR I=1 TO 4 :V(I)=II(I) :NEXT I
      │720 PRINT :PRINT "Due only to I: II="; :GOSUB 5980
      └730 PRINT "which is";V(3);"at an angle of";V(4);"degrees"
       740 END
      4000 REM Module to convert from rectangular to polar form
      4010 REM Enter with V(1),V(2) - return with V(3),V(4)
      4020 V(3)=SQR(V(1)^2+V(2)^2)
      4030 IF V(1)>0 THEN V(4)=(180/3.14159)*ATN(V(2)/V(1))
      4040 IF V(1)<0 THEN V(4)=180*SGN(V(2))+(180/3.14159)*ATN(V(2)/V(1))
      4050 IF V(1)=0 THEN V(4)=90*SGN(V(2))
      4060 IF V(2)=0 THEN IF V(1)<0 THEN V(4)=180
      4070 RETURN
      4100 REM Conversion from polar to rectangular form
      4110 REM Enter with V(3),V(4) - return with V(1),V(2)
      4120 V(1)=V(3)*COS(V(4)*3.14159/180)
      4130 V(2)=V(3)*SIN(V(4)*3.14159/180)
      4140 RETURN
      4400 REM Module to Add or Subtract VC=VA + or - VB
      4410 REM Enter with VA,VB - return with VC
      4420 VC(1)=VA(1)+VB(1)
      4430 VC(2)=VA(2)+VB(2)
      4440 V(1)=VC(1)  :V(2)=VC(2)
      4450 GOSUB 4000 :REM Convert to polar form
      4460 VC(3)=V(3)  :VC(4)=V(4)
      4470 RETURN
      4500 REM Enter here for subtraction operation
      4510 REM VC=VA-VB
      4520 VC(1)=VA(1)-VB(1)
      4530 VC(2)=VA(2)-VB(2)
      4540 GOTO 4440 :REM Convert to polar form and return
      4600 REM Module to multiply and divide VA by VB
      4610 REM VC=VA*VB
      4620 VC(3)=VA(3)*VB(3)
      4630 VC(4)=VA(4)+VB(4)
      4640 V(3)=VC(3)  :V(4)=VC(4)
      4650 GOSUB 4100 :REM Convert to rectangular form
      4660 VC(1)=V(1)  :VC(2)=V(2)
      4670 RETURN :REM Return from both multiply and divide
      4700 REM Enter here for divide
      4710 REM VC=VA/VB
      4720 VC(4)=VA(4)-VB(4)
      4730 IF VB(3)=0 THEN VC(3)=1E10 :REM VC(3) very large
      4740 IF VB(3)<>0 THEN VC(3)=VA(3)/VB(3)
      4750 GOTO 4640 :REM Convert to rectangular form and return
      5500 REM Module to accept vector input
      5510 REM (1) in Rectangular form, (2) in Polar form
      5520 PRINT "(1) Enter rectangular form, or (2) Enter polar form:";
      5530 INPUT C
      5540 IF C=0 THEN END
      5550 IF C>2 THEN GOTO 5520
      5560 ON C GOSUB 5800,5900
      5570 PRINT
      5580 RETURN
      5800 REM Module to accept input in rectangular form
```

```
5810 REM Accept input of V(1),V(2), and convert to V(3),V(4)
5820 INPUT "X=";V(1)
5830 INPUT "Y=";V(2)
5840 GOSUB 4000 :REM Convert to the polar form
5850 RETURN
5900 REM Module to accept input in polar form
5910 REM Accept input of V(3),V(4), and convert to V(1),V(2)
5920 INPUT "Magnitude=";V(3)
5930 INPUT "at angle";V(4)
5940 GOSUB 4100 :REM Convert to the rectangular form
5950 RETURN
5980 PRINT V(1); :IF V(2)>=0 THEN PRINT "+j";V(2) :RETURN
5990 IF V(2)<0 THEN PRINT "-j";ABS(V(2)) :RETURN
```

READY

RUN

```
*
This program determines the current I
for the network of Figure 18-1 using
the Superposition theorem.

Input the following network information:
Z1=
(1) Enter rectangular form, or (2) Enter polar form:? 1
X=? 0
Y=? 6


Z2=
(1) Enter rectangular form, or (2) Enter polar form:? 1
X=? 6
Y=? -8


E1=
(1) Enter rectangular form, or (2) Enter polar form:? 2
Magnitude=? 20
at angle? 30


I1=
(1) Enter rectangular form, or (2) Enter polar form:? 2
Magnitude=? 2
at angle? 0



The current through Z2 is 1.4981 +j 4.166
which is 4.4272 at an angle of 70.2218 degrees

This total current resulting from the following:

Due only to E: IE= 2.0981 +j 2.366
which is 3.1623 at an angle of 48.435 degrees

Due only to I: II=-.6 +j 1.8
which is 1.8974 at an angle of 108.435 degrees


READY
```

*Example 18.2, ICA

RUN

```
This program determines the current I
for the network of Figure 18-1 using
the Superposition theorem.

Input the following network information:
Z1=
(1) Enter rectangular form, or (2) Enter polar form:? 1
X=? 0
Y=? -5

Z2=
(1) Enter rectangular form, or (2) Enter polar form:? 1
X=? 0
Y=? 8

E1=
(1) Enter rectangular form, or (2) Enter polar form:? 2
Magnitude=? 10
at angle? 0

I1=
(1) Enter rectangular form, or (2) Enter polar form:? 2
Magnitude=? 0.3
at angle? 60


The current through Z2 is-.25 -j 3.7663
which is 3.7746 at an angle of-93.7974 degrees

This total current resulting from the following:

Due only to E: IE= 4.4226E-06 -j 3.3333
which is 3.3333 at an angle of-90 degrees

Due only to I: II=-.25 -j .433
which is .5 at an angle of-120 degrees
```

READY

The first run will analyze the network of Figure 18.4, and the second run will determine the current **I** for the network of Figure 18.5a. Note in Figure 18.5b that the redrawn network

using block impedances looks different from Figure 18.1. However, reversing the position of I_1 and Z_2 will result in the proper format and a correct solution for **I**.

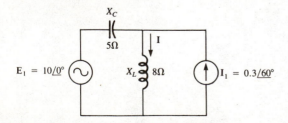

FIGURE 18.4

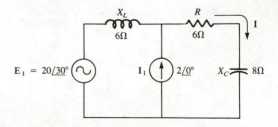

FIGURE 18.5a

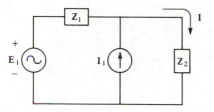

FIGURE 18.5b

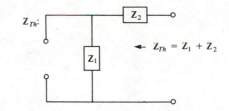

FIGURE 18.7

18.3 THEVENIN'S THEOREM

Thevenin's theorem will now be applied to the network of Figure 18.6 to determine the Thevenin impedance and source for the network to the left of the load Z_L.

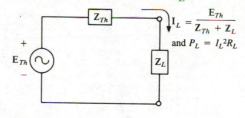

FIGURE 18.8

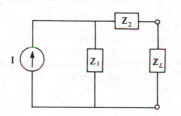

FIGURE 18.6

The Thevenin impedance is defined by Figure 18.7 and the Thevenin voltage by Figure 18.8.

The Thevenin network appears in Figure 18.9 with the return of the load Z_L.

FIGURE 18.9

In Program 18-2, the network is analyzed by lines 320 through 560. The statements before

```
10 REM ***** PROGRAM 18-2 *****
20 REM *******************************************
30 REM Program analyzes network of Figure 18-6
40 REM using Thevenin's theorem.
50 REM *******************************************
60 REM
100 DIM V(4),VA(4),VB(4),VC(4)
110 DIM Z1(4),Z2(4),ZL(4),I(4),N(4),D(4)
120 DIM ZT(4),ET(4),IL(4)
130 DIM IE(4),II(4),IT(4)
140 PRINT "This program solves for Zth and"
150 PRINT "Eth, and uses Thevenin's theorem"
160 PRINT "to solve for the load current of Figure 18-6."
170 PRINT
180 PRINT "Input the following circuit information:"
190 PRINT "Z1=" :GOSUB 5500
200 FOR I=1 TO 4
210 Z1(I)=V(I)
220 NEXT I
```

(Continued)

```
       ┌ 230  PRINT "Z2=" :GOSUB 5500
       │ 240  FOR I=1 TO 4
 Z₂    │ 250  Z2(I)=V(I)
       └ 260  NEXT I
       ┌ 270  PRINT "ZL=" :GOSUB 5500
       │ 280  FOR I=1 TO 4
 Zʟ    │ 290  ZL(I)=V(I)
       └ 300  NEXT I
  I      310  PRINT "I=" :GOSUB 5500
       ┌ 320  FOR I=1 TO 4 :I(I)=V(I) :NEXT I
       │ 330  REM Determine Zth=Z1+Z2
       │ 340  FOR I=1 TO 4
 Zₜₕ   │ 350  VA(I)=Z1(I)
       │ 360  VB(I)=Z2(I)
       │ 370  NEXT I
       └ 380  GOSUB 4400
       ┌ 390  REM Determine Eth=I*Z1
       │ 400  FOR I=1 TO 4
 Eₜₕ   │ 410  ZT(I)=VC(I)
       │ 420  VA(I)=I(I) :VB(I)=Z1(I)
       │ 430  NEXT I
       └ 440  GOSUB 4600
       ┌ 450  REM Determine Zth+ZL
       │ 460  FOR I=1 TO 4
Zₜₕ+Zʟ │ 470  ET(I)=VC(I)
       │ 480  VA(I)=ZT(I) :VB(I)=ZL(I)
       │ 490  NEXT I
       └ 500  GOSUB 4400
       ┌ 510  REM Determine IL=Eth/(Zth+ZL)
       │ 520  FOR I=1 TO 4
       │ 530  D(I)=VC(I)
 Iʟ    │ 540  VA(I)=ET(I) :VB(I)=D(I)
       │ 550  NEXT I
       └ 560  GOSUB 4700
       ┌ 570  REM Prepare to print IL
       │ 580  FOR I=1 TO 4
       │ 590  IL(I)=VC(I)
       │ 600  NEXT I
Print  │ 610  PRINT
 Iʟ    │ 620  PRINT "The current through ZL is";
       │ 630  FOR I=1 TO 4
       │ 640  V(I)=IL(I)
       │ 650  NEXT I
       └ 660  GOSUB 5980:PRINT "which is";V(3);"at an angle of";V(4);"degrees"
       ┌ 670  PRINT :PRINT "with Thevenin circuit values of:"
       │ 680  PRINT
       │ 690  FOR I=1 TO 4
       │ 700  V(I)=ZT(I)
Print  │ 710  NEXT I
 Th    │ 720  PRINT "Thevenin impedance, Zth=";
Values │ 730  GOSUB 5980:PRINT "which is";V(3);"at an angle of";V(4);"degrees"
       │ 740  FOR I=1 TO 4
       │ 750  V(I)=ET(I)
       │ 760  NEXT I
       │ 770  PRINT:PRINT "Thevenin voltage, Eth=";
       └ 780  GOSUB 5980:PRINT "which is";V(3);"at an angle of";V(4);"degrees"
         790  END
            ┌ 4000-5990
            │ as in Program 18-1
            │
            │
            ↓
```

```
RUN
.
This program solves for Zth and
Eth, and uses Thevenin's theorem
to solve for the load current of Figure 18-6.

Input the following circuit information:
Z1=
(1) Enter rectangular form, or (2) Enter polar form:? 1
X=? 20
Y=? 0

Z2=
(1) Enter rectangular form, or (2) Enter polar form:? 1
X=? 10
Y=? 0

ZL=
(1) Enter rectangular form, or (2) Enter polar form:? 1
X=? 10
Y=? 0

I=
(1) Enter rectangular form, or (2) Enter polar form:? 1
X=? 5
Y=? 0


The current through ZL is 2.5 +j 0
which is 2.5 at an angle of 0 degrees
.
with Thevenin circuit values of:

Thevenin impedance, Zth= 30 +j 0
which is 30 at an angle of 0 degrees

Thevenin voltage, Eth= 100 +j 0
which is 100 at an angle of 0 degrees

READY

RUN

This program solves for Zth and
Eth, and uses Thevenin's theorem
to solve for the load current of Figure 18-6.

Input the following circuit information:
Z1=
(1) Enter rectangular form, or (2) Enter polar form:? 1
X=? 3.077
Y=? 0.615

Z2=
(1) Enter rectangular form, or (2) Enter polar form:? 1
X=? 0
Y=? 0
```

*Problem 18.8b, ICA

```
ZL=
(1) Enter rectangular form, or (2) Enter polar form:? 1
X=? 0
Y=? -6

I=
(1) Enter rectangular form, or (2) Enter polar form:? 2
Magnitude=? 0.1
at angle? 0


The current through ZL is .016 +j .048
which is .051 at an angle of 71.5591 degrees

with Thevenin circuit values of:

Thevenin impedance, Zth= 3.077 +j .615
which is 3.1379 at an angle of 11.3028 degrees

Thevenin voltage, Eth= .3077 +j .061
which is .3138 at an angle of 11.3028 degrees


READY
```

and after these lines simply ask for the network data, printout the results, and include the subroutines introduced earlier. Note that lines 4000 through 5990 were omitted to save space. The result of the $\mathbf{Z}_1 + \mathbf{Z}_2$ operation on line 380 is stored as $VC(I)$ until line 410. There, $ZT(I)$ is defined so that $VC(I)$ can act as the temporary storage of the $\mathbf{I} \cdot \mathbf{Z}_1$ product. The operation of line 410 is included under the heading of the $\mathbf{E}_{Th}$ subroutine to use the I loop of line 400. Thus, an additional I loop for the line 410 operation need not be written. The same process continues for $\mathbf{E}_{Th}$ on lines 470 and $\mathbf{I}_L$ on line 590.

On line 640 the values of $V(1)$, $V(2)$, $V(3)$, and $V(4)$ are defined to permit a printout of the results using the format statements of lines 5980, 5990, and 660. Since $\mathbf{Z}_{Th}$ and $\mathbf{E}_{Th}$ were defined by $ZT(I)$ and $ET(I)$ for temporary storage within the program they can each be printed out once the corresponding values of $V(I)$ are defined.

The first run analyzes the network of Figure 18.10, which is essentially a dc system since none of the elements are frequency-dependent. The Thevenin resistance is 30 ohms

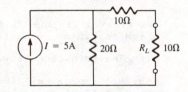

FIGURE 18.10

and $E_{Th} = 5 \times 20 = 100V$, as is shown by the printout. The load current is determined by $I_L = E_{Th}/(R_L + R_{Th}) = 100/40 = 2.5A$.

The second run examines the network of Figure 18.11, which initially does not look like

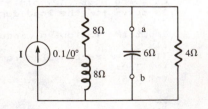

FIGURE 18.11

Figure 18.6. However, when redrawn as shown in Figure 18.12 the similarities become more apparent.

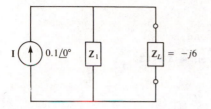

FIGURE 18.12

$\mathbf{Z}_1$ and $\mathbf{Z}_L$ are as defined below:

$$\mathbf{Z}_1 = 4\underline{/0°}\|(8 + j8) = 3.077 + j0.615$$
$$= R + jX_L$$
$$\mathbf{Z}_L = 6\underline{/-90°} = -j6$$

For Figure 18.12, however, $\mathbf{Z}_2 = 0$ ohms (as defined in Figure 18.6) in response to the request for $\mathbf{Z}_2$ in the second run. $\mathbf{Z}_{Th}$ is now equal to $\mathbf{Z}_1$ as appears in the results, and $\mathbf{I}_L = \mathbf{E}_{Th}/(\mathbf{Z}_1 + \mathbf{Z}_L)$. The second run clearly shows how a program written for a standard format can be used to solve a variety of network problems.

The last program (18-3) of Chapter 18 will tabulate the power to the load of Figure 18.3 for a range of values for R_L and X_L. The results show that maximum power is transferred to the load when $R_L = R_{Th}$ and $X_L = X_C$. In the first part of the run, X_L is set equal to X_C, and the power to the load is tabulated for values of R_L that range from 0 to $2R_{Th}$ in steps of $R_{Th}/10$. In the provided run, $R_{Th} = 20$ ohms and R_L will extend from 0 to 40 ohms in steps of two ohms. When $X_L = X_C$ in Figure 18.13:

```
 10 REM ***** PROGRAM 18-3 *****
 20 REM ***************************************
 30 REM Program to determine and tabulate PL vs RL
 40 REM for a Thevenin circuit and load as shown
 50 REM in Figure 18-13.
 60 REM ***************************************
 70 REM
100 PRINT "This program will tabulate power vs load"
110 PRINT "resistance over a range of load values"
120 PRINT "for a Thevenin circuit as shown in Figure 18-13."
130 PRINT
140 PRINT "Enter the following data:"
150 INPUT "Eth=";ET
160 INPUT "Rth=";RT
170 INPUT "Xc=";XC
180 PRINT
190 REM Tabulate PL vs RL
200 PRINT "Over a range of RL values(XL=Xc):"
210 PRINT
220 F$="      ###.###      ###.###"
230 PRINT "        RL            PL"
240 PRINT "      (ohms)        (watts)"
250 FOR RL=0 TO 2*RT STEP RT/10
260 PL=(ET^2*RL)/((RT+RL)^2)
270 PRINT USING F$,RL,PL
280 NEXT RL
290 REM Tabulate PL vs XL
300 PRINT:PRINT
310 PRINT "Over a range of XL values(Rth=RL):"
320 PRINT:PRINT "        XL            PL"
330 PRINT "      (ohms)        (watts)"
340 FOR XL=0 TO 2*XC STEP XC/10
350 PL=(ET^2*RL)/((RT+RL)^2+(XL-XC)^2)
360 PRINT USING F$,XL,PL
370 NEXT XL
380 END
```

Labels at left of program lines:
- Lines 150–170: Input Data
- Lines 190–240: P_L vs. R_L
- Lines 250–280: Range of R_L
- Lines 290–330: P_L vs. X_L
- Lines 340–370: Range of X_L

READY

RUN

This program will tabulate power vs load
resistance over a range of load values
for a Thevenin circuit as shown in Figure 18-13.

Enter the following data:
Eth=? 10
Rth=? 25
Xc=? 50

Over a range of RL values(XL=Xc):

RL (ohms)	PL (watts)
0.000	0.000
2.500	0.331
5.000	0.556
7.500	0.710
10.000	0.816
12.500	0.889
15.000	0.938
17.500	0.969
20.000	0.988
22.500	0.997
25.000	1.000
27.500	0.998
30.000	0.992
32.500	0.983
35.000	0.972
37.500	0.960
40.000	0.947
42.500	0.933
45.000	0.918
47.500	0.904
50.000	0.889

Over a range of XL values(Rth=RL):

XL (ohms)	PL (watts)
0.000	0.615
5.000	0.654
10.000	0.692
15.000	0.730
20.000	0.766
25.000	0.800
30.000	0.830
35.000	0.855
40.000	0.873
45.000	0.885
50.000	0.889
55.000	0.885
60.000	0.873
65.000	0.855
70.000	0.830
75.000	0.800
80.000	0.766
85.000	0.730
90.000	0.692
95.000	0.654
100.000	0.615

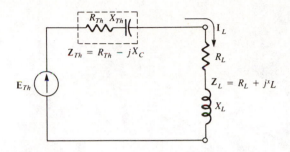

FIGURE 18.13

$$I_L = \frac{E_{Th}}{Z_{Th} + Z_L} = \frac{E_{Th}}{(R_{Th} - X_C) + (R_L + X_L)}$$

$$= \frac{E_{Th}}{R_{Th} + R_L}$$

and $P_L = I_L^2 R_L = \frac{E_{Th}^2}{(R_{Th} + R_L)^2} \cdot R_L$

as appears on line 260. The levels of P_L will then be printed out in a table, as defined by lines 220 through 240, for a range of values of R_L and fixed values of E_{Th} and R_{Th}.

In the case of $R_{Th} = R_L$, the range of X_L is defined by X_C on line 340. The magnitude of the current and power to the load is determined by:

$$I_L = \frac{E_{Th}}{\sqrt{(R_{Th} + R_L)^2 + (X_L - X_C)^2}}$$

and $P_L = I_L^2 R_L = \frac{E_{Th}^2 R_L}{(R_{Th} + R_L)^2 + (X_L - X_C)^2}$

Note in the printout that maximum power is delivered when $R_{Th} = R_L$ and $X_L = X_C$. If X_{Th} in Figure 18.13 had been an inductor, the load reactance would have to be capacitive to satisfy maximum power transfer conditions.

EXERCISES

Write a program to perform the following tasks.

1. Using superposition, determine the voltage **V** of Figure 18.14. For the first run, use $\mathbf{E}_1$

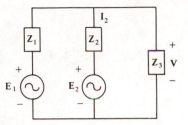

FIGURE 18.14

$= 30\underline{/30°}$, $\mathbf{E}_2 = 60\underline{/10°}$, $\mathbf{Z}_1 = 6$, $\mathbf{Z}_2 = -j6$, and $\mathbf{Z}_3 = +j8$.

2. Find the Thevenin equivalent network for the elements external to $\mathbf{Z}_3$ in Figure 18.14. For the first run, use the same values in problem 18-1 for all the elements except $\mathbf{Z}_3$.

3. Using the results of program 18-2, determine the rectangular components of $\mathbf{Z}_3$ needed to establish maximum power transfer to $\mathbf{Z}_3$. Determine the maximum power to $\mathbf{Z}_3$.

4. Using the results of problem 18-3, plot P_{Z3} versus the magnitude of $\mathbf{Z}_{Th}$ and $\mathbf{Z}_3$ for $+ jX_3 = - jX_{Th}$ and R_3 in the range $R_{Th/4}$ to $2R_{Th}$ in increments of $R_{Th/4}$. For the first run, use the values in problem 18-2.

5. Find the Norton equivalent network for the elements external to $\mathbf{Z}_3$ in Figure 18.14. For the first run use the values in problem 18-1 for all the elements except $\mathbf{Z}_3$.

6. Find the Thevenin equivalent network for all elements external to the series source $\mathbf{E}_2$ and $\mathbf{Z}_2$. Then determine the current $\mathbf{I}_2$. Use the values of problem 18-1.

7. Find the Norton equivalent network for all the elements external to the impedance $\mathbf{Z}_2$ in Figure 18.14. For the first run, use the values in problem 18-1.

8. Replace $\mathbf{Z}_L$ of Figure 18.6 by $\mathbf{Z}_3$ and place a current source $\mathbf{I}_2$ in parallel with $\mathbf{Z}_3$, supplying current to the junction of $\mathbf{Z}_2$ and $\mathbf{Z}_3$. Determine the current through $\mathbf{Z}_2$ using superposition. For the first run, use $\mathbf{I} = 10\underline{/0°}$, $\mathbf{I}_2 = 50\underline{/90°}$, $\mathbf{Z}_1 = 10$, $\mathbf{Z}_2 = +j20$, and $\mathbf{Z}_3 = -j30$.

Power-AC

19

19.1 INTRODUCTION

The content of this chapter lends itself beautifully to computer techniques. For any number of elements, the total real, reactive, and apparent power of a network can be determined using a series of straightforward mathematical relationships. The individual calculations are not difficult, but, because there are so many, they can be quite time-consuming. Once the program is written, it is particularly pleasant to see the long list of results printed out after you have taken a few seconds to provide the input data.

19.2 TOTAL REAL, REACTIVE, AND APPARENT POWER

The total real, reactive, and apparent power for the network of Figure 19.1 will be determined by Program 19-1. Note that five individual impedances are included, with the real and reactive power levels indicated for each load. Lines 210 through 250 will request the data for each impedance, while lines 180 and 190 will request the input data for the voltage E. The I loop will extend through 280 before the next

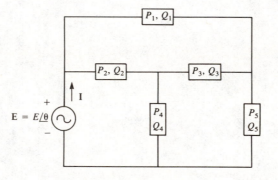

FIGURE 19.1

```
 10 REM *****  PROGRAM 19-1  *****
 20 REM ************************************************
 30 REM This program calculates the total real,
 40 REM reactive and apparent power of a network
 50 REM with five individual loads.
 60 REM ************************************************
 70 REM
100 DIM P(5),Q(5),S(5)
110 PRINT "This program calculates the total real,"
120 PRINT "reactive and apparent power of a network"
130 PRINT "with five individual loads."
140 PRINT
150 PRINT "Input the following data:"
160 PRINT "(use negative sign for capacitive vars)"
170 PRINT
180 INPUT "E=";E
190 INPUT "at an angle=";EA
200 PRINT
210 FOR I=1 TO 5
220 PRINT "For";I;"   ";
230 INPUT "P(watts)=";P(I)
240 PRINT TAB(8);
250 INPUT "Q(vars)=";Q(I)
260 PT=PT+P(I)
270 QT=QT+Q(I)
```

Input — 180, 190

Input — 210–250

P,P_q — 260, 270

```
      280 NEXT I
      290 PRINT
    ┌ 300 PRINT "The apparent power associated with each load"
    │ 310 PRINT "is the following:"
    │ 320 PRINT
P_a │ 330 FOR I=1 TO 5
    │ 340 S(I)=SQR(P(I)^2+Q(I)^2)
    │ 350 PRINT "S";I;"=";S(I)
    │ 360 NEXT I
    └ 370 ST=SQR(PT^2+QT^2)
      380 PRINT:PRINT
    ┌ 390 PRINT "Total real power, PT=";PT;"watts"
Power│ 400 PRINT
Output│ 410 PRINT "Total reactive power, QT=";QT;"vars"
    │ 420 PRINT
    └ 430 PRINT "Total apparent power, ST=";ST;"VA"
    ┌ 440 FP=PT/ST
    │ 450 TH=-57.296*ATN(QT/PT)
    │ 460 IF QT>0 THEN IA=EA-TH
    │ 470 IF QT<0 THEN IA=EA+TH
F_p │ 480 PRINT
    │ 490 PRINT "Power factor angle=";IA;"degrees"
    │ 500 PRINT
    │ 510 PRINT "Power factor=";FP;
    │ 520 IF QT>0 THEN PRINT "(lagging)"
    └ 530 IF QT<0 THEN PRINT "(leading)"
    ┌ 540 I=ST/E
I_T │ 550 PRINT
    └ 560 PRINT "Input current:";I;"at an angle of";IA;"degrees"
      570 END
```

<u>RUN</u>

*
This program calculates the total real,
reactive and apparent power of a network
with five individual loads.

Input the following data:
(use negative sign for capacitive vars)

E=? <u>50</u>
at an angle=? <u>60</u>

```
For 1    P(watts)=? 200
         Q(vars)=? 100
For 2    P(watts)=? 200
         Q(vars)=? 100
For 3    P(watts)=? 100
         Q(vars)=? -200
For 4    P(watts)=? 100
         Q(vars)=? -200
For 5    P(watts)=? 0
         Q(vars)=? 0
```

The apparent power associated with each load
is the following:

```
S 1 = 223.6068
S 2 = 223.6068
S 3 = 223.6068
S 4 = 223.6068
S 5 = 0
```

*Problem 19.4, ICA

```
Total real power, PT= 600 watts

Total reactive power, QT=-200 vars

Total apparent power, ST= 632.4555 VA

Power factor angle= 78.435 degrees

Power factor= .9487 (leading)

Input current: 12.6491 at an angle of 78.435 degrees


READY

RUN
.
This program calculates the total real,
reactive and apparent power of a network
with five individual loads.

Input the following data:
(use negative sign for capacitive vars)

E=? 100
at an angle=? 90

For 1    P(watts)=? 0
         Q(vars)=? 200
For 2    P(watts)=? 0
         Q(vars)=? -600
For 3    P(watts)=? 300
         Q(vars)=? 0
For 4    P(watts)=? 0
         Q(vars)=? 0
For 5    P(watts)=? 0
         Q(vars)=? 0

The apparent power associated with each load
is the following:

S 1 = 200
S 2 = 600
S 3 = 300
S 4 = 0
S 5 = 0


Total real power, PT= 300 watts

Total reactive power, QT=-400 vars

Total apparent power, ST= 500 VA

Power factor angle= 143.1303 degrees

Power factor= .6 (leading)

Input current: 5 at an angle of 143.1303 degrees


READY
```

*Problem 19.2, ICA

value of *I* is considered, to permit a determination of the total real and reactive power by lines 260 and 270 respectively. Note that line 270 will only provide the correct total reactive power if capacitive vars are entered as negative values. Line 260 reveals that the total power is simply the sum of the individual power losses of each impedance, while the total reactive power is the algebraic sum of the reactive power levels of each load. The apparent power for each load is then determined by line 340, using $S = \sqrt{P^2 + Q^2}$, and printed out by line 350. The total apparent power is determined by line 370, and the results printed out by lines 390 through 430. Finally, the decimal value of the power factor is determined by line 440, and the angle in degrees calculated on line 450. The results for theta and the power factor are then printed out by lines 490 and 510 respectively, before the magnitude of the *I* of Figure 19.1 is determined from $I = S_T/E$ and printed out with the associated theta by line 560.

The results of the first run indicate that the network has a very high power factor, due to $P_T > Q_T$, and the network is capacitive, as revealed by the positive sign for the angle of **I**. Recall that **I** leads **E** for capacitive networks, and **E** leads **I** for inductive networks. In this case, since $\mathbf{E} = E\underline{/0°}$, **I** leads **E** by 2.862°.

19.3 *R-L-C* SERIES CIRCUIT

Program 19-2 will determine the various powers and the power factor for the series *R-L-C* network of Figure 19.2. The input data is requested on lines 130 through 170 and the calculations

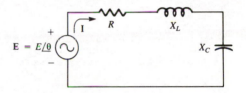

FIGURE 19.2

```
10 REM *****  PROGRAM 19-2  *****
20 REM ************************************************
30 REM Program to analyze the series R-L-C circuit
40 REM of Figure 19-2.
50 REM ************************************************
60 REM
100 PRINT "For the circuit of Figure 19-2, enter"
110 PRINT "the following data:"
120 PRINT
130 INPUT "Supply voltage = ";EM        ┐
140 INPUT "at angle";EA                  │
150 INPUT "R=";R                    Input│
160 INPUT "XL=";XL                       │
170 INPUT "XC=";XC                       ┘
180 PRINT
190 REM Now do calculations              ┐
200 XT=XL-XC                             │
210 ZM=SQR(R^2+XT^2)                     │
220 ZA=57.296*ATN(XT/R)                  │
230 IE=EM/ZM                             │
240 IA=EA-ZA                        Calc.│
250 PR=IE^2*R                            │
260 QL=IE^2*XL                           │
270 QC=-IE^2*XC                          │
280 PT=PR                                │
290 QT=QL-QC                             │
300 ST=SQR(PT^2+QT^2)                    │
310 FP=PT/ST                             ┘
```

(Continued)

```
┌320 REM Now print results
│330 PRINT "The input impedance is";ZM;"at an angle of";ZA;"degrees"
│340 PRINT "and the circuit current is";IE;"at an angle of";IA;"degrees"
│350 PRINT
│360 PRINT "PT=";PR;"watts"
│370 PRINT "QL=";QL;"vars"
│380 PRINT "QC=";QC;"vars"
│390 PRINT "QT=";QT;"vars"
│400 PRINT "ST=";ST;"VA"
└410 PRINT "FP=";FP
 420 END
```

Print
Results

READY

RUN
*
For the circuit of Figure 19-2, enter
the following data:

Supply voltage = ? 50
at angle? 0
R=? 3
XL=? 9
XC=? 5

The input impedance is 5 at an angle of 53.1303 degrees
and the circuit current is 10 at an angle of-53.1303 degrees

PT= 300 watts
QL= 900 vars
QC= 500 vars
QT= 400 vars
ST= 500 VA
FP= .6

READY

RUN

For the circuit of Figure 19-2, enter
the following data:

Supply voltage = ? 20.51
at angle? 30.65
R=? 2.2E3
XL=? 8.4E3
XC=? 3.3E3

The input impedance is 5554.2776 at an angle of 66.6662 degrees
and the circuit current is 3.6926E-03 at an angle of-36.0162 degrees

PT= .03 watts
QL= .1145 vars
QC= .045 vars
QT= .07 vars
ST= .076 VA
FP= .3961

READY

*Problem 19.1, ICA

performed by lines 200 through 310. The remainder of the program prints out the results. Line 210 will determine the magnitude of the total impedance using $Z_T = \sqrt{R^2 + (X_L - X_C)^2}$, and line 230 will then determine the effective value of the current I using Ohm's law: $I = E/Z_T$. The associated angle is determined for the total impedance and current **I** on lines 220 and 240 respectively. On lines 250 through 270, the individual powers are determined and, on lines 280 through 300, the total real, reactive, and apparent power is determined. Line 310 determines the power factor before printing out the results starting on line 320.

The input values chosen for the first run are fairly standard for instructional purposes, while the chosen values for the second run are more abstract. In either case, however, note the clarity and accuracy of the result, and, if you

run the program, the undetectable difference in time to perform the calculations.

19.4 SYSTEM ANALYSIS

The following program (19-3) will determine those series internal components of the "black box" of Figure 19.3 that will satisfy the specified terminal values. Note the angle θ associated with the input voltage to establish a reference angle for the input impedance and current **I**. As indicated in the program, the values of E, S_T, and F_p are provided, along with an indication of whether the system is leading or lagging. The program will then determine the magnitude of the input current using $I = S_T/E$ on line 170 before calculating the phase angle of the input current using the function defined on

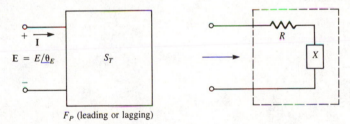

F_P (leading or lagging)

FIGURE 19.3

```
        10 REM *****   PROGRAM 19-3   *****
        20 REM ****************************************************
        30 REM Program designed to calculate the series components
        40 REM of a system (Fig. 19-3) based on the terminal data.
        50 REM ****************************************************
        60 REM
       100 DEF FNC(X)=-ATN(X/SQR(1-X^2))+1.5708
       110 PRINT "For the network of Fig. 19-3"
      ┌120 INPUT "   ST=";ST
       130 INPUT "Supply voltage, E=";EE
Input │140 INPUT "at angle ";EA
Data  │150 INPUT "with Fp=";FP
      └160 INPUT "(1) lead, or (2) lag";L
      ┌170 IE=ST/EE
       180 IA=57.296*FNC(FP)
       190 IF L=2 THEN IA=-IA
Calc. │200 ZM=EE/IE
       210 ZA=EA-IA
       220 R=ZM*COS(ZA/57.296)
      └230 X=ZM*SIN(ZA/57.296)
```

(Continued)

```
240 REM Print results
250 PRINT
260 PRINT "Input current is";IE;"at an angle of";IA;"degrees"
270 PRINT "with an input impedance of";ZM;"at an angle of";ZA;"degrees"
280 PRINT
290 PRINT "Series circuit equivalent is:"
300 PRINT "R=";R;"ohms, and"
310 IF L=1 THEN PRINT "XC="; ELSE PRINT "XL=";
320 PRINT ABS(X);"ohms"
330 END
```

Print
Results

```
READY

RUN
*
For the network of Fig. 19-3
   ST=? 5000
Supply voltage, E=? 100
at angle ? 0
with Fp=? 0.6
(1) lead, or (2) lag? 2

Input current is 50 at an angle of-53.1305 degrees
with an input impedance of 2 at an angle of 53.1305 degrees

Series circuit equivalent is:
R= 1.2 ohms, and
XL= 1.6 ohms

READY

RUN

For the network of Fig. 19-3
   ST=? 1.5E3
Supply voltage, E=? 20
at angle ? 30
with Fp=? 0.3
(1) lead, or (2) lag? 1

Input current is 75 at an angle of 72.5429 degrees
with an input impedance of .2667 at an angle of-42.5429 degrees

Series circuit equivalent is:
R= .1965 ohms, and
XC= .1803 ohms

READY
```

*Example 19.4, ICA

line 100. Since $\cos^{-1}x$ is not provided in computer memory, the following trigonometric identity is employed, since $\tan^{-1}x$ is typically available:

$$\theta = \cos^{-1}x = -\tan^{-1}\left(\frac{x}{\sqrt{1 - x^2}}\right) + 90°$$

However, since θ must be measured in radians, the 90° is replaced by its radian measure of $\pi/2 = 1.5708$. Line 180 converts the radian solution to degrees by multiplying by 57.296, since 1 radian equals 57.296°. Line 190 will insure that the proper sign is associated with the phase angle of the input current, as determined by whether the system is inductive (lagging) or capacitive (leading). If the system is inductive ($L = 2$), the current lags the voltage and the angle must be negative.

The magnitude and angle of the input impedance can then be determined from $\mathbf{Z}_T$

$$= Z_T\underline{/\theta_Z} = E\underline{/\theta_E}/I\underline{/\theta} = E/I\underline{/\theta_E} - \theta$$

$$= EE/IE\underline{/EA} - IA.$$ as appearing on lines 200 and 210. Once the polar form for the impedance is known, the magnitude of the resistive and reactive components can be determined from $X = Z_T\cos\theta_Z$ and $Y = Z_T\sin\theta_Z$

The results are then printed out by lines 260 through 320. Note in the printouts how inductive and capacitive networks affect the sign of the angle associated with the input impedance and current.

19.5 F_p AND S_T VERSUS P_T AND Q_T

Program 19-4 is an interesting tabulation of the effect of an increasing level of real power dis-sipation on the total apparent power and power factor of a system. Note in the provided printout that the levels of reactive power for each component of the system remain the same, but the real power for each increases by a factor of 1, 2, 3, and so on. Note also how the power factor changes from one that indicates a highly reactive system, since Q_T is greater than P_T ($400 > 150$), to one that is highly resistive, since $P_T > Q_T$ ($1500 > 400$). The total real power will increase linearly, since $P_T = NP_1 + NP_2 + NP_3 = N(P_1 + P_2 + P_3)$, but the total apparent power will increase non-linearly, since S_T is related to P_T and Q_T by the square and square root operations. It is particularly interesting to note how S_T will approach P_T as it continues to grow, since, eventually, $P_T >> Q_T$ and $S_T = \sqrt{P_T^2 + Q_T^2} \simeq \sqrt{P_T^2} = P_T$.

The program is actually quite straightforward, with the initial value of P_T determined on line 190 and the incremented value determined by line 390. For each pass of the I loop beginning at line 370, the value of P_T will increase by the value I, before the new value of S_T and F_p is determined on lines 400 and 410. The F$ format

```
        10 REM *****   PROGRAM 19-4  *****
        20 REM ************************************************
        30 REM Program to provide a table of power factor and
        40 REM total apparent power versus the total real and
        50 REM reactive power for increasing levels
        60 REM of real power.
        70 REM ************************************************
        80 REM
       100 PRINT "For a network of three blocks, enter"
       110 PRINT "(use - sign for capacitive vars)"
      ┌120 INPUT "P1=";P1
       │130 INPUT "Q1=";Q1
Input │140 INPUT "P2=";P2
       │150 INPUT "Q2=";Q2
       │160 INPUT "P3=";P3
      └170 INPUT "Q3=";Q3
       180 PRINT
      ┌190 PT=P1+P2+P3
       │200 QT=Q1+Q2+Q3
Calc. │210 ST=SQR(PT^2+QT^2)
      └220 FP=PT/ST
      ┌230 PRINT "Initially,"
       │240 PRINT "PT=";PT;"watts"
Info. │250 PRINT "QT=";QT;"vars"
       │260 PRINT "ST=";ST;"va"
      └270 PRINT "and, FP=";FP
```

(Continued)

```
        280 PRINT
Format ┌290 F$="#### #### #### #### #### #### "
       └300 G$="##### #### #### #.###"
       ┌310 PRINT "For QT fixed and a range of real power"
Heading│320 PRINT "from 1 to 10 times its initial value"
       └330 PRINT "the following results were obtained:"
        340 PRINT
       ┌350 REM Now print table
        360 PRINT "  P1    Q1    P2    Q2    P3    Q3    PT    QT    ST      FP"
        370 FOR I=1 TO 10
Print   380 PRINT USING F$, I*P1;Q1;I*P2;Q2;I*P3;Q3;
Table   390 PT=I*(P1+P2+P3)
        400 ST=SQR(PT^2+QT^2)
        410 FP=PT/ST
        420 PRINT USING G$, PT;QT;ST;FP
       └430 NEXT I
        440 END
```

READY

RUN

```
For a network of three blocks, enter
(use - sign for capacitive vars)
P1=? 50
Q1=? 100
P2=? 25
Q2=? -200
P3=? 75
Q3=? -300


Initially,
PT= 150 watts
QT=-400 vars
ST= 427.2002 va
and, FP= .3511

For QT fixed and a range of real power
from 1 to 10 times its initial value
the following results were obtained:
```

P1	Q1	P2	Q2	P3	Q3	PT	QT	ST	FP
50	100	25	-200	75	-300	150	-400	427	0.351
100	100	50	-200	150	-300	300	-400	500	0.600
150	100	75	-200	225	-300	450	-400	602	0.747
200	100	100	-200	300	-300	600	-400	721	0.832
250	100	125	-200	375	-300	750	-400	850	0.882
300	100	150	-200	450	-300	900	-400	985	0.914
350	100	175	-200	525	-300	1050	-400	1124	0.934
400	100	200	-200	600	-300	1200	-400	1265	0.949
450	100	225	-200	675	-300	1350	-400	1408	0.959
500	100	250	-200	750	-300	1500	-400	1552	0.966

READY

of line 290 is for the individual values of power as indicated by line 380, while $G\$$ is reserved for the calculated values of line 420. Note that there is no requirement for the format statement of lines 290 and 300 to be on the same line. The same is true for the PRINT statements of lines 380 and 420.

19.6   TEST ROUTINE

The last program (19-5) of the chapter will test the user's ability to find the solution to the system of Figure 19.1 for two to five components. The function appearing on line 110 is required to calculate the power factor angle determined by $\theta = \cos^{-1}F_p$ on line 490. The program will first randomly generate the values of P and Q using the subroutine starting on line 650. The values of P_T and Q_T are then calculated on lines 210 and 220 and checked against the user values on lines 280, 340, and 350. On line 410, the value of S_T is calculated and compared against the provided value on line 420. The angle and power factor are then calculated and compared on lines 500, 510, 580, and 590.

Note the number of elements generated in each run and the fact that the program will check the sign of the reactive power components.

```
10 REM *****  PROGRAM 19-5  *****
20 REM *********************************
30 REM Program to test power calculations
40 REM for the network of Figure 19-1.
50 REM *********************************
60 REM
100 DIM P(6),Q(6)
110 DEF FNC(X)=-ATN(X/SQR(1-X^2))+1.5708
120 N=INT(2+5*RND(XV))  :REM For 2 to 5 individual loads
130 PRINT "For the network of Figure 19-1"
140 FOR I=1 TO N
150 GOSUB 650 :REM Select P and Q
160 P(I)=P
170 Q(I)=Q
180 PRINT "P(";I;")=";P(I);"watts"
190 PRINT "Q(";I;")=";Q(I);"vars"
200 PRINT
210 PT=PT+P(I)
220 QT=QT+Q(I)
230 NEXT I
240 PRINT
250 REM Now ask questions
260 PRINT
270 INPUT "PT=";PA
280 IF ABS(PT-PA)>ABS(0.01*PT) THEN GOTO 310
290 PRINT "Correct"
300 GOTO 320
310 PRINT "No, it's";PT;"watts"
320 PRINT
330 INPUT "QT=";QA
340 IF SGN(QT)<>SGN(QA) THEN GOTO 380
350 IF ABS(QT-QA)>ABS(0.01*QT) THEN GOTO 380
360 PRINT "Correct"
370 GOTO 390
380 PRINT "No, it's";QT;"vars"
390 PRINT
400 INPUT "ST=";SA
410 ST=SQR(PT^2+QT^2)
420 IF ABS(ST-SA)>ABS(0.01*ST) THEN GOTO 450
430 PRINT "Correct"
440 GOTO 460
```

(Continued)

```
      450 PRINT "No, its";ST;"VA"
      460 PRINT
      470 INPUT "Theta(degrees)=";TA
      480 FP=PT/ST
      490 TH=57.296*FNC(FP)
      500 IF SGN(TA)<>SGN(TH) THEN GOTO 540
      510 IF ABS(TH-TA)>ABS(0.01*TH) THEN GOTO 540
      520 PRINT "Correct"
      530 GOTO 550
      540 PRINT "No, it's";TH;"degrees"
      550 PRINT
      560 INPUT "FP=";FA
      570 FP=PT/ST
      580 IF SGN(FA)<>SGN(FP) THEN GOTO 620
      590 IF ABS(FA-FP)>ABS(0.01*FP) THEN GOTO 620
      600 PRINT "Correct"
      610 GOTO 630
      620 PRINT "No, it's";FP
      630 END
      650 REM Module to select P and Q
      660 SG=+1
      670 S=INT(1+3*RND(XV))
      680 IF S=2 THEN SG=-1  :REM select - sign one of 3 times
      690 P=10*INT(1+50*RND(XV)) :REM 10<P<500
      700 Q=10*SG*INT(1+50*RND(XV)) :REM SG=+ :10<Q<500, SG=- : -500<Q<-10
      710 RETURN
```

READY

RUN

```
For the network of Figure 19-1
P( 1 )= 180 watts
Q( 1 )=-200 vars

P( 2 )= 450 watts
Q( 2 )= 280 vars

P( 3 )= 370 watts
Q( 3 )= 110 vars

P( 4 )= 100 watts
Q( 4 )= 240 vars

P( 5 )= 170 watts
Q( 5 )= 240 vars


PT=? 1270
Correct

QT=? 650
No, it's 670 vars

ST=? 1436
Correct

Theta(degrees)=? 27.8
Correct

FP=? 0.885
Correct
```

READY

EXERCISES

Write a program to perform the following tasks.

1. Expand Program 19-1 to include any number of elements N.

2. Given the voltage across a coil in the form $v = V_m \sin\omega t$ and the current of the coil in the form $i = I_m\sin(\omega t - 90°)$, tabulate the power determined by the product $p = v \cdot i$ at $\omega t = 0°$ through $360°$ in increments of $10°$. Be sure to preserve the sign resulting from each product. For the first run, use $V_m = 10V$, $\omega = 100$ and $I_m = 2A$.

3. Use the results of problem 19-2 to plot the curve of the power to the coil versus ωt for one full cycle of the applied voltage.

4. Given S_T, F_p, E_T, P_1, Q_1, P_2, and Q_2 for the network of Figure 19.4, determine P_3 and Q_3. For the first run, use $S_T = 4000VA$, $F_p = 0.8$, $E_T = 200V$, $P_1 = 2000W$, $Q_1 = 400vars(Ind.)$, $P_2 = 1000W$, and $Q_2 = 500vars(Cap.)$.

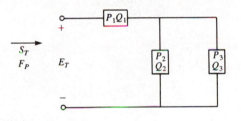

FIGURE 19.4

5. Repeat Program 19-2 for a parallel R-L-C network. That is, given the applied voltage $\mathbf{E}$ $= E\underline{/\theta}$, R, X_L and X_C, print out P_T, Q_L, Q_C, Q_T, S_T, and F_p.

6. Repeat Program 19-3 for the internal parallel components. That is, given the applied voltage $\mathbf{E} = E\underline{/\theta}$, the total apparent power S_T, the power factor F_p, and an indication of whether the network is leading or lagging, determine the internal parallel components.

7. Tabulate N, P_T, Q_T, S_T, and F_p for an increasing number of loads which have a dissipation and reactive power component directly related to the number of loads. For instance, if only one load is present, the total real and reactive power is P_1 and Q_1 respectively; if two loads, the first remains P_1 and Q_1, while the second is $2P_1$ and $2Q_1$. For the first run, use $N = 10$, $P_1 = 20W$, and $Q_1 = 40W$. Repeat for $N = 10$, $P_1 = 40W$, and $Q_1 = 20W$ and note any differences.

8. Use the results of problem 19-7 to plot S_T versus N. Label both axes.

9. For the series R-L-C circuit analyzed by Program 19-2, tabulate P_T, Q_T, S_T, and F_p versus frequency for a frequency range extending from 0Hz to 20,000Hz in increments of 1000Hz. For the first run, use $R = 100$ ohms, $L = 5mH$, and $C = 0.1\mu F$.

10. Using the results of problem 19-9, plot F_p versus frequency for the indicated range. Label both axes.

Resonance

20

20.1 INTRODUCTION

There are a number of sequential calculations performed in the analysis of series and parallel resonant circuits that are appropriate for computer analysis. In particular is the testing of the value of Q before choosing the suitable equations for the parallel resonance circuit. The resonance curves are also an excellent opportunity to test our newly developed plotting skills.

20.2 SERIES RESONANCE

A computer analysis of the series resonant circuit of Figure 20.1 will be performed by Program 20-1. The resonant frequency is determined on line 200 using $f_s = 1/2\pi\sqrt{LC}$, followed by $X_L = \omega L$ and $X_C = 1/\omega C$ on lines 220 and 230. The current and various voltages are then determined before the bandwidth and cut-off fre-

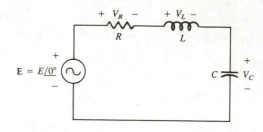

FIGURE 20.1

quencies are determined by lines 290 through 310. The maximum power is calculated on line 320, with the power at the band frequencies determined on line 330. The remainder of the program prints out the results. The program is quite easy to write but the wealth of information it provides is apparent from the two printouts. Note in each printout that $V_L = V_C$ at resonance, and also how an increase in Q will result in a reduced bandwidth relative to the resonant frequency.

```
10 REM *****  PROGRAM 20-1  *****
20 REM ******************************
30 REM Program to analyze the series
40 REM resonant circuit.
50 REM ******************************
60 REM
100 PRINT "This program analyzes the"
110 PRINT "series resonant circuit of Figure 20-1."
120 PRINT
130 PRINT "Enter the following data:"
140 INPUT "The supply voltage(rms), E=";EE
150 INPUT "R=";R
160 INPUT "L=";L
170 INPUT "C=";C
180 PRINT
190 REM Perform calculations
200 FS=1/(2*3.14159*SQR(L*C))
210 WS=2*3.14159*FS
220 XL=WS*L
230 XC=1/(WS*C)
240 Q=XL/R
250 IE=EE/R
260 VR=IE*R
270 VL=IE*XL
280 VC=IE*XC
290 BW=FS/Q
300 F2=FS+(BW/2)
310 F1=FS-(BW/2)
320 PM=IE^2*R
330 P=PM/2
```

Input (lines 140–170)
Calc. (lines 190–330)

```
┌ 340 REM Now print results
│ 350 PRINT "The resonant frequency, f=";FS;"hertz, or w=";WS;"r/s"
│ 360 PRINT "XL=";XL;"ohms, and XC=";XC;"ohms"
│ 370 PRINT "with circuit Q=";Q
│ 380 PRINT
│ 390 PRINT "The current, I=";IE;"amps"
│ 400 PRINT "and voltages are:"
│ 410 PRINT "VR=";VR;"volts"
│ 420 PRINT "VL=";VL;"volts"
│ 430 PRINT "VC=";VC;"volts"
│ 440 PRINT
│ 450 PRINT "The bandwidth, BW=";BW;"hertz"
│ 460 PRINT "with, f1=";F1;"hertz"
│ 470 PRINT "and,   f2=";F2;"hertz"
│ 480 PRINT
│ 490 PRINT "Maximum power dissipated, Pmax=";PM;"watts"
│ 500 PRINT TAB(22);
└ 510 PRINT "with P(HPF)=";P;"watts"
  520 END
```

Print Results (brace label for lines 410–510)

READY

RUN

This program analyzes the
series resonant circuit of Figure 20-1.

Enter the following data:
The supply voltage(rms), E=? 50
R=? 5
L=? 1E-3
C=? 20E-6

The resonant frequency, f= 1125.3963 hertz, or w= 7071.0678 r/s
XL= 7.0711 ohms, and XC= 7.0711 ohms
with circuit Q= 1.4142

The current, I= 10 amps
and voltages are:
VR= 50 volts
VL= 70.7107 volts
VC= 70.7107 volts

The bandwidth, BW= 795.7754 hertz
with, f1= 727.5087 hertz
and, f2= 1523.284 hertz

Maximum power dissipated, Pmax= 500 watts
                    with P(HPF)= 250 watts

READY

This program analyzes the
series resonant circuit of Figure 20-1.

Enter the following data:
The supply voltage(rms), E=? 1.2
R=? 200
L=? 10E-3
C=? 0.01E-6

```
The resonant frequency, f= 1.5916E+04 hertz, or w= 1E+05 r/s
XL= 1000 ohms, and XC= 1000 ohms
with circuit Q= 5

The current, I= 6E-03 amps
and voltages are:
VR= 1.2 volts
VL= 6 volts
VC= 6 volts

The bandwidth, BW= 3183.1016 hertz
with, f1= 1.4324E+04 hertz
and,  f2= 1.7507E+04 hertz

Maximum power dissipated, Pmax= 7.2E-03 watts
                    with P(HPF)= 3.6E-03 watts


READY
```

In Program 20-2, the resonant frequency, bandwidth, and value of C are given and the value of R and L determined. The quality factor is determined on line 160, followed by L on line 170, using the fact that $X_L = X_C$ or $L = 1/4\pi^2 f^2 C$. The magnitude of ω_s and $X_L = \omega_s L$ can then be determined before R is determined from $R = X_L/Q_S$. Note in each printout the use of a 10^3 multiplier on line 250 to permit the use of mH as a unit of measure.

```
 10 REM *****  PROGRAM 20-2  *****
 20 REM ******************************************
 30 REM Program does design calculations for the
 40 REM series resonant circuit of (Figure 20-1).
 50 REM ******************************************
 60 REM
100 PRINT "Enter the following data for the circuit of Figure 20-1."
110 PRINT
```
Input
```
120 INPUT "fs=";FS
130 INPUT "BW=";BW
140 INPUT "C=";C
```
Design
```
150 REM Determine components
160 QS=FS/BW
170 L=1/(4*(3.14159*FS)^2*C)
180 WS=2*3.14159*FS
190 XL=WS*L
200 R=XL/QS
```
Output
```
210 REM Print results
220 PRINT
230 PRINT "With Qs=";QS
240 PRINT "at fs=";FS;"hertz"
250 PRINT "use L=";1000*L;"mH"
260 PRINT "with XL=";XL;"ohms"
270 PRINT "and  R=";R;"ohms"
280 END
```

```
READY
```

```
RUN
*
Enter the following data for the circuit of Figure 20-1.

fs=? 2800
BW=? 200
C=? 0.1E-6

With Qs= 14
at fs= 2800 hertz
use L= 32.3091 mH
with XL= 568.411 ohms
and  R= 40.6008 ohms


READY

RUN

Enter the following data for the circuit of Figure 20-1.

fs=? 100E3
BW=? 4E3
C=? 1000E-12

With Qs= 25
at fs= 1E+05 hertz
use L= 2.533 mH
with XL= 1591.5508 ohms
and  R= 63.662 ohms


READY


*Example 20.4, ICA
```

20.3 PARALLEL RESONANCE

Analyzing the parallel resonant network of Figure 20.2 is always more lengthy and difficult

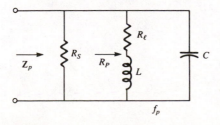

FIGURE 20.2

than analyzing the series resonant circuit of Figure 20.1. However, once the various equations are properly entered in the program, the analysis will always be performed properly and accurately.

With the computer, the user can analyze the network using the full equation for any value of Q without a measurable loss in time and energy. However, the test of $Q \geq 10$ will appear as an exercise in including a decision operation—one set of equations will be used if $Q \geq 10$ and another if $Q < 10$.

Once the network values and resonant frequency are provided on lines 130 through 180 of Program 20-3, the value of Q will be deter-

```
 10 REM *****  PROGRAM 20-3  *****
 20 REM *****************************************************
 30 REM Program to analyze the parallel resonant network
 40 REM of Figure 20-2.
 50 REM *****************************************************
 60 REM
100 PRINT "This program analyzes the"
110 PRINT "parallel resonant circuit of Figure 20-2."
120 PRINT
130 PRINT "Enter the following component data:"
140 PRINT
150 INPUT "Rs(source resistance)=";RS
160 INPUT "L=";L
170 INPUT "with Rl(coil resistance)=";RL
180 INPUT "fp(parallel resonant frequency)=";FP
190 REM Perform calculations
200 WP=6.28318*FP
210 XL=WP*L
220 Q=XL/RL
230 IF Q>=10 THEN GOTO 300
240 REM Calculations for Q<10
250 RP=(RL^2+XL^2)/RL
260 ZP=RS*RP/(RS+RP)
270 C=(1/(WP^2*L))*(Q^2/(1+Q^2))
280 QP=WP*C*ZP
290 GOTO 360
300 REM Calculations for Q>=10
310 RP=Q^2*RL
320 ZP=RS*RP/(RS+RP)
330 C=1/(WP^2*L)
340 QP=ZP/(WP*L)
350 REM Complete calculations and print results
360 BW=FP/QP
370 PRINT
380 PRINT "For Q=";Q;"    ";
390 IF Q>=10 THEN PRINT "(Q>=10)"
400 IF Q<10 THEN PRINT "(Q<10)"
410 PRINT "Rp=";RP;"ohms"
420 PRINT "ZTp=";ZP;"ohms"
430 PRINT "and at the desired resonant frequency, C=";1E6*C;"uF"
440 PRINT
450 PRINT "The resulting Qp=";QP
460 PRINT "with a bandwidth, BW=";BW/1000;"kHz"
470 END
```

Input — lines 150–180
Calc. — lines 190–340 ($Q < 10$: lines 250–280; $Q > 10$: lines 310–340)
Calc. BW & Print Results — lines 350–460

```
READY

*
This program analyzes the
parallel resonant circuit of Figure 20-2.

Enter the following component data:

Rs(source resistance)=? 40E3
L=? 1E-3
with Rl(coil resistance)=? 10
fp(parallel resonant frequency)=? 40E3
```

*Example 20.7, ICA

```
For  Q= 25.1327      (Q>=10)
Rp= 6316.5361 ohms
ZTp= 5455.1024 ohms
and at the desired resonant frequency, C= .016 uF

The resulting Qp= 21.7052
with a bandwidth, BW= 1.8429 kHz


READY


This program analyzes the
parallel resonant circuit of Figure 20-2.

Enter the following component data:

Rs(source resistance)=? 30E3
L=? 0.01E-3
with Rl(coil resistance)=? 4Ω
fp(parallel resonant frequency)=? 2E6

For  Q= 3.1416      (Q<10)
Rp= 434.7835 ohms
ZTp= 428.5723 ohms
and at the desired resonant frequency, C= 5.75E-04 uF

The resulting Qp= 3.0967
with a bandwidth, BW= 645.8467 kHz


READY
```

mined from $Q = X_L/R_\ell$ on line 220. If $Q \geq 10$, the program will move to line 300, where the equations used to determine the various quantities of interest will be in a reduced, simpler form. For instance, the magnitude of R_p (representing the input resistance of the tank circuit of Figure 20.2) is determined by $Q^2 R_\ell$ on line 310 and $(R_\ell{}^2 + X_L{}^2)/R_\ell$ on line 250. The total impedance Z_p is then the parallel combination of R_p and R_s, as in Figure 20.2, and determined on line 260. The value of C is determined from $C = 1/\omega_p^2 L$ $\left(\text{from } \omega_p = \dfrac{1}{\sqrt{LC}}\right)$ on line 330, the network quality factor Q_p by $Q_p = Z_p/X_L = Z_p/\omega_p L$, and the bandwidth $BW = f_p/Q_p$ on line 360.

If $Q \leq 10$, the equations are more complex and $R_p = (R_\ell^2 + X_L^2)/R_\ell$, with $C = \dfrac{1}{\omega_p^2 L}\left(\dfrac{Q^2}{1 + Q^2}\right)$. The results for *both* levels of

Q are printed out on lines 380 through 460. Note the division by 1000 on line 460 to permit the use of the kHz unit of measure. Note also in the printout the reduced value of the quality factor (Q_p versus Q), due primarily to the source resistance R_s. In addition, note the relative range (compared to the resonant frequency) of the bandwidth, with a low (second run) and high Q (first run), and the reduced value of Z_{T_p} as compared to R_p, due to the parallel combination of elements.

In Program 20-4 the network of Figure 20.3 is designed according to the specifications provided for R_s, R_ℓ, L, f_p, BW, and $V_{C\text{max}}$. Essentially, the design requires that the value of R (the resistor added to establish a particular Q_p) and C (to establish f_p), and the magnitude of I (as limited by $V_{C\text{max}}$) be determined.

The first step is to determine the network quality factor from $Q_p = f_p/BW$, as on line 260.

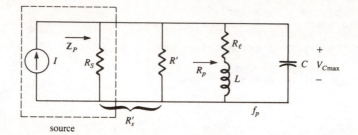

FIGURE 20.3

```
10 REM *****  PROGRAM 20-4  *****
20 REM ************************************
30 REM Program to design the parallel
40 REM resonant circuit of Figure 20-3.
50 REM ************************************
60 REM
100 PRINT "This program selects components for the"
110 PRINT "parallel resonant circuit of Figure 20-3."
120 PRINT
130 PRINT "Enter the following data:"
140 INPUT "Rs(source resistance)=";RS
150 PRINT
160 INPUT "Coil, L=";L
170 INPUT "with Rl(coil resistance)=";RL
180 PRINT
190 INPUT "fp(parallel resonant frequency)=";FP
200 INPUT "BW(bandwidth)=";BW
210 PRINT
220 INPUT "Vc(max)=";VM
230 PRINT
240 REM Now do calculations
250 WP=6.28318*FP
260 QP=FP/BW
270 XL=WP*L
280 Q=XL/RL
290 IF Q>=10 THEN RP=Q^2*RL
300 IF Q<10 THEN RP=(RL^2+XL^2)/RL
310 R1=(RP*QP*XL)/(RP-XL*QP)   :REM R1 is Rs'
320 R2=(R1*RS)/(RS-R1)   :REM R2 is R'
330 IF Q>=10 THEN C=1/(WP*XL)
340 IF Q<10 THEN C=XL/(WP*(RL^2+XL^2))
350 IP=(VM*(R1+RP))/(R1*RP)
360 REM Now print results
370 PRINT "Results are:"
380 PRINT:PRINT "For wp=";WP;"r/s"
390 PRINT "Qp=";QP
400 PRINT "XL=";XL;"ohms"
410 PRINT "and Q=";Q
420 PRINT "Rp=";RP/1000;"kilohms"
430 PRINT "Rs=";RS/1000;"kilohms"
440 PRINT "and R'=";R2/1000;"kilohms"
450 PRINT
460 PRINT "For the desired resonance condition, C=";1E6*C;"uF"
470 PRINT "and the peak current, IP=";IP*1000;"mA"
480 END
```

The program is annotated on the left margin with: **Input** (lines 160–220), **Calc.** (lines 240–350), **Print Results** (lines 360–470).

READY

```
RUN
*
This program selects components for the
parallel resonant circuit of Figure 20-3.

Enter the following data:
Rs(source resistance)=? 40E3

Coil, L=? 1E-3
with Rl (coil resistance)=? 10

fp(parallel resonant frequency)=? 50E3
BW(bandwidth)=? 2500

Vc(max)=? 10

Results are:

For wp- 3.1416E+05 r/s
Qp= 20
XL= 314.159 ohms
and Q= 31.4159
Rp= 9.8696 kilohms
Rs= 40 kilohms
and R'= 30.4565 kilohms

For the desired resonance condition, C= .01 uF
and the peak current, IP= 1.5916 mA


READY
```

*Example 20.10, ICA

The value of Q is determined from $Q = X_L/R_\ell$ = $\omega_p L/R_\ell$ before being tested on lines 290 and 300, where the appropriate equations are used to determine R_p, as in Figure 20.3. The equation for Q_p is then written as

$$Q_p = \frac{R'_s \| R_p}{X_L} = \frac{R'_s R_p/(R'_s + R_L)}{X_L}$$
$$= \frac{R'_s R_p}{X_L(R'_s + R_L)}$$

or $R'_s R_p = Q X_L(R'_s + R_L)$
$$= Q X_L R'_s + Q X_L R_L$$

and $R'_s = \dfrac{R_p Q_p X_L}{R_p - X_L Q_p}$

as on line 310. However,

$$R'_s = R_s \| R' = \frac{R_s R'}{R_s + R'}$$

and

$$R' = \frac{R'_s R_s}{R_s - R'_s}$$

as on line 320.

The value of Q will then determine which equations will be used to calculate C on lines 330 and 340. The input impedance of the network, as shown in Figure 20.4 is determined by $R'_s \| R_p$ and

$$I = \frac{V_{Cmax}}{Z_p} = \frac{V_{Cmax}}{R'_s \| R_p}$$
$$= \frac{V_{Cmax}}{\dfrac{R'_s \cdot R_p}{R'_s + R_p}} = \frac{V_{Cmax}(R'_s + R_p)}{R'_s \cdot R_p}$$

as on line 350.

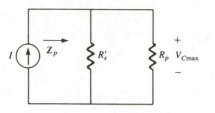

FIGURE 20.4

The results are then printed out on lines 360 through 470. Note the value of R', compared to the source resistance, to result in the desired bandwidth and the calculated quality factor.

20.4 PLOTTING ROUTINES

The next two programs will plot the familiar series and parallel resonance curves. In both cases, the plotting routine employed in earlier chapters (line 3100 through 3290) will be employed. A detailed review of the subroutine appears in those chapters and will not be repeated here.

Once the network parameters are provided in Program 20-5 for the series resonant circuit (lines 130 through 160), various important quantities will be calculated and printed out with the resulting plot. The data required for the plot is gathered by the subroutine at line 350. For a range of frequencies extending from $f_s/10$ to $2f_s$ (in increments of $f_s/20$), the impedance Z will be determined from $Z = \sqrt{R^2 + (X_L - X_C)^2}$, and the current determined from $I = E/Z$. The values of $D(1)$, $D(2)$, etc. will then be set equal to the resulting current level. Note the F (frequency) loop extending from 360 to 440, with the value of J incremented by line 370 for the next value of $D(J)$. The result is that the value of I for each increment in frequency will be stored as $D(1)$, $D(2)$, $D(3)$, etc. for the routine at 3100 where $D(1)$ appears on lines 3130, 3150 and 3230. Note on the provided printout the value of I_{max} printed by line 3210, where FM has the largest value in

```
              10 REM *****   PROGRAM 20-5   *****
              20 REM ************************************************
              30 REM Program ro provide the frequency plot for"
              40 REM the current of a series resonant circuit.
              50 REM ************************************************
              60 REM
             100 PRINT "For the series resonant circuit"
             110 PRINT "of Figure 20-1, enter:"
             120 DIM D(100)
            ┌130 INPUT "R=";R
             140 INPUT "L=";L
     Input  │150 INPUT "C=";C
             160 INPUT "Supply voltage(rms), E=";EE
            └170 PRINT
            ┌180 FS=1/(6.28318*SQR(L*C))
             190 WS=6.28318*FS
             200 XL=WS*L
  fs,ωs     │210 XC=1/(WS*C)
  Q          220 Q=XL/R
             230 PRINT "At fs=";FS/1000;"kHz"
            └240 PRINT "with Q=";Q
            ┌250 BW=FS/Q
             260 F1=FS-BW/2
  BW        │270 F2=FS+BW/2
  f1,f2      280 PRINT "Bandwidth is, BW=";BW/1000;"kHz"
             290 PRINT "with f1=";F1/1000;"kHz, and f2=";F2/1000;"kHz"
            └300 PRINT
```

```
/ peak  ┌310  IP=EE/R
        └320  PRINT "The peak current, IP=";IP;"amps"
         330  PRINT
         340  PRINT
        ┌350  REM Calculate data for plot
         360  FOR F=FS/10 TO 2*FS STEP FS/20
         370  J=J+1
         380  W=6.28318*F
         390  X1=W*L
Calc.    400  X2=1/(W*C)
Plot     410  Z=SQR(R^2+(X1-X2)^2)
Points   420  I=EE/Z
         430  D(J)=I
         440  NEXT F
         450  N=J
         460  PRINT "For";N;"entries"
         470  PRINT
        └480  GOSUB 3100
         490  END
        ┌3100  REM Module to plot data in array D(I) with N entries
         3110  FS=64
Plot     3120  CP=0
Module   3130  FM=ABS(D(I))
         3140  FOR I=2 TO N
         3150  IF ABS(D(I))>FM THEN FM=ABS(D(I))
         3160  NEXT I
         3170  REM FM now has largest value in array
         3180  FOR I=1 TO FS
         3190  PRINT "-";
         3200  NEXT I
         3210  PRINT INT(FM+0.5)
         3220  FOR I= 1 TO N
         3230  P=INT(CP+FS*D(I)/FM+0.5)
         3240  IF P<CP THEN PRINT TAB(P);"*";TAB(CP);"I"
         3250  IF P=CP THEN PRINT TAB(CP);"I"
         3260  IF P>CP THEN PRINT TAB(CP);"I";TAB(P);"*"
         3270  NEXT I
         3280  PRINT:PRINT
         3290  RETURN


READY


For the series resonant circuit
of Figure 20-1, enter:
R=? 10
L=? 4E-3
C=? 0.4E-6
Supply voltage(rms), E=? 100

At fs= 3.9789 kHz
with Q= 10
Bandwidth is, BW= .3979 kHz
with f1= 3.7799 kHz, and f2= 4.1778 kHz

The peak current, IP= 10 amps
```

For 38 entries

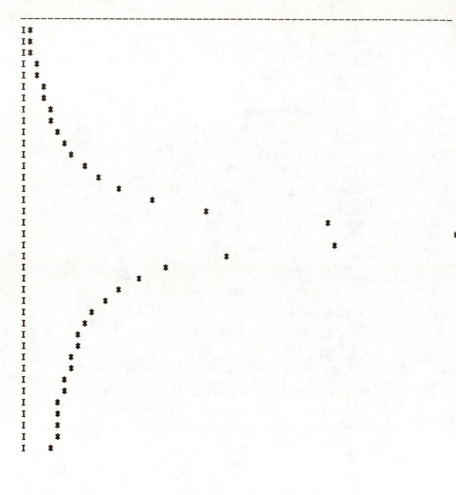

READY

the array, as noted by line 3170. The various discontinuities you see on the graph are necessary because intermediate locations on the horizontal printout are lacking.

Program 20-6 is very similar to Program 20-5, except that the resulting equations are measurably more complex. The network employed in defining the necessary parameters appears in Figure 20.5. Note the change in

equations, defined by the value of Q and appearing on lines 260, 270, 400, and 410. The network of Figure 20.5 is terminally (at the source) equivalent to the network of Figure 20.4. It was developed through a series of manipulations of the total admittance equation for the network of Figure 20.4. Lines 390 through 460 will determine the total admittance and impedance of the network in a manner best

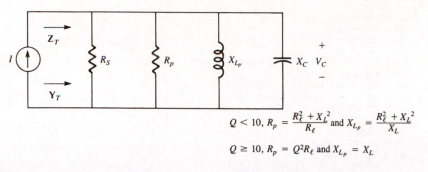

$$Q < 10, \; R_p = \frac{R_\ell^2 + X_L^2}{R_\ell} \; \text{and} \; X_{L_p} = \frac{R_\ell^2 + X_L^2}{X_L}$$

$$Q \geq 10, \; R_p = Q^2 R_\ell \; \text{and} \; X_{L_p} = X_L$$

FIGURE 20.5

described by Figure 20.6. The magnitude of the total admittance is determined by $Y_T = \sqrt{G_T^2 + B_T^2}$, and the total impedance from $Z_T = 1/Y_T$. The magnitude of the voltage across the capacitor for the frequency of interest is then determined by $V_C = IZ_T$ and stored as $D(1)$ for the first pass on line 480. The module at

3100 will provide the plot appearing in the printout, which, you will note, has a much higher Q than that resulting from the chosen series resonance circuit. Note the maximum whole number value of the voltage V_C on the graph, provided by line 3210 of the plotting module.

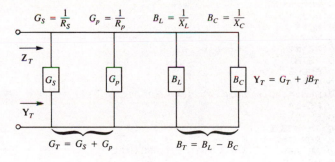

FIGURE 20.6

```
10 REM *****   PROGRAM 20-6   *****
20 REM ***********************************
30 REM Program to provide a plot of
40 REM the voltage versus frequency
50 REM for a parallel resonant network.
60 REM ***********************************
70 REM
100 PRINT "For the parallel resonant circuit"
110 PRINT "of Figure 20-2, enter:"
120 DIM D(100)
130 INPUT "Rs=";RS
140 INPUT "L=";L
150 INPUT "Rl=";RL
160 INPUT "C=";C
170 INPUT "Supply current(rms), I=";IE
180 PRINT
190 FP=1/(6.28318*SQR(L*C))
```

(Continued)

```
 200 WP=6.28318*FP
 210 XL=WP*L
 220 XC=1/(WP*C)
 230 Q=XL/RL
 240 PRINT "At fp=";FP/1000;"kHz"
 250 PRINT "with Q=";Q
 260 IF Q<10 THEN RP=(RL^2+XL^2)/RL
 270 IF Q>=10 THEN RP=Q^2*RL
 280 ZP=RS*RP/(RS+RP)
 290 QP=ZP/XL
 300 BW=FP/QP
 310 F1=FP-BW/2
 320 F2=FP+BW/2
 330 PRINT "BW=";BW/1000;"kHz"
 340 PRINT "with f1=";F1/1000;"kHz, and f2=";F2/1000;"kHz"
 350 PRINT
 360 REM Calculate data for plot
 370 FOR F=F1/4 TO 2*F2 STEP FP/20
 380 J=J+1
 390 W=6.28318*F
 400 IF Q<10 THEN X1=(RL^2+(W*L)^2)/XL
 410 IF Q>=10 THEN X1=W*L
 420 X2=1/(W*C)
 430 GT=1/RS+1/RP
 440 BT=(1/X1-1/X2)
 450 YT=SQR(GT^2+BT^2)
 460 ZT=1/YT
 470 VC=IE*ZT
 480 D(J)=VC
 490 NEXT F
 500 N=J
 510 PRINT "For";N;"entries"
 520 PRINT
 530 GOSUB 3100
 540 END
3100 REM Module to plot data in array D(I) with N entries
3110 FS=64
3120 CP=0
3130 FM=ABS(D(I))
3140 FOR I=2 TO N
3150 IF ABS(D(I))>FM THEN FM=ABS(D(I))
3160 NEXT I
3170 REM FM now has largest value in array
3180 FOR I=1 TO FS
3190 PRINT "-";
3200 NEXT I
3210 PRINT INT(FM+0.5)
3220 FOR I= 1 TO N
3230 P=INT(CP+FS*D(I)/FM+0.5)
3240 IF P<CP THEN PRINT TAB(P);"*";TAB(CP);"I"
3250 IF P=CP THEN PRINT TAB(CP);"I"
3260 IF P>CP THEN PRINT TAB(CP);"I";TAB(P);"*"
3270 NEXT I
3280 PRINT:PRINT
3290 RETURN


READY
```

```
For the parallel resonant circuit
of Figure 20-2, enter:
Rs=? 40E3
L=? 1E-3
R1=? 10
C=? 0.02E-6
Supply current(rms), I=? 10E-3

At fp= 35.5882 kHz
with Q= 22.3607
BW= 1.7905 kHz
with f1= 34.6929 kHz, and f2= 36.4834 kHz

For 37 entries

---------------------------------------------------------- 43
I*
I*
I*
I *
I *
I *
I  *
I  *
I   *
I   *
I    *
I     *
I       *
I         *
I            *
I            *
I              *
I                *
I        *
I      *
I      *
I    *
I    *
I   *
I   *
I   *
I   *
I  *
I  *
I  *
I  *
I  *
I *
I *
I *
I *


READY
```

20.5 DOUBLE-TUNED FILTER

The next Program (20-7) will examine the network of Figure 20.7, which is designed to reject one frequency and accept another. At increasing frequencies (from zero), the components were chosen to first establish a parallel resonance condition with C and L_p. Within the corresponding frequency range, the magnitude of X_{L_s} can be ignored compared to X_C. The impedance Z (X in the program) will then be very large at the parallel resonance frequency, resulting in a low level for V_L, as determined by the voltage-divider-rule. The result is that the resonant frequency for the parallel resonant circuit is the "rejected" frequency at the output.

As the frequency continues to rise, the magnitude of X_{L_p} will rise to the point where it can be ignored and approximated by an open-circuit compared to the net reactance of the parallel L-C branch. The magnitude of X_C will approach that of X_L, and a series resonance condition established between L_s, C, and R_L.

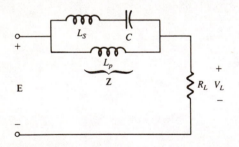

FIGURE 20.7

When resonance occurs, the impedance Z will be at a minimum, and the major portion of E will appear across R_L, identifying the "accepted" frequency.

The resonant frequency for the series resonance can be determined from $f_s = 1/2\pi\sqrt{L_s C}$ (line 200), and the parallel resonant frequency from $f_p = 1/2\pi\sqrt{L_p C}$ (line 210). The range of frequencies on line 290 was chosen to extend from one-fourth the lower (rejected) frequency to twice the higher (accepted) frequency.

In general, the voltage across the load R_L is determined by

$$\mathbf{V}_L = \frac{\mathbf{R_L E}}{\mathbf{R_L + Z}} \qquad \text{(line 380)}$$

with $\mathbf{Z} = X_{L_p} \| X'$ and $X' = X_{L_s} - X_C$ (line 340)

$$= \frac{X_{L_p} \cdot X'}{X_{L_p} + X'} \qquad \text{(line 360)}$$

Note in both printouts that the net impedance of Z is very low at the series resonance condition, to establish $V_L \cong E$. In the second run, the quality factor of the parallel resonant circuit is higher, as identified by the sharper negative peak, while the quality factor for the series resonant circuit is lower as noted by the slowly decaying curve. In fact, for the series resonant circuit,

$$Q = \frac{X_{L_s}}{R} = \frac{2\pi f_s L_s}{R}$$

$$= \frac{(6.28)(599.4 \times 10^3)(141 \times 10^{-3})}{2000} \cong 0.27$$

```
10 REM *****  PROGRAM 20-7  *****
20 REM ***************************************
30 REM Program to provide a plot of
40 REM current over a range of frequency
50 REM around the resonant frequency
60 REM in a doubly tuned circuit.
70 REM ***************************************
80 REM
100 PRINT "For the doubly-tuned circuit of Figure 20-4"
110 PRINT "enter the following data:"
120 DIM D(150)
```

```
130 INPUT "Capacitor, C=";C
140 INPUT "E=";E
150 INPUT "Lp=";LP
160 INPUT "Ls=";LS
170 INPUT "RL=";RL
180 PRINT
190 WS=6.28318*FS :WR=6.28318*FR
200 F1=1/(3.14159*2*SQR(LS*C))
210 F2=1/(2*3.14159*SQR(LP*C))
220 PRINT "For a calculated rejected frequency, F2=";F2/1000;"kHz"
230 PRINT "and a calculated accepted frequency, F1=";F1/1000;"kHz"
240 PRINT
250 PRINT "Over a range of frequency from";(F2/4)/1000;"kHz";
260 PRINT " to";2*F1/1000;"kHz"
270 PRINT
280 REM Calculate data for plot
290 FOR F=F2/4 TO 2*F1 STEP F2/8
300 J=J+1
310 W=6.28318*F
320 XL=W*LS
330 XC=1/(W*C)
340 XT=XL-XC
350 XP=W*LP
360 X=XP*XT/(XP+XT)
370 DE=SQR(RL^2+X^2)
380 VL=RL*E/DE
390 D(J)=VL
400 NEXT F
410 N=J
420 PRINT "For";N;"entries"
430 PRINT
440 GOSUB 3100
450 END
3100 REM Module to plot data in array D(I) with N entries
3110 FS=64
3120 CP=0
3130 FM=ABS(D(I))
3140 FOR I=2 TO N
3150 IF ABS(D(I))>FM THEN FM=ABS(D(I))
3160 NEXT I
3170 REM FM now has largest value in array
3180 FOR I=1 TO FS
3190 PRINT "-";
3200 NEXT I
3210 PRINT INT(FM+0.5)
3220 FOR I= 1 TO N
3230 P=INT(CP+FS*D(I)/FM+0.5)
3240 IF P<CP THEN PRINT TAB(P);"*";TAB(CP);"I"
3250 IF P=CP THEN PRINT TAB(CP);"I"
3260 IF P>CP THEN PRINT TAB(CP);"I";TAB(P);"*"
3270 NEXT I
3280 PRINT:PRINT
3290 RETURN
```

READY

RUN

```
For the doubly-tuned circuit of Figure 20-4
enter the following data:
Capacitor, C=? 10E-12
E=? 5
```

```
Lp=? 63.4E-3
Ls=? 7E-3
RL=? 10E3
```

For a calculated rejected frequency, F2= 199.883 kHz
and a calculated accepted frequency, F1= 601.5497 kHz

Over a range of frequency from 49.9708 kHz to 1203.0993 kHz

For 47 entries

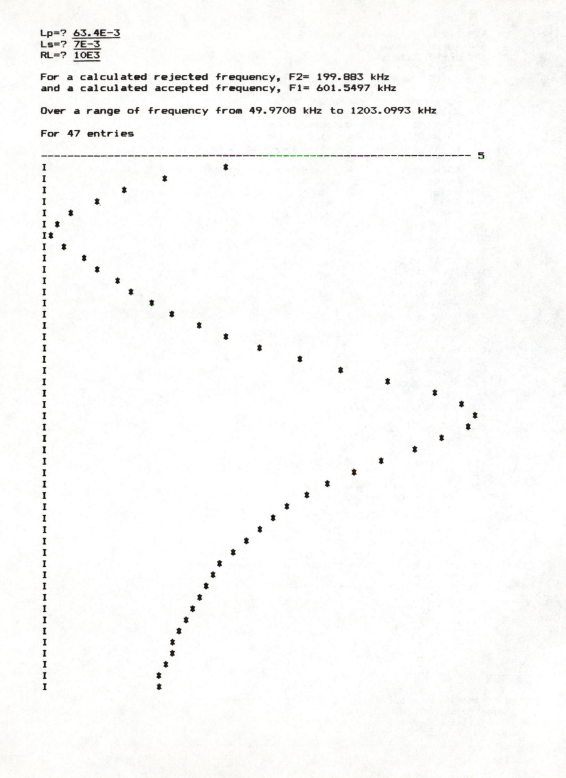

For the doubly-tuned circuit of Figure 20-4
enter the following data:
Capacitor, C=? <u>500E-12</u>
E=? <u>20</u>
Lp=? <u>1.12E-3</u>
Ls=? <u>141E-6</u>
RL=? <u>2E3</u>

For a calculated rejected frequency, F2= 212.6799 kHz
and a calculated accepted frequency, F1= 599.4127 kHz

Over a range of frequency from 53.17 kHz to 1198.8254 kHz

For 44 entries

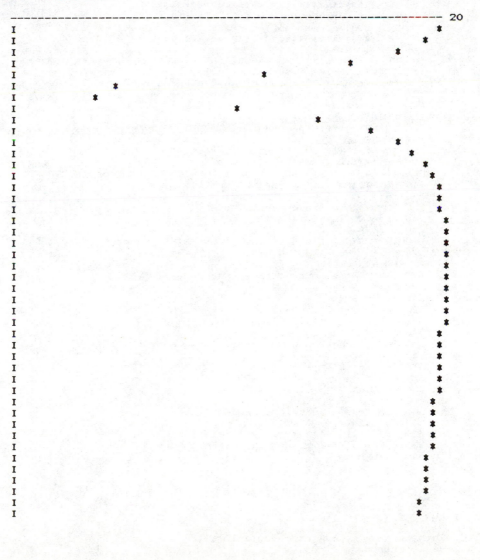

while, for the parallel resonant circuit, R_ℓ is assumed to be zero, resulting in a very high Q.

20.6 TESTING ROUTINE

The last program (20-8) of the chapter will test your ability to analyze the series resonant cir-cuit of Figure 20.1. The values of R, L, and C are randomly generated and the resonant fre-quency, quality factor, and cut-off frequencies requested. Note, in particular, lines 520 and 530, which use multiplying factors for L and C to permit applying the units of mH and μF.

```
10 REM *****  PROGRAM 20-8  *****
20 REM ***********************************
30 REM Program to test calculations for a
40 REM series resonant circuit.
50 REM ***********************************
60 REM
100 PRINT "This program tests the calculations for a"
110 PRINT "series resonant circuit."
120 PRINT:PRINT "For the circuit of Figure 20-1:"
130 GOSUB 500 :REM Randomly select component values
140 PRINT:PRINT "R=";R;"ohms"
150 PRINT "L=";1000*L;"mH"
160 PRINT "C=";1E6*C;"uF"
170 PRINT
180 INPUT "What is the value of fs";FA
190 FS=1/(6.28318*SQR(L*C))
200 IF ABS(FS-FA)<ABS(0.01*FS) THEN PRINT "Correct":GOTO 220
210 PRINT "No, it's";FS/1000;"kHz"
220 PRINT
230 INPUT "What is circuit Q";QA
240 XL=6.28318*FS*L
250 Q=XL/R
260 IF ABS(Q-QA)<ABS(0.01*Q) THEN PRINT "Correct":GOTO 280
270 PRINT "No, it's";Q
280 PRINT
290 INPUT "The lower cutoff frequency f1=";FA
300 BW=FS/Q
310 F1=FS-BW/2
320 IF ABS(F1-FA)<ABS(0.01*F1) THEN PRINT "Correct":GOTO 340
330 PRINT "No, it's";F1/1000;"kHz"
340 PRINT
350 INPUT "and the upper cutoff frequency f2=";FA
360 F2=FS+BW/2
370 IF ABS(F2-FA)<ABS(0.01*F2) THEN PRINT "Correct":GOTO 390
380 PRINT "No, it's";F2/1000;"kHz"
390 PRINT
400 END
500 REM Module to select component values
510 R=10*INT(1+10*RND(XV))  :REM 10<R<100
520 L=1E-3*INT(1+25*RND(XV))  :REM 1mH<L<25mH
530 C=1E-6*INT(1+25*RND(XV))  :REM 1uF<C<25uF
540 RETURN


READY


RUN

This program tests the calculations for a
series resonant circuit.
```

```
For the circuit of Figure 20-1:

R= 20 ohms
L= 17 mH
C= 15 uF

What is the value of fs? 315
Correct

What is circuit Q? 1.5
No, it's 1.6833

The lower cutoff frequency f1=? 222
Correct

and the upper cutoff frequency f2=? 409
Correct


READY

RUN

This program tests the calculations for a
series resonant circuit.

For the circuit of Figure 20-1:

R= 20 ohms
L= 18 mH
C= 2 uF

What is the value of fs? 839
Correct

What is circuit Q? 4.7
Correct

The lower cutoff frequency f1=? 750
Correct

and the upper cutoff frequency f2=? 927
Correct


READY
```

EXERCISES

Write a program to perform each of the following tasks.

1. Given BW, f_s, and R, determine Q_s, X_L, X_C, L, and C for a series resonant circuit. For the first run, use $BW = 200Hz$, $f_s = 2000Hz$, and $R = 2$ ohms.

2. Given f_1, f_2, Q_s, and R, determine BW, f_s, X_L, X_C, L, and C for a series resonant circuit. For the first run, use $f_1 = 5400Hz$, $f_2 = 6000Hz$, $Q_s = 9.5$, and $R = 2$ ohms.

3. Given $\mathbf{E}$, I_{peak}, BW, and f_s, determine L, C, and the cut-off frequencies. For the first run, use $\mathbf{E} = 5\underline{/0°}$, $I_{peak} = 500mA$, $BW = 120Hz$, and $f_s = 8400Hz$.

4. Plot the magnitude of $\mathbf{Z}_T$ versus frequency for a series resonant circuit. For the first run, use the same input as in the provided run of Program 20-5. Compare the results.

5. Plot the magnitude of $\mathbf{V}_L$ versus frequency for a series resonant circuit. For the first run, use the same input as in the provided run of Program 20-5. Compare the locations of the peak values of the curves and their general shapes.

6. Given f_p, BW, L, and $Q\left(= \dfrac{X_L}{R_\ell}\right)$, determine R_s and C for the parallel resonant network of Figure 20.2. For the first run, use $f_p = 100\text{kHz}$, $BW = 2500\text{Hz}$, $L = 2\text{mH}$, and $Q = 80$.

7. Given $R_s = \infty\Omega$, BW, Q, $V_{C\text{max}}$, and an applied current source $\mathbf{I}$ for the parallel resonant network of Figure 20.2, determine L, R_ℓ, and C. For the first run, use $BW = 500\text{Hz}$, $Q = 30$, $V_{C\text{max}} = 1.8\text{V}$, and $\mathbf{I} = 20 \times 10^{-3}\underline{/0°}$.

8. Plot I_L versus frequency for the parallel resonant network of Figure 20.2 due to an applied current source $\mathbf{I}$. Assume $R_s \cong \infty\Omega$. Label both axes.

9. Given the accepted and rejected frequencies and the capacitor C for the network of Figure 20.7, determine L_s and L_p. For the first run, use $C = 500\text{E} - 12$, with a rejected frequency of 200kHz and an accepted frequency of 600kHz. Compare L_s and L_p with the values used in the provided runs of Program 20-7.

10. Develop a test routine for the parallel resonant network of Figure 20.2 that is similar in content to Program 20-8.

Three-Phase Networks

21

21.1   INTRODUCTION

For balanced loads (equal phase impendances), the Y-Y and Δ-Δ configurations and a combination Δ-Y load will be examined in detail. For the unbalanced situation, the general solution for the phase currents of a Y-connected load will be developed.

21.2 Y-Y CONFIGURATION

The first program (21-1) will perform a complete analysis of the Y-Y connected, three-phase system of Figure 21.1. The phase voltage E_ϕ, and phase impedance components R_ϕ and X_ϕ are provided, and the program will calculate Z_ϕ, I_ϕ, V_ϕ, P_T, Q_T, S_T, E_L, I_L, and the power factor. The program is titled, and the input data provided on line 100 through 160. The line voltage is determined by $E_L = \sqrt{3}E_\phi$

(line 190), and the load phase voltage by the fact that $V_\phi = E_\phi$ (line 210). Line 220 will then determine the impedance of each phase through $Z_\phi = \sqrt{R_\phi^2 + X_\phi^2}$, followed by the phase and line currents, using $I_\phi = V_\phi/Z_\phi$ on line 230.

The power dissipated in each phase is determined by line 270, using $P_\phi = I_\phi^2 R_\phi$, and the total power on line 280, using $P_T = 3P_\phi$. The reactive power for each phase is determined on line 330, using $Q_\phi = I_\phi^2 X_\phi$, with $Q_T = 3Q_\phi$ on line 340. The net apparent power is then $S_T = \sqrt{P_T^2 + Q_T^2}$ (line 370), with the power factor determined by $F_p = P_T/S_T$ (line 380).

Note that the performed calculations do not require knowing whether X_ϕ is capacitive or inductive. A switch in reactive elements will simply change the sign of each resulting phase angle and require that the power factor be changed from leading to lagging (or vice versa).

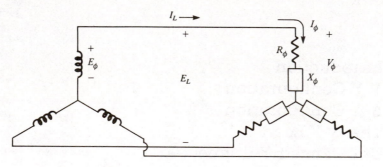

FIGURE 21.1

```
10 REM *****  PROGRAM 21-1  *****
20 REM *********************************************
30 REM Program to analyze the Y-Y connected
40 REM 3-phase network.
50 REM *********************************************
60 REM
100 PRINT "This program analyzes the"
110 PRINT "3-phase, Y-Y connected network."
120 PRINT
130 PRINT "Enter the following network information:"
140 INPUT "Phase voltage, E=";EP
150 INPUT "and for a Y-connected series load, R(phase)=";RP
160 INPUT "with X(phase)=";XP
170 REM Now do calculations and print results
180 PRINT
```

```
E_L    ┌ 190 EL=SQR(3)*EP
V_∅    │ 200 PRINT "The line voltage E(line)=";EL;"volts"
       └ 210 PRINT "and the load phase voltage=";EP;"volts"
       ┌ 220 ZP=SQR(RP^2+XP^2)
Z_∅    │ 230 IP=EP/ZP
I_∅,I_L│ 240 PRINT
       │ 250 PRINT "The individual phase impedance is";ZP;"ohms"
       └ 260 PRINT "with phase and line currents of";IP;"amps"
       ┌ 270 PP=IP^2*RP
P_∅    │ 280 PT=3*PP
P_T    │ 290 PRINT
       │ 300 PRINT "The power for one phase is";PP;"watts"
       └ 310 PRINT "for a total of";PT;"watts"
       ┌ 320 PRINT
       │ 330 QP=IP^2*XP
Q_∅    │ 340 QT=3*QP
Q_T    │ 350 PRINT "The reactive power is";QP;"vars per phase"
       └ 360 PRINT "for a total of";QT;"vars"
       ┌ 370 ST=SQR(PT^2+QT^2)
       │ 380 FP=PT/ST
S_T    │ 390 PRINT
F_p    │ 400 PRINT "For the network, ST=";ST;"VA"
       └ 410 PRINT "and Fp=";FP
         420 END
```

READY

RUN
*
This program analyzes the
3-phase, Y-Y connected network.

Enter the following network information:
Phase voltage, E=? <u>120</u>
and for a Y-connected series load, R(phase)=? <u>3</u>
with X(phase)=? <u>4</u>

The line voltage E(line)= 207.8461 volts
and the load phase voltage= 120 volts

The individual phase impedance is 5 ohms
with phase and line currents of 24 amps

The power for one phase is 1728 watts
for a total of 5184 watts

The reactive power is 2304 vars per phase
for a total of 6912 vars

For the network, ST= 8640 VA
and Fp= .6

READY

RUN

This program analyzes the
3-phase, Y-Y connected network.

*Example 21.1, ICA

```
Enter the following network information:
Phase voltage, E=? 110
and for a Y-connected series load, R(phase)=? 100
with X(phase)=? 40

The line voltage E(line)= 190.5256 volts
and the load phase voltage= 110 volts

The individual phase impedance is 107.7033 ohms
with phase and line currents of 1.0213 amps

The power for one phase is 104.3103 watts
for a total of 312.931 watts

The reactive power is 41.7241 vars per phase
for a total of 125.1724 vars

For the network, ST= 337.037 VA
and Fp= .9285

READY
```

21.3 Δ - Δ CONFIGURATION

Program 21-2 will provide the magnitude of all the quantities of interest for the network of Figure 21.2. The program is very similar to Program 21-1, but uses the following pertinent equations:

$$V_\phi = E_L = E_\phi, \; I_L = \sqrt{3}I_\phi, \; Z_\phi = R_\phi \| X_\phi$$
$$\text{and } P_\phi = V_\phi{}^2/R_\phi$$

Otherwise, the sequence of steps parallels that

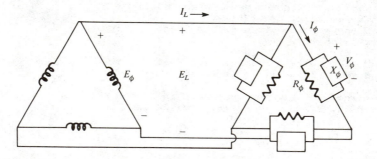

FIGURE 21.2

```
10 REM ***** PROGRAM 21-2 *****
20 REM *************************************************
30 REM Program to analyze the delta-delta
40 REM connected 3-phase network.
50 REM *************************************************
60 REM
100 PRINT "This program analyzes the"
110 PRINT "3-phase, delta-delta connected network."
120 PRINT
130 PRINT "Enter the following circuit information:"
140 INPUT "Phase voltage, E=";EP
150 INPUT "and for a Delta-connected parallel load, R(phase)=";RP
160 INPUT "with X(phase)=";XP
```

Input

```
        170 REM Now do calculations and print results
        180 PRINT
       ┌190 EL=EP
E_L    │200 VP=EL
V_∅    │210 PRINT "The line voltage, E(line)=";EL;"volts"
       └220 PRINT "and load phase voltage=";EP;"volts"
       ┌230 ZP=RP*XP/SQR(RP^2+XP^2)
       │240 IP=EP/ZP
Z_∅    │250 PRINT
I_∅,I_L│260 PRINT "The individual phase impedance is";ZP;"ohms"
       │270 IL=SQR(3)*IP
       │280 PRINT "with a phase current=";IP;"amps"
       └290 PRINT "and line current =";IL;"amps"
       ┌300 PP=VP^2/RP
P_∅    │310 PT=3*PP
P_T    │320 PRINT
       │330 PRINT "The power for one phase is";PP;"watts"
       └340 PRINT "for a total of";PT;"watts"
        350 PRINT
       ┌360 QP=VP^2/XP
Q_∅    │370 QT=3*QP
Q_T    │380 PRINT "The reactive power is";QP;"vars per phase"
       └390 PRINT "for a total of";QT;"vars"
       ┌400 ST=SQR(PT^2+QT^2)
S_T    │410 FP=PT/ST
F_p    │420 PRINT
       │430 PRINT "For the network, ST=";ST;"VA"
       └440 PRINT "and Fp=";FP
        450 END
```

READY

<u>**RUN**</u>

```
*
This program analyzes the
3-phase, delta-delta connected network.

Enter the following circuit information:
Phase voltage, E=? 440
and for a Delta-connected parallel load, R(phase)=? 6
with X(phase)=? 6

The line voltage, E(line)= 440 volts
and load phase voltage= 440 volts

The individual phase impedance is 4.2426 ohms
with a phase current= 103.709 amps
and line current = 179.6292 amps

The power for one phase is 3.2267E+04 watts
for a total of 9.68E+04 watts

The reactive power is 3.2267E+04 vars per phase
for a total of 9.68E+04 vars

For the network, ST= 1.369E+05 VA
and Fp= .7071
```

READY

*Problem 21.24, ICA

of Program 21-1. Be sure to analyze the system of Figure 21.1 or 21.2 by longhand before using the computer to fully appreciate the time and energy saved (in addition to the high level of consistency and accuracy).

21.4   THE Δ-Y LOAD

A complete analysis of the Δ-Y load of Figure 21.3 can be a drawn out affair, which will often produce incorrect results due to the loss of sign or an improperly placed decimal point. Given E_L and the impendance components, Program 21-3 will determine the phase currents, power to each phase, total real, reactive, and apparent powers, and the power factor. The sequential list of calculations required is provided below for review purposes. Note, on lines 340 and 350, that the sign must be provided with the input because two reactive loads are present and there is no requirement that they be the same.

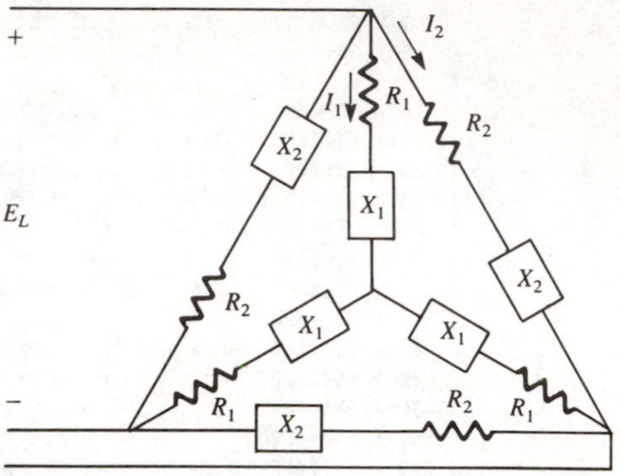

$$Y: I_\phi = I_1 = V_\phi/Z_\phi = \frac{E_L/\sqrt{3}}{\sqrt{R_1^2 + X_1^2}}$$

$$P_{T1} = 3I_\phi^2 R_\phi = 3I_1^2 R_1,$$

FIGURE 21.3

$$Q_{T1} = 3I_\phi^2 X_\phi = 3I_1^2 X_1$$

$$S_T = 3V_\phi I_\phi = 3(E_L/\sqrt{3})(I_1)$$

$$\Delta: I_\phi = I_2 = V_\phi/Z_\phi = \frac{E_2}{\sqrt{R_2^2 + X_2^2}}$$

$$P_{T2} = 3I_\phi^2 R_\phi = 3I_2^2 R_2,$$

$$Q_{T2} = 3I_\phi^2 X_\phi = 3I_2^2 X_2$$

$$S_T = 3V_\phi I_\phi = 3E_L I_L$$

```
10 REM *****  PROGRAM 21-3  *****
20 REM ********************************
30 REM Program to analyze a 3-phase
40 REM delta-wye load.
50 REM ********************************
60 REM
100 PRINT "This program analyzes the 3-phase"
110 PRINT "delta-wye load."
120 PRINT
130 PRINT "Enter the following network information:"
140 PRINT
150 INPUT "Line voltage, E=";EL
160 PRINT "For the series-connected wye load:"
170 INPUT "R=";R1
180 INPUT "X=";X1 :REM Enter negative sign if capacitive
190 PRINT
200 PRINT "And for the series-connected delta load:"
210 INPUT "R=";R2
220 INPUT "X=";X2 :REM Enter negative sign if capacitive
230 REM Now do calculations and print results
240 PRINT:PRINT
```

```
      ┌ 250  Z1=SQR(R1^2+X1^2)
      │ 260  I1=EL/(SQR(3)*Z1)
      │ 270  P1=3*I1^2*R1
      │ 280  Q1=3*I1^2*X1
      │ 290  S1=3*EL*I1/SQR(3)
   Y  │ 300  PRINT "For the wye connection:"
      │ 310  PRINT "The phase current I=";I1;"amps"
      │ 320  PRINT "The total power dissipated is";P1;"watts"
      │ 330  PRINT "With a net reactive power of Q=";ABS(Q1);"vars";
      │ 340  IF SGN(X1)=-1 THEN PRINT "(cap.)"
      │ 350  IF SGN(X1)=1 THEN PRINT "(ind.)"
      └ 360  PRINT "and apparent power of:";S1;"VA"
        370  PRINT:PRINT
      ┌ 380  Z2=SQR(R2^2+X2^2)
      │ 390  I2=EL/Z2
      │ 400  P2=3*I2^2*R2
      │ 410  Q2=3*I2^2*X2
      │ 420  S2=3*EL*I2
   Δ  │ 430  PRINT "For the delta connected load:"
      │ 440  PRINT "The phase current is, I=";I2;"amps"
      │ 450  PRINT "The total power dissipated is";P2;"watts"
      │ 460  PRINT "with a net reactive power of Q=";ABS(Q2);"vars";
      │ 470  IF SGN(X2)=-1 THEN PRINT "(cap.)"
      │ 480  IF SGN(X2)=1 THEN PRINT "(ind.)"
      └ 490  PRINT "and apparent power of:";S2;"VA"
        500  PRINT:PRINT
      ┌ 510  PT=P1+P2
      │ 520  QT=Q1+Q2
 PT   │ 530  PRINT "For the combined system:"
      │ 540  PRINT "The total power dissipated is";PT;"watts"
 QT   │ 550  PRINT "and the net reactive power QT=";ABS(QT);"vars";
      │ 560  IF SGN(QT)=-1 THEN PRINT "(cap.)"
      └ 570  IF SGN(QT)=1 THEN PRINT "(ind.)"
      ┌ 580  ST=SQR(PT^2+QT^2)
      │ 590  FP=PT/ST
 ST   │ 600  PRINT:PRINT "The total apparent power ST=";ST;"VA"
 Fp   │ 610  PRINT "with a network power factor Fp=";FP;
      │ 620  IF SGN(QT)=-1 THEN PRINT "(leading)"
      └ 630  IF SGN(QT)=1 THEN PRINT "(lagging)"
        640  END


READY


RUN
*
This program analyzes the 3-phase
delta-wye load.

Enter the following network information:

Line voltage, E=? 200
For the series-connected wye load:
R=? 4
X=? 3

And for the series-connected delta load:
R=? 6
X=? -8
```

*Example 21.6, ICA

```
For the wye connection:
The phase current I= 23.094 amps
The total power dissipated is 6400 watts
With a net reactive power of Q= 4800 vars(ind.)
and apparent power of: 8000 VA


For the delta connected load:
The phase current is, I= 20 amps
The total power dissipated is 7200 watts
with a net reactive power of Q= 9600 vars(cap.)
and apparent power of: 1.2E+04 VA


For the combined system:
The total power dissipated is 1.36E+04 watts
and the net reactive power QT= 4800 vars(cap.)

The total apparent power ST= 1.4422E+04 VA
with a network power factor Fp= .943 (leading)


READY
```

Total: $P_T = P_{T1} + P_{T2}$

$Q_T = \pm Q_{T1} \pm Q_{T2}$

$S_T = \sqrt{P_T^2 + Q_T^2}$

$F_p = \dfrac{P_T}{S_T}$

Note in the printout that lines 620 and 630 pick up on the sign and identify whether the network has a leading or lagging power factor.

21.5 UNBALANCED *Y*-CONNECTED LOAD

The analysis of an unbalanced *Y*-connected network, such as in Figure 21.4, requires that the input and load phase angles be included throughout the analysis. The result is the re-introduction of the routines provided earlier that perform the $+$, $-$, $\div$, and $\times$ operations with vectors in either the polar or rectangular

$$I_1 = I_{an} = \frac{E_{BA}Z_3 - E_{AC}Z_2}{\triangle_z}$$

$$I_2 = I_{bn} = \frac{E_{CB}Z_1 - E_{BA}Z_3}{\triangle_z}$$

$$I_3 = I_{cn} = \frac{E_{AC}Z_2 - E_{CB}Z_1}{\triangle_z}$$

$$\triangle_z = Z_1Z_2 + Z_1Z_3 + Z_2Z_3$$

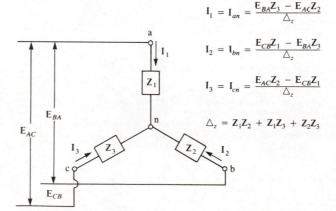

FIGURE 21.4

form. The applicable routines from 4000 through 5990 are not included here to save space.

As indicated by the printout, lines 160 through 270 of Program 21-4 simply request the applied line voltage and the impedances Z_1, Z_2, and Z_3. The REM statements on lines 300, 320, and 340 reveal that the denominator (which is the same for each phase current) will be deter-

mind first. Lines 370 through 420 determine the numerator of I_1: lines 430 through 480 the numerator of I_2, and lines 490 through 540 the numerator of I_3. Lines 550 through 610 determine the individual currents and the remainder of the program prints out the results. Note in the printout the resulting four-place accuracy for the long series of calculations.

```
10 REM *****  PROGRAM 21-4  *****
20 REM ************************************************
30 REM Program to calculate the phase currents of
40 REM an unbalanced 3-phase, 3-wire Y-connected load.
50 REM ************************************************
60 REM
100 DIM V(4),VA(4),VB(4),VC(4)
110 DIM Z1(4),Z2(4),Z3(4),N1(4),N2(4),N3(4),D(4)
120 DIM E1(4),E2(4),E3(4),I1(4),I2(4),I3(4)
130 PRINT "This program calculates the phase currents"
140 PRINT "of an unbalanced 3-phase, 3-wire Y-connected load."
150 PRINT
160 PRINT "Line voltage, Eba=" :GOSUB 5500
170 FOR I=1 TO 4 :E1(I)=V(I) :NEXT I
180 PRINT "Line voltage, Eac=" :GOSUB 5500
190 FOR I=1 TO 4 :E2(I)=V(I) :NEXT I
200 PRINT "Line voltage, Ecb=" :GOSUB 5500
210 FOR I=1 TO 4 :E3(I)=V(I) :NEXT I
220 PRINT "Load impedance, Z1=" :GOSUB 5500
230 FOR I=1 TO 4 :Z1(I)=V(I) :NEXT I
240 PRINT "Load impedance, Z2=" :GOSUB 5500
250 FOR I=1 TO 4 :Z2(I)=V(I) :NEXT I
260 PRINT "Load impedance, Z3=" :GOSUB 5500
270 FOR I=1 TO 4 :Z3(I)=V(I) :NEXT I
280 PRINT:PRINT
290 FOR I=1 TO 4 :VA(I)=Z1(I) :VB(I)=Z2(I) :NEXT I
300 GOSUB 4600 :REM Z1*Z2
310 FOR I=1 TO 4 :D(I)=VC(I) :VB(I)=Z3(I) :NEXT I
320 GOSUB 4600 :REM Z1*Z3
330 FOR I=1 TO 4 :D(I)=D(I)+VC(I) :VA(I)=Z2(I) :NEXT I
340 GOSUB 4600 :REM Z2*Z3
350 FOR I=1 TO 4 :D(I)=D(I)+VC(I) :VA(I)=E1(I) :V(I)=D(I) :NEXT I
360 GOSUB 4000 :REM Convert to the polar form
370 D(3)=V(3) :D(4)=V(4)
380 GOSUB 4600 :REM E1*Z3
390 FOR I=1 TO 4 :N1(I)=VC(I) :VA(I)=E2(I) :VB(I)=Z2(I) :NEXT I
400 GOSUB 4600 :REM E2*Z2
410 FOR I=1 TO 4 :VA(I)=N1(I) :VB(I)=VC(I) :NEXT I
420 GOSUB 4500 :REM N1=E1*Z3-E2*Z2
430 FOR I=1 TO 4 :N1(I)=VC(I) :VA(I)=E3(I) :VB(I)=Z1(I) :NEXT I
440 GOSUB 4600 :REM E3*Z1
450 FOR I=1 TO 4 :N2(I)=VC(I) :VA(I)=E1(I) :VB(I)=Z3(I) :NEXT I
460 GOSUB 4600 :REM E1*Z3
470 FOR I=1 TO 4 :VA(I)=N2(I) :VB(I)=VC(I) :NEXT I
480 GOSUB 4500 :REM N2=E3*Z1-E1*Z3
490 FOR I=1 TO 4 :N2(I)=VC(I) :VA(I)=E2(I) :VB(I)=Z2(I) :NEXT I
500 GOSUB 4600 :REM E2*Z2
510 FOR I=1 TO 4 :N3(I)=VC(I) :VA(I)=E3(I) :VB(I)=Z1(I) :NEXT I
```

(Continued)

```
520 GOSUB 4600 :REM E3*Z1
530 FOR I=1 TO 4 :VA(I)=N3(I) :VB(I)=VC(I) :NEXT I
540 GOSUB 4500 :REM N3=E2*Z2-E3*Z1
550 FOR I=1 TO 4 :N3(I)=VC(I) :VA(I)=N1(I) :VB(I)=D(I) :NEXT I
560 GOSUB 4700 :REM N1/D
570 FOR I=1 TO 4 :I1(I)=VC(I) :VA(I)=N2(I) :NEXT I
580 GOSUB 4700 :REM N2/D
590 FOR I=1 TO 4 :I2(I)=VC(I) :VA(I)=N3(I) :NEXT I
600 GOSUB 4700 :REM N3/D
610 FOR I=1 TO 4 :I3(I)=VC(I) :NEXT I
620 REM Calculations now complete
630 PRINT "The phase currents are "
640 PRINT "Ian=";
650 FOR I=1 TO 4 :V(I)=I1(I) :NEXT I
660 GOSUB 5960 :REM Print answer in rectangular and polar form
670 PRINT:PRINT "Ibn=";
680 FOR I=1 TO 4 :V(I)=I2(I) :NEXT I
690 GOSUB 5960
700 PRINT:PRINT "Icn=";
710 FOR I=1 TO 4 :V(I)=I3(I) :NEXT I
720 GOSUB 5960
730 END
```

```
    ┌─ 4000-5990
    │  as in earlier programs
    │
    │
    │
    ▼
```

<u>RUN</u>

*

This program calculates the phase currents
of an unbalanced 3-phase, 3-wire Y-connected load.

Line voltage, Eba=
(1) Enter rectangular form, or (2) Enter polar form:? <u>2</u>
Magnitude=? <u>200</u>
at angle? <u>0</u>

Line voltage, Eac=
(1) Enter rectangular form, or (2) Enter polar form:? <u>2</u>
Magnitude=? <u>200</u>
at angle? <u>120</u>

Line voltage, Ecb=
(1) Enter rectangular form, or (2) Enter polar form:? <u>2</u>
Magnitude=? <u>200</u>
at angle? <u>-120</u>

Load impedance, Z1=
(1) Enter rectangular form, or (2) Enter polar form:? <u>2</u>
Magnitude=? <u>166</u>
at angle? <u>-90</u>

Load impedance, Z2=
(1) Enter rectangular form, or (2) Enter polar form:? <u>2</u>
Magnitude=? <u>200</u>
at angle? <u>0</u>

*Example 21.7, ICA

```
Load impedance, Z3=
(1) Enter rectangular form, or (2) Enter polar form:? 2
Magnitude=? 200
at angle? 0



The phase currents are
Ian= .8938 at an angle of 28.9348 degrees
which is .7822 +j .4324

Ibn= .9124 at an angle of 225.3608 degrees
which is-.6411 -j .6492

Icn= .2587 at an angle of 123.0558 degrees
which is-.1411 +j .2168


READY
```

21.6 TESTING ROUTINE

The last program (21 5) of the chapter will test your ability to perform the standard calculations for *Y-Y*-connected or Δ-Δ-connected balanced loads of Figures 21.1 and 21.2. Lines 170 through 190 will determine which system will be analyzed, while the subroutine at 700 and 800 will perform all the calculations required for the *Y-Y* and Δ-Δ systems respectively. The three required quantities of E_ϕ, R_ϕ, and X_ϕ are determined by lines 600 through 670, before each solution is tested using the standard routine introduced in earlier chapters.

```
10 REM ***** PROGRAM 21-5 *****
20 REM ******************************************
30 REM Program to test the calculations
40 REM of a Y-Y or Delta-Delta connected system.
50 REM ******************************************
60 REM
100 PRINT "This program tests the calculations of"
110 PRINT "a Y-Y or delta-delta connection with series-connected loads."
120 PRINT
130 GOSUB 600 :REM Randomly select system values
140 PRINT "For E(phase)=";EP;"volts"
150 PRINT "      R(phase)=";RP;"ohms"
160 PRINT "and,X(phase)=";XP;"ohms"
170 C=INT(1+2*RND(XV))
180 IF C=1 THEN PRINT "in a Y-Y connected system"
190 IF C=2 THEN PRINT "in a delta-delta connected system"
200 PRINT:PRINT "Determine the following values:"
210 PRINT
220 ON C GOSUB 700,800
230 INPUT "E(line)=";EA
240 IF SGN(EA)<>SGN(EL) THEN GOTO 260
250 IF ABS(EA-EL)<ABS(0.01*EL) THEN PRINT "Correct" :GOTO 270
260 PRINT "No, it's";EL;"volts"
270 PRINT
280 INPUT "Phase impedance, Z(phase)=";ZA
290 IF SGN(ZA)<>SGN(ZP) THEN GOTO 310
300 IF ABS(ZA-ZP)<ABS(0.01*ZP) THEN PRINT "Correct" :GOTO 320
310 PRINT "No, it's";ZP;"ohms"
```

(Continued)

```
320 PRINT
330 INPUT "Impedance phase angle, Theta=";TA
340 IF SGN(TH)<>SGN(TA) THEN GOTO 360
350 IF ABS(TA-TH)<ABS(0.01*TH) THEN PRINT "Correct" :GOTO 370
360 PRINT "No. it's";TH;"degrees"
370 PRINT
380 INPUT "Line current, I(line)=";IA
390 IF SGN(IA)<>SGN(IL) THEN GOTO 410
400 IF ABS(IA-IL)<ABS(0.01*IL) THEN PRINT "Correct" :GOTO 420
410 PRINT "No, it's";IL;"amps"
420 PRINT
430 INPUT "Phase current, I(phase)=";IA
440 IF SGN(IA)<>SGN(IP) THEN GOTO 460
450 IF ABS(IA-IP)<ABS(0.01*IP) THEN PRINT "Correct" :GOTO 470
460 PRINT "No, it's";IP;"amps"
470 PRINT
480 INPUT "and power per phase =";PA
490 IF SGN(PA)<>SGN(PP) THEN GOTO 510
500 IF ABS(PA-PP)<ABS(0.01*PP) THEN PRINT "Correct" :GOTO 520
510 PRINT "No, it's";PP;"watts"
520 PRINT
530 END
600 REM Randomly select component values
610 EP=10*INT(10+15*RND(XV)) :REM 100<EP<250
620 RP=INT(1+25*RND(XV)) :REM 1<RP<25
630 SG=+1
640 S=INT(1+4*RND(XV)) :REM Capacitive load 1 out of 4 times
650 IF S=3 THEN SG=-1
660 XP=SG*INT(1+25*RND(XV))
670 RETURN
700 REM Do calculations for Wye-Wye system
710 EL=SQR(3)*EP
720 VP=EL
730 ZP=SQR(RP^2+XP^2)
740 TH=(180/3.14159)*ATN(XP/RP)
750 IP=VP/ZP
760 IL=IP
770 PP=IP^2*RP
780 RETURN
800 REM Do calculations for Delta-Delta system
810 EL=EP
820 VP=EL
830 ZP=ABS(RP*XP/(SQR(RP^2+XP^2)))
840 TH=(180/3.14159)*ATN(XP/RP)
850 IP=VP/ZP
860 IL=SQR(3)*IP
870 PP=VP^2/RP
880 RETURN
```

READY

<u>RUN</u>

```
This program tests the calculations of
a Y-Y or delta-delta connection with series-connected loads.

For E(phase)= 240 volts
    R(phase)= 6 ohms
and,X(phase)=-1 ohms
in a Y-Y connected system
```

Determine the following values:

E(line)=? <u>10</u>
No, it's 415.6922 volts

Phase impedance, Z(phase)=? <u>5</u>
No, it's 6.0828 ohms

Impedance phase angle, Theta=? <u>3</u>
No. it's-9.4623 degrees

Line current, I(line)=? <u>3</u>
No, it's 68.3394 amps

Phase current, I(phase)=? <u>68</u>
Correct

and power per phase =? <u>28E3</u>
Correct

READY

<u>RUN</u>

This program tests the calculations of
a Y-Y or delta-delta connection with series-connected loads.

For E(phase)= 190 volts
    R(phase)= 18 ohms
and,X(phase)=-8 ohms
in a delta-delta connected system

Determine the following values:

E(line)=? <u>190</u>
Correct

Phase impedance, Z(phase)=? <u>7.3</u>
Correct

Impedance phase angle, Theta=? <u>24</u>
No. it's-23.9625 degrees

Line current, I(line)=? <u>45</u>
Correct

Phase current, I(phase)=? <u>26</u>
Correct

and power per phase =? <u>2005</u>
Correct

READY

EXERCISES

Write a program to perform each of the following tasks.

1. Given a Y-Δ system with a balanced load, determine the load phase voltage, individual phase impedance and current, real and reactive power to each phase, the total apparent power, and the power factor of the load. For the first run, use E_L = 208V and $R - jX = 100 - j100$ as the phase impedance

2. Repeat problem 21-1 for a balanced $\Delta - Y$ system. For the first run, use E_L = 120V, and $R - jX = 15 - j20$ as the impedance of each phase.

3. Modify Program 21-1 to analyze a balanced $Y - Y$ system with a parallel load in each phase. For the first run, use $\mathbf{Z}_\phi = \mathbf{R} \parallel \mathbf{X}_C = 6\underline{/0°} \parallel 8\underline{/-90°}$ and E_L = 120V.

4. Modify Program 21-2 to analyze a balanced $\Delta - \Delta$ system with a parallel load in each phase. For the first run, use $\mathbf{Z}_\phi = \mathbf{R} \parallel \mathbf{X}_C = 20\underline{/0°} \parallel 20\underline{/-90°}$ and E_L = 100V.

5. Develop a general solution for the unbalanced 4-wire, Y-connected load for a series impedance in each phase.

6. Develop a general solution for the unbalanced 3-wire Δ-connected load with a series impedance in each phase.

7. Develop a test routine for the Y-Δ or Δ-Y systems with balanced loads. Randomly generate which system will be analyzed and then randomly generate the line voltage and the series impedance elements of each phase.

Non-Sinusoidal Signals

22

22.1 INTRODUCTION

The analysis of networks with non-sinusoidal inputs will provide a fine opportunity to demonstrate the versatility and value of computer analysis techniques. Applying a non-sinusoidal signal to the simplest of networks can result in a long series of calculations that is lengthy, if not impossible, to perform by hand.

The effective value of non-sinusoidal voltages and currents will be determined, and the sum of two non-sinusoidal signals calculated, with a plot of the solution. The chapter will also contain a complete solution of an R-C network with an applied non-sinusoidal signal and a routine for determining the power delivered to a load by a non-sinusoidal input.

22.2 EFFECTIVE VALUE

The first three programs will calculate the effective value of non-sinusoidal signals and the power delivered to a resistive load. In Program 22-1, a nonsinusoidal current of the form:

$$i(t) = I_0 + I_1 \sin \omega t + I_2 \sin 2\omega t + I_3 \sin 3\omega t$$

is applied to a resistor R, and the effective value of the current and power are determined using the following equations:

$$I_{\text{eff}} = \sqrt{I_o{}^2 + \frac{(I_1{}^2 + I_2{}^2 + I_3{}^2)}{2}}$$

$$P = (I_{\text{eff}})^2 \cdot R$$

As indicated by the printout, the magnitudes of the peak values of the individual com-

```
10 REM *****  PROGRAM 22-1  *****
20 REM ******************************************************
30 REM Program calculates the effective value
40 REM of a non-sinusoidal current and the power
50 REM to a resistive load.
60 REM ******************************************************
70 REM
100 PRINT "For a non-sinusoidal current, with the following format:"
110 PRINT "      i(t)=I0+I1sin(wt)+I2sin(2wt)+I3sin(3wt)"
120 PRINT "enter the following signal information:"
130 INPUT "I0=";I0
140 INPUT "I1=";I1
150 INPUT "I2=";I2
160 INPUT "I3=";I3
170 INPUT "Circuit resistance, R=";R
180 REM Calculate the effective current, IE
190 IE=SQR(I0^2+(I1^2+I2^2+I3^2)/2)
200 REM Determine total power dissipated by R
210 PT=IE^2*R
220 PRINT "Effective current, Ieff=";IE;"amps"
230 PRINT "with a total power distribution of P=";PT;"watts"
240 END
```

Input — lines 130–170
I_{eff} — lines 180–190
P — lines 200–210
Output — lines 220–230

```
READY

RUN

For a non-sinusoidal current, with the following format:
     i(t)=I0+I1sin(wt)+I2sin(2wt)+I3sin(3wt)
enter the following signal information:
I0=? 5
I1=? 12
I2=? 0
I3=? 6
```

```
Circuit resistance, R=? 5
Effective current, Ieff= 10.7238 amps
with a total power distribution of P= 575 watts


READY


RUN


For a non-sinusoidal current, with the following format:
    i(t)=I0+I1sin(wt)+I2sin(2wt)+I3sin(3wt)
enter the following signal information:
I0=? 12E-3
I1=? 0
I2=? 42.5E-3
I3=? 16.8E-3
Circuit resistance, R=? 2.2E3
Effective current, Ieff= .034 amps
with a total power distribution of P= 2.6141 watts


READY
```

ponents of the above equations are requested on lines 130 through 160. Once the value of R is provided on line 170, the effective value of the current is determined, along with the dissipated power. Note in the printout that the format of the current is provided to insure that the frequency of each component is apparent. Note

also that all the terms do not have to be present to apply the program.

In Program 22-2, a square-wave signal is applied to a resistive network, and the effective value and power is calculated for a particular number of terms. For an infinite number of terms, the effective value of a square wave is

```
        10 REM *****  PROGRAM 22-2  *****
        20 REM *************************************************
        30 REM Program to calculate the effective value of a
        40 REM square wave signal from its Fourier components
        50 REM and determines the power to a resistive load.
        60 REM *************************************************
        70 REM
       ┌100 PRINT "This program calculates the effective value"
Title  │110 PRINT "of a square wave signal from it's Fourier"
       └120 PRINT "components and determines the power to a resistive load."
        130 PRINT
        140 PRINT "For the square wave signal, what is the peak"
        150 PRINT "voltage, Vm";
       ┌160 INPUT VM
Input  └170 INPUT "Circuit resistance, R=";R
        180 REM Calculate Veff
        190 INPUT "How many terms of the Fourier series should be used";NT
       ┌200 FOR N=0 TO NT-1
  S    │210 S=S+(1/(2*N+1))^2
       └220 NEXT N
Veff   ┌230 VE=((4*VM)/(SQR(2)*3.14159))*SQR(S)
  P    └240 PT=VE^2/R
        250 PRINT
       ┌260 PRINT "The effective value for";NT;"terms is";VE;"volts"
Output └270 PRINT "with a total power dissipation of";PT;"watts"
        280 END


READY
```

```
RUN

This program calculates the effective value
of a square wave signal from its Fourier
components and determines the power to a resistive load.

For the square wave signal, what is the peak
voltage, Vm? 20
Circuit resistance, R=? 5
How many terms of the Fourier series should be used? 3

The effective value for 3 terms is 19.319 volts
with a total power dissipation of 74.6446 watts


READY

RUN

This program calculates the effective value
of a square wave signal from its Fourier
components and determines the power to a resistive load.

For the square wave signal, what is the peak
voltage, Vm? 20
Circuit resistance, R=? 5
How many terms of the Fourier series should be used? 25

The effective value for 25 terms is 19.9188 volts
with a total power dissipation of 79.3518 watts


READY
```

equal to the peak value of the signal ($V_{eff} = V_{peak}$), and the power will approach $P = (V_{peak})^2/R$. In fact, the number of terms in the two examples was chosen to demonstrate this fact.

The Fourier series expansion for a square-wave voltage is:

$$V = \frac{4}{\pi} Vm \left[\sin \omega t + \frac{1}{3} \sin 3\omega t + \frac{1}{5} \sin 5\omega t \right.$$

$$\left. + \cdots + \frac{1}{n} \sin n\omega t \right]$$

Applying the general equation for the effective value will result in:

$$V_{eff} =$$

$$\sqrt{0^2 + \left(\frac{4Vm}{\sqrt{2}\pi}\right)^2 + \left(\frac{4Vm}{\sqrt{2}\pi 3}\right)^2 + \left(\frac{4Vm}{\sqrt{2}\pi 5}\right)^2}$$

$$= \frac{4Vm}{\sqrt{2}\pi} \sqrt{1 + \left(\frac{1}{3}\right)^2 + \left(\frac{1}{5}\right)^2 + \left(\frac{1}{7}\right)^2 + \left(\frac{1}{9}\right)^2}$$

Each term within the square root can be determined from the following equation:

$$S = \left(\frac{1}{2N + 1}\right)^2$$

by substituting the proper value of N. For instance, if $NT = 5$ on line 190, then $NT - 1 = 4$ (line 200) and $S = 1$; for $N = 0$, $S = \left(\frac{1}{3}\right)^2$; for $N = 1$, $S = \left(\frac{1}{5}\right)^2$; for $N = 2$, $S = \left(\frac{1}{7}\right)^2$; for $N = 3$, and $S = \left(\frac{1}{9}\right)^2$; for $N = 4$. Of course $S = S + \left(\frac{1}{2N + 1}\right)^2$ on line 210 will determine the sum of the terms for use in equation 230, where

$$V_{eff} = \frac{4 \ Vm}{\sqrt{2}\pi} \sqrt{S}$$

The power delivered is then determined from $P = (V_{eff})^2/R$, and the results printed out. Note in the second run that 25 terms can be applied without expanding the program, because the equation for S appears above.

The last program (22-3) of the section will tabulate the values of V_{eff} and P for Program 22-2 for an increasing number of terms. Note that the K loop extends from line 200 through 300, while the N loop begins on line 220 and

```
10 REM ***** PROGRAM 22-3 *****
20 REM ************************************************
30 REM Program to tabulate the effective voltage
40 REM and dissipated power for a square wave input
50 REM expressed by a range of terms of a Fourier series.
60 REM ************************************************
70 REM
100 PRINT "This program tabulates the effective voltage"
110 PRINT "and power for a square-wave input"
120 PRINT "expressed by a range of terms of a Fourier series."
130 PRINT
140 PRINT "For an input square wave, what is the peak"
150 PRINT "voltage, Vm";
160 INPUT VM                                          ] Input
170 INPUT "Circuit resistance, R=";R
180 PRINT:PRINT " N    Veff    %D      Pdis"          ] Heading
200 FOR K=1 TO 10
210 S=0
220 FOR N=0 TO K-1
230 S=S+(1/(2*N+1))^2      ] S
240 NEXT N
250 VE=((4*VM)/(SQR(2)*3.14159))*SQR(S)
260 PT=VE^2/R                                  ] Calc.
270 PD=ABS(100*(VM-VE)/VE)
280 F$="## ###.### ##.###   ##.##"   ] Output format
290 PRINT USING F$,K,VE,PD,PT
300 NEXT K
310 END
```

Labels at left of listing: Input (160–170), Heading (180), K Loop / # terms (200–300), S (220–240), Calc. (250–270), Output format (280–290).

```
READY


RUN


This program tabulates the effective voltage
and power for a square-wave input
expressed by a range of terms of a Fourier series.

For an input square wave, what is the peak
voltage, Vm? 20
Circuit resistance, R=? 5

 N   Veff    %D    Pdis
 1  18.006 11.072  64.85
 2  18.980  5.372  72.05
 3  19.319  3.525  74.64
 4  19.489  2.619  75.97
 5  19.592  2.083  76.77
 6  19.660  1.729  77.30
 7  19.709  1.477  77.69
 8  19.745  1.289  77.98
 9  19.774  1.144  78.20
10  19.797  1.028  78.38


READY
```

ends on line 240. For each value of K, the sum of the series of terms within the square root sign will be determined by the N loop, and the effective voltage and power determined by lines 250 and 260. Line 270 will calculate the percent difference between the peak value (which V_{eff} will approach with an increasing number of terms) and the calculated effective value. The format for the printout is defined by line 280, and the values to be printed out in that format are provided on line 290. Note how quickly the percent difference falls within two percent. The maximum power dissipation is 80 watts, as determined by $P_{max} = V_m^2/R$.

22.3 *R-C* CIRCUIT

The analysis of the *R-C* circuit of Figure 22.1 for a non-sinusoidal voltage *e(t)* requires that the

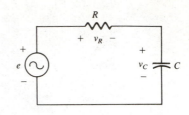

FIGURE 22.1

network be analyzed for each term of the Fourier series expansion of the applied voltage. In Program 22-4, the input signal is limited to a single dc and ac component. However, the analysis can be expanded to any number of terms by simply repeating the process defined for the sinusoidal term and using the correct frequency and magnitude. For the dc portion of *e(t)*, the network of Figure 22.1 can be redrawn, as shown in Figure 22.2. Recall that a capacitor can be replaced on an approximate basis by an

```
 10 REM *****  PROGRAM 22-4  *****
 20 REM *********************************************************
 30 REM Program to calculate the non-sinusoidal expressions
 40 REM for the currents and voltages of a series R-C circuit
 50 REM for a non-sinusoidal input.
 60 REM *********************************************************
 70 REM
100 PRINT "For a series R-C circuit with a non-sinusoidal"
110 PRINT "input voltage, enter the following data:"
120 PRINT
130 INPUT "R=";R
140 INPUT "C=";C
150 PRINT
160 PRINT "For an input voltage of form:"
170 PRINT "     e(t)=E0+E1sin(wt)"
180 PRINT
190 INPUT "E0=";E0
200 INPUT "E1=";E1
210 INPUT "w=";W
220 PRINT
230 REM Perform calculations and print results
240 XC=1/(W*C)
250 ZT=SQR(R^2+XC^2)
260 ZA=-57.296*ATN(XC/R)
270 IE=E1/(SQR(2)*ZT)
280 IA=-ZA
290 VR=IE*R
300 VC=IE*XC
310 RA=IA :REM Angle of voltage across R
320 CA=IA-90 :REM Angle of voltage across C
```

Input _(bracket spanning lines 130–210)_

Calc. _(bracket spanning lines 230–320)_

```
        ┌ 330 PRINT "Resulting signals are:"
 i(t)   │ 340 PRINT
 vᵣ(t)  │ 350 A0=0 :A1=IE :A2=IA :PRINT "i(t)="; :GOSUB 500
 v_C(t) │ 360 A0=0 :A1=VR :A2=RA :PRINT "Vr(t)="; :GOSUB 500
        │ 370 A0=E0 :A1=VC :A2=CA :PRINT "Vc(t)="; :GOSUB 500
        └ 380 PRINT
        ┌ 390 PRINT "I(effective)=";IE;"amps"
 E_ff   │ 400 PRINT "Vr(effective)=";VR;"volts"
 Values │ 410 PRINT "Vc(effective)=";SQR(E0^2+VC^2);"volts"
        └ 420 PRINT
   P    ┌ 430 P=IE^2*R
        └ 440 PRINT "and power dissipated is";P;"watts"
          450 END
        ┌ 500 REM Module to print sinusoidal expression
Sinu-   │ 510 PRINT A0;"+";A1*SQR(2);"sin(";W;"t";
soidal  │ 520 IF A2>=0 THEN PRINT "+";A2; ELSE PRINT A2;
Format  │ 530 PRINT ")"
        └ 540 RETURN
```

READY

RUN

*
For a series R-C circuit with a non-sinusoidal
input voltage, enter the following data:

R=? 3
C=? 0.125

For an input voltage of form:
      e(t)=E0+E1sin(wt)

E0=? 12
E1=? 10
w=? 2

Resulting signals are:

i(t)= 0 + 2 sin(2 t+ 53.1303)
Vr(t)= 0 + 6 sin(2 t+ 53.1303)
Vc(t)= 12 + 8 sin(2 t-36.8697)

I(effective)= 1.4142 amps
Vr(effective)= 4.2426 volts
Vc(effective)= 13.2665 volts

and power dissipated is 6 watts

READY

RUN

For a series R-C circuit with a non-sinusoidal
input voltage, enter the following data:

R=? 6E3
C=? 0.2E-7

*Example 23.4, ICA

```
For an input voltage of form:
    e(t)=E0+E1sin(wt)

E0=? 4E-3
E1=? 1E-3
w=? 1E4

Resulting signals are:

i(t)= 0 + 1.2804E-07 sin( 1E+04 t+ 39.8057 )
Vr(t)= 0 + 7.6822E-04 sin( 1E+04 t+ 39.8057 )
Vc(t)= 4E-03 + 6.4018E-04 sin( 1E+04 t-50.1943 )


I(effective)= 9.0536E-08 amps
Vr(effective)= 5.4321E-04 volts
Vc(effective)= 4.0255E-03 volts

and power dissipated is 4.918E-11 watts


READY
```

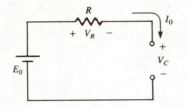

FIGURE 22.2

open-circuit once the transient phase has passed. The current I is zero and voltage $V_C = E$, and since $I = O$, $V_R = I_o R = OV$. For any value of E, the answers will remain the same, resulting in the values of AO appearing on lines 350 through 370. Line 350 is reserved for $i(t)$, while 360 and 370 define $v_R(t)$ and $v_C(t)$ respectively.

The second term of $e(t)$ requires that we analyze the network of Figure 22.3. The value of X_C must be determined for the applied fre-

quency (line 240), and the total impendance from $Z_T = \sqrt{R^2 + X_C^2}$, to permit calculating the magnitude of the effective value of the current on line 270 using $I_{eff} = \dfrac{E_1/\sqrt{2}}{Z_T}$. The phase angle (in degrees) for the total impedance is determined on line 260, which incorporates a negative sign because of the capacitive reactance. The angle for the current I and the magnitude of the effective value of the voltages across the elements is then determined on lines 280 through 300. Line 310 reveals that the angle associated with the voltage across the resistor is the same as that of the current **I**. However, the angle associated with the voltage across the capacitor must lag the current associated with the capacitor by 90° (line 320).

On lines 350 through 370, $A1$ is the effective value and $A2$ the associated angle. A_0, A_1, and A_2 are all provided, in preparation for the standard output statement of lines 510 through 530. Note that the peak value is determined by multiplying each value of $A1$ by $\sqrt{2}$, and that a plus sign was added for positive angles, since a negative sign would automatically appear if $A2 < 0$.

Since the current and the voltage across the resistor will only have a sinusoidal component, the effective value equals the effective

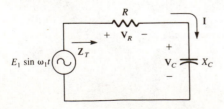

FIGURE 22.3

value determined solely by the sinusoidal component of the input, appearing on lines 390 and 400. The effective value of the voltage across the capacitor is determined by $V_{eff} = \sqrt{E_o^2 + V_C^2}$, where V_C is the effective value. The power delivered is determined on line 430 and printed on line 440. Note in the provided runs of the program that, even through the peak values and frequency may change significantly, the dc components of the current and $v_R(t)$ are still 0, and the angle associated with the current and with $\mathbf{V}_R$ is the same.

22.4 ADDITION OF NON-SINUSOIDAL SIGNALS

The addition of two non-sinusoidal functions in their Fourier series expansion requires that the dc terms and each harmonic (or frequency) be considered individually. That is, the dc terms are first added algebraically, followed by an addition of the terms of the same frequency using vector algebra.

In Program 22-5 the general format for the non-sinusoidal function appears on line 130. The dc terms for each are requested on lines 240 and 250, and the harmonics on lines 260 through 340. Note on line 190 that the number of terms N includes the dc term and the number of harmonics.

The algebraic addition of the dc terms occurs on line 370, while the sum of the sinusoidal terms is obtained by lines 380 through 440, using the indicated I loop. As an example, consider the case of finding the sum of two fundamental frequency components for the provided printout. Line 390 will define $A1 = A1(1) = 30$ and $A2 = A2(1) = 40$. Line 400 will then determine the *sum* of the real and imaginary compo-

```
10 REM *****  PROGRAM 22-5  *****
20 REM *******************************************
30 REM Program to add two non-sinusoidal voltages.
40 REM *******************************************
50 REM
100 DIM V1(15),A1(15),V2(15),A2(15),VS(15),VA(15)
110 PRINT "This program add two non-sinusoidal voltages"
120 PRINT "of the form:"
130 PRINT "v(t)=V0 + V1sin(wt+TH1) + V2sin(2wt+TH2) + V3sin(3wt+TH3) + ..."
140 PRINT
150 PRINT "Enter the following signal information:"
160 PRINT
170 INPUT "Radian frequency, w=";W
180 PRINT :PRINT "Enter the number of terms for the voltage"
190 PRINT "where N=dc + harmonics, i.e. N=1 + 4 = 5 for up "
200 PRINT "to and including the 4th harmonic."
210 PRINT:INPUT "Number of terms, N=";N
220 PRINT:PRINT "For the following, the number in the parentheses"
230 PRINT "refers to the term in the Fourier series expansion."
240 PRINT:INPUT "V1(0)=";V1(0)
250 INPUT "V2(0)=";V2(0)
260 FOR I=1 TO N-1
270 PRINT
280 PRINT "V1(";I;")=";
290 INPUT V1(I)
300 INPUT "at angle";A1(I)
310 PRINT "V2(";I;")=";
320 INPUT V2(I)
330 INPUT "at angle";A2(I)
340 NEXT I
350 PRINT
360 REM Calculate the sum terms
370 VS(0)=V1(0)+V2(0)
380 FOR I=1 TO N-1
```

(Continued)

```
390 A1=A1(I) :A2=A2(I)
400 XS=V1(I)*COS(A1/57.296)+V2(I)*COS(A2/57.296)
410 YS=V1(I)*SIN(A1/57.296)+V2(I)*SIN(A2/57.296)
420 VS(I)=SQR(XS^2+YS^2)
425 IF XS=O THEN VA(I)=90 :GOTO 440
430 VA(I)=57.296*ATN(YS/XS)
440 NEXT I
450 PRINT
460 PRINT "Equation for the sum of the voltages is:"
470 PRINT "v(t)=v1(t) + v2(t) =";VS(0);
480 FOR I=1 TO N-1
490 PRINT TAB(21);
500 PRINT "+";VS(I);"sin(";W*I;"t";
510 IF VA(I)>=0 THEN PRINT "+";
520 PRINT VA(I);")"
530 NEXT I
540 END
```

READY

RUN

```
*
This program add two non-sinusoidal voltages
of the form:
v(t)=V0 + V1sin(wt+TH1) + V2sin(2wt+TH2) + V3sin(3wt+TH3) + ...

Enter the following signal information:

Radian frequency, w=? 60

Enter the number of terms for the voltage
where N=dc + harmonics, i.e. N=1 + 4 = 5 for up
to and including the 4th harmonic.

Number of terms, N=? 4

For the following, the number in the parentheses
refers to the term in the Fourier series expansion.

V1(0)=? 60
V2(0)=? 20

V1( 1 )=? 70
at angle? 0
V2( 1 )=? 30
at angle? 0

V1( 2 )=? 20
at angle? 90
V2( 2 )=? -20
at angle? 90

V1( 3 )=? 10
at angle? 60
V2( 3 )=? 5
at angle? 90
```

*Problem 23.15a, ICA

```
Equation for the sum of the voltages is:
v(t)=v1(t) + v2(t) = 80 + 100 sin( 60 t+ 0 )
                   + 0 sin( 120 t+ 90 )
                   + 14.5466 sin( 180 t+ 69.8961 )


READY
```

nents using the right triangle relationships of Figure 22.4. That is,

$$XS = X1 + X2 = V1(1) \cos \theta_1 + V2(1) \cos \theta_2$$
$$= V1(1) \cos A1/57.296$$
$$+ V2(1) \cos A2/57.296$$
$$= 2 \cos (30/57.296)$$
$$+ 5\cos (40/57.296)$$

and similarly:

$$YS = Y1 + Y2 = V1(1) \sin \theta_1 + V2(1) \sin \theta_2$$
$$= 2\sin (30/57.296)$$
$$+ 5\sin (40/57.296)$$

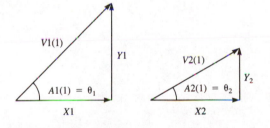

FIGURE 22.4

The magnitude and angle of the polar form of the *sum* is then determined by lines 420 and 430 respectively and stored as $VS(1)$ and $VA(1)$. The same routine will then be repeated for each set of terms with the same frequency.

When the I loop is complete, the results will be printed out by lines 460 through 530, using an additional I loop from $I = 1$ to $N - 1$.

Note in the above program that the effective values are not employed in the calculations because they were not to be printed out, and that the results are in the sinusoidal domain. The printout will also indicate the proper value of ω for each term and include each harmonic that appears in either or both of the non-sinusoidal signals.

Program 22-6 is essentially a repeat of Program 22-5, except for the plotting routines that begin at line 540. The horizontal location of the plot points is determined by line 620, while the time interval is determined by $t = 0$ to $t = 2T$ (where T is the period of the wave form),

```
10 REM *****  PROGRAM 22-6  *****
20 REM ***********************************************
30 REM Program to add two non-sinusoidal voltages
40 REM and then plot the resultant.
50 REM ***********************************************
60 REM
100 DIM V1(15),V2(15),VS(15),VA(15)
110 DIM A1(15),A2(15),D(150)
120 PRINT "This program finds the sum of two non-sinusoidal voltages"
130 PRINT "and then plots resulting voltage."
140 PRINT "Enter the following signal information:"
150 PRINT
160 INPUT "Radian frequency, w=";W
170 PRINT "Enter the number of terms for the voltage"
180 PRINT "where N = dc + harmonics, i.e., N=1 + 4 =5 for"
190 PRINT "up to and including the 4th harmonic."
200 PRINT
210 INPUT "Number of terms=";N
220 PRINT :N=N-1 :REM Reduce N by 1 (N for sinusoidal terms only)
```

(Continued)

```
230 INPUT "V1(0)=";V1(0)
240 INPUT "V2(0)=";V2(0)
250 FOR I=1 TO N
260 PRINT
270 PRINT "V1(";I;")=";
280 INPUT V1(I)
290 INPUT "at angle";A1(I)
300 PRINT "V2(";I;")=";
310 INPUT V2(I)
320 INPUT "at angle";A2(I)
330 NEXT I
340 PRINT
350 REM Calculate the sum terms
360 VS(0)=V1(0)+V2(0)
370 FOR I=1 TO N
380 A1=A1(I)  :A2=A2(I)
390 XS=V1(I)*COS(A1/57.296)+V2(I)*COS(A2/57.296)
400 YS=V1(I)*SIN(A1/57.296)+V2(I)*SIN(A2/57.296)
410 VS(I)=SQR(XS^2+YS^2)
420 IF XS=0 THEN VA(I)=90 :GOTO 440
430 VA(I)=57.296*ATN(YS/XS)
440 NEXT I
450 PRINT
460 PRINT "Equation for the sum of the voltages is:"
470 PRINT "V(t)=V1(t)+V2(t)=";VS(0);
480 FOR I=1 TO N
490 PRINT TAB(21);
500 PRINT "+";VS(I);"sin(";W*I;"t";
510 IF VA(I)>=0 THEN PRINT "+";
520 PRINT VA(I);")"
530 NEXT I
540 REM Obtain data for plot of sum voltage
550 PRINT "Plot follows ";
560 J=0
570 FOR T=0 TO 12.56637/W STEP .1256637/W
580 J=J+1
590 WT=W*T
600 D(J)=VS(0)
610 FOR I=1 TO N
620 D(J)=D(J)+VS(I)*SIN(WT+VA(I)/57.296)
630 NEXT I
640 NEXT T
650 REM Perform plot
660 N=J
670 PRINT " for";N;"points.":PRINT:PRINT
680 GOSUB 3100
690 END
3100 REM Module to plot data in array D(I) with N entries
3110 FS=32 :REM + or - 32 in this plot
3120 CP=32
3130 FM=ABS(D(I))
3140 FOR I=2 TO N
3150 IF ABS(D(I))>FM THEN FM=ABS(D(I))
3160 NEXT I
3170 REM FM now has largest value in array
3180 FOR I=1 TO 2*FS
3190 PRINT "-";
3200 NEXT I
3210 PRINT INT(FM+0.5)
3220 FOR I= 1 TO N
3230 P=INT(CP+FS*D(I)/FM+0.5)
3240 IF P<CP THEN PRINT TAB(P);"*";TAB(CP);"I"
```

```
3250 IF P=CP THEN PRINT TAB(CP);"I"
3260 IF P>CP THEN PRINT TAB(CP);"I";TAB(P);"*"
3270 NEXT I
3280 PRINT:PRINT
3290 RETURN
```

READY

RUN

This program finds the sum of two non-sinusoidal voltages
and then plots resulting voltage.
Enter the following signal information:

Radian frequency, w=? 30
Enter the number of terms for the voltage
where N = dc + harmonics, i.e., N=1 + 4 =5 for
up to and including the 4th harmonic.

Number of terms=? 3

V1(0)=? 5
V2(0)=? -4

V1(1)=? 4
at angle? 30
V2(1)=? 5
at angle? -30

V1(2)=? 3
at angle? 45
V2(2)=? -3
at angle? -45

Equation for the sum of the voltages is:
V(t)=V1(t)+V2(t)= 1   + 7.8103 sin(30 t-3.6705)
                    + 4.2426 sin(60 t+ 90)
Plot follows for 101 points.

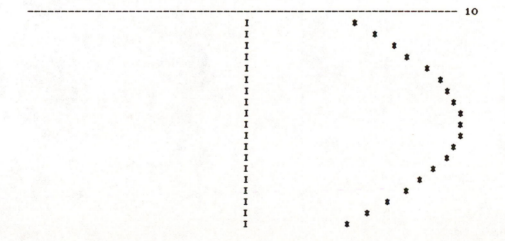

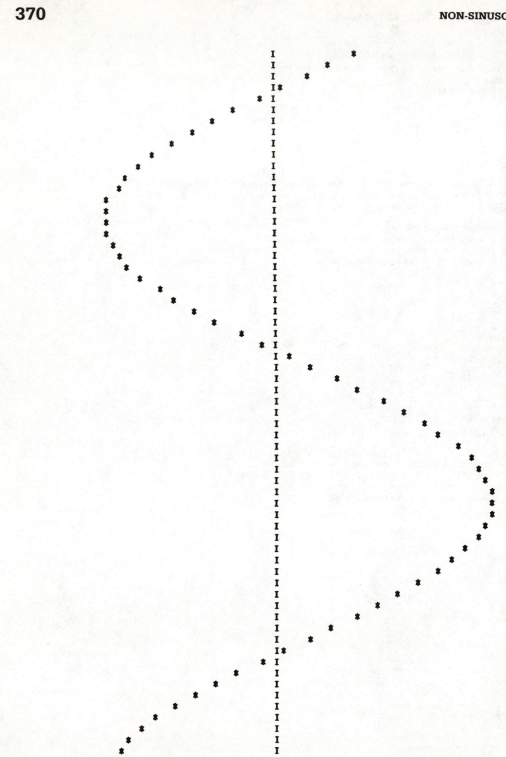

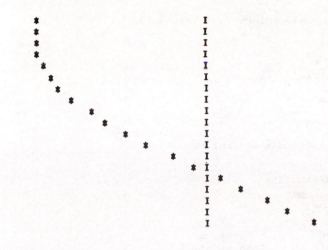

READY

or $t = 0$ to $t = 2T = 2\left(\dfrac{2\pi}{\omega}\right) = \dfrac{4\pi}{\omega} = \dfrac{12.56637}{\omega}$,

since $\omega = 2\pi f = 2\pi/T$. The time interval between plot points is $1/100$ of $2T$ or $2T/100 = 0.1256637/\omega$. The plot point $D(J) = D(0) = VS(0)$ is the algebraic sum of the dc terms, while lines 610 through 630 define the values of $D(1)$, $D(2)$ etc., through $D(101)$, as instructed by line 570. Once the values of $D(I)$ are established, the routine of line 3100 can be employed to plot the results. Note that the printout has a sinusoidal appearance, but with a major portion of the plot above the provided axis and a maximum integer value of 10.

22.5   TESTING ROUTINE

The last program (22-7) of the chapter will test your ability to determine the effective value of the current for the R-C network of Figure 22.1. The values of R and C and the components of the sinusoidal expression for $v(t)$ are all randomly generated within the boundaries spec-

```
10 REM *****  PROGRAM 22-7  *****
20 REM **********************************************************
30 REM Program to test the calculations of the
40 REM effective value of the non-sinusoidal current
50 REM of a series R-C circuit.
60 REM **********************************************************
70 REM
100 PRINT "For a series R-C circuit with a non-sinusoidal input"
110 PRINT "voltage, calculate the effective value of the current"
120 PRINT "for the following:"
```

(Continued)

```
130 GOSUB 600 :REM Randomly select data values
140 PRINT
150 PRINT "R=";R;"ohms"
160 PRINT "C=";C;"farads"
170 PRINT
180 PRINT "and an applied voltage of:"
190 PRINT
200 REM Print equation
210 PRINT "e(t)=";E0;
220 IF SGN(E1)=1 THEN PRINT "+";
230 PRINT E1;"sin(";W;"t)"
240 REM Perform calculations and print results
250 XC=1/(W*C)
260 ZT=SQR(R^2+XC^2)
270 ZA=-57.296*ATN(XC/R)
280 IE=E1/(SQR(2)*ZT)
290 IA=-ZA
300 VR=IE*R
310 VC=IE*XC
320 RA=IA :REM Angle of voltage across R
330 CA=IA-90 :REM Angle of voltage across C
340 PRINT
350 PRINT "The effective value of the non-sinusoidal current(amps) is";
360 INPUT IA
370 IF SGN(IA)<>SGN(IE) THEN GOTO 400
380 IF ABS(IA-IE)>ABS(0.01*IE) THEN GOTO 400
390 PRINT "Correct" :END
400 PRINT "No, it's";IE;"amps"
410 END
600 REM Select circuit and signal values
610 R=INT(1+20*RND(XV)) :REM 1<R<20
620 C=0.01*INT(1+100*RND(XV))   :REM 0.01<C<1
630 GOSUB 700 :REM Get sign
640 E0=SG*INT(1+25*RND(XV))
650 GOSUB 700
660 E1=SG*INT(1+25*RND(XV))
670 W=10*INT(1+25*RND(XV))
680 RETURN
700 REM Select sign (+ or -)
710 SG=1
720 S=INT(1+4*RND(XV))
730 IF S=3 THEN SG=-1
740 RETURN
```

READY

RUN

```
For a series R-C circuit with a non-sinusoidal input
voltage, calculate the effective value of the current
for the following:

R= 16 ohms
C= .61 farads

and an applied voltage of:

e(t)= 14 -22 sin( 250 t)

The effective value of the non-sinusoidal current(amps) is? 0.8
No, it's-.9723 amps
```

READY

For a series R-C circuit with a non-sinusoidal input
voltage, calculate the effective value of the current
for the following:

R= 1 ohms
C= .02 farads

and an applied voltage of:

e(t)=-5 + 12 sin(70 t)

The effective value of the non-sinusoidal current(amps) is? _6.9_
Correct

ified by lines 610, 620, 640, 660, 670, and 720.
Note in the provided runs the wide variations of
randomly generated values within the specified
limits.

EXERCISES

Write a program to perform each of the following tasks.

1. For a non-sinusoidal voltage of the following form:

$$v = V_o + V_1 \sin\omega t + V_2 \sin2\omega t$$
$$+ V_3 \sin3\omega t + \cdots + V_N \sin N\omega t$$

determine the effective value of the voltage
and the power delivered to a resistor R. For the
first run, use $V_o = 10V$, $V_1 = 5V$, $V_2 = 4V$, V_3
$= 3V$, $V_4 = 2V$ and $V_5 = 1V$.

2. The half-wave rectified signal has the following Fourier series expansion:

$$v = 0.318V_m + 0.500\ V_m \sin\omega t$$
$$- 0.212\ V_m \sin(2\omega t + 90°)$$
$$- 0.0424\ V_m \sin(4\omega t + 90°) + \cdots$$

Tabulate the effective value with just the dc
term, then with the first harmonic, the second
harmonic, and finally the fourth harmonic. That
is, list N vs. V_{eff} with $N = 0$ for just the dc term,
$N = 1$ for the dc term and the first harmonic, N

$= 2$ for the dc term plus both the first and
second harmonic, and $N = 3$ for all the terms.
For the first run, use $V_m = 10V$.

3. Plot the function of problem 22-2 for just
the dc term and the first harmonic.

4. Repeat the problem 22-3 for the dc term
and the first two harmonics. Compare the results with the results of problem 22-3.

5. Repeat problem 22-3 for all the terms appearing in problem 22-2 and compare your results with the curves obtained for problems 22-3
and 22-4.

6. Repeat Program 22-4 for a series R-L network. For the first run, use $R = 6\Omega$, $L = 0.02$H
and $e = 18 + 30 \sin400t$.

7. Plot the non-sinusoidal voltage e and the
resulting current i of problem 22-6 on the same
graph.

8. Repeat problem 22-7 for the voltage v_R
and current i_R.

9. Add a dc term of $\dfrac{-V_m}{2}$ to the half-wave
rectified signal of problem 22-2 and plot the
results. What is the impact on the resulting
plot?

10. Develop a test routine for the R-L network
of problem 22-6 similar in content to Program
22-7.

Transformers

23

23.1 INTRODUCTION

The general solution to a number of typical transformer problems will be examined in detail in this chapter. The ideal, iron-core, and air-core transformers will all be included in the analysis, along with the concept of impedance matching and the operation of multiple-load transformers. Three of the programs will use the vector analysis subroutines.

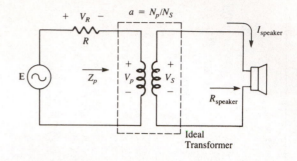

FIGURE 23.1

23.2 IMPEDANCE MATCHING

The impedance matching situation of Figure 23.1 will be examined in detail by Program 23-1. The magnitude of E and the resistor values of R and $R_{speaker}$ are provided and the program will determine the value of a for maximum power transfer to the load and the magnitude of the various currents, voltages, and power levels of the network. For an ideal transformer, the primary impedance Z_p of Figure 23.1 can be determined from $Z_p = a^2 Z_s$. For maximum power transfer, the internal resistance of the source

(R) must equal Z_p or $a = \sqrt{Z_p/Z_s} = \sqrt{R/Z_s}$. In this case, $Z_s = R_{speaker} = R_s$ and $a = \sqrt{R/R_s}$ for maximum power transfer, as on line 180.

If $R = Z_p$, then $V_R = V_p$ and $V_R = V_p = E/2$, as defined by line 210. The secondary voltage is determined from $V_p/V_s = N_p/N_s = a$ or $V_s = V_p/a$, appearing on line 220. The speaker current is determined from $I_s = V_s/R_s$, and the primary current from $I_p = V_p/Z_p = V_p/R$ (lines 270 and 280). Finally, the power delivered by the source is determined by $P_{source} = E \cdot I_p$, and the power to the load by $P_{speaker} = I_s^2 R_s$, as calculated by lines 320 and 330. For maximum power transfer conditions, the per-

```
10 REM ***** PROGRAM 23-1 *****
20 REM ***********************************************
30 REM Program to perform the basic calculations
40 REM associated with an impedance matching situation.
50 REM ***********************************************
60 REM
100 PRINT "To perform the basic calculations associated"
110 PRINT "with an impedance matching transformer"
120 PRINT "enter the following data:"
130 PRINT
140 INPUT "Supply voltage, E=";E
150 INPUT "Source resistance, R=";R
160 INPUT "Speaker resistance, Rs=";RS
170 REM Perform calculations
180 A=SQR(R/RS)
190 PRINT
200 PRINT "The transformer turns ratio is";A
210 VP=E/2
220 VS=VP/A
230 PRINT:PRINT "For maximum power transfer conditions"
240 PRINT "the speaker voltage will be";VS;"volts"
250 PRINT "and primary voltage";VP;"volts"
260 PRINT
```

Input (lines 140–160)
a (lines 180–200)
V (lines 210–260)

```
     ┌270 IS=VS/RS
     │280 IP=VP/R
   I │290 PRINT "The speaker current is";IS;"amps"
     └300 PRINT "and source current is";IP;"amps"
      310 PRINT
     ┌320 PP=E*IP
   P │330 PS=IS^2*RS
     │340 PRINT "The speaker power is";PS;"watts"
     └350 PRINT "and the source power is";PP;"watts"
      360 PRINT
     ┌370 N=PS/PP
  η% │380 PRINT "The transfer efficiency=";N
     └390 PRINT "which is";N*100;"%"
      400 END
```

READY

RUN

To perform the basic calculations associated
with an impedance matching transformer
enter the following data:

Supply voltage, E=? <u>10</u>
Source resistance, R=? <u>648</u>
Speaker resistance, Rs=? <u>8</u>

The transformer turns ratio is 9

For maximum power transfer conditions
the speaker voltage will be .5556 volts
and primary voltage 5 volts

The speaker current is .069 amps
and source current is 7.716E-03 amps

The speaker power is .039 watts
and the source power is .077 watts

The transfer efficiency= .5
which is 50 %

READY

RUN
*
To perform the basic calculations associated
with an impedance matching transformer
enter the following data:

Supply voltage, E=? <u>20</u>
Source resistance, R=? <u>36</u>
Speaker resistance, Rs=? <u>4</u>

The transformer turns ratio is 3

For maximum power transfer conditions
the speaker voltage will be 3.3333 volts
and primary voltage 10 volts

*Problem 24.17, ICA

```
The speaker current is .8333 amps
and source current is .2778 amps

The speaker power is 2.7778 watts
and the source power is 5.5556 watts

The transfer efficiency= .5
which is 50 %


        READY
```

cent efficiency, determined by lines 370 and
printed out by lines 380 and 390, will always be
50%—note the results of the provided run of the
program.

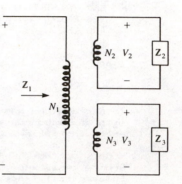

FIGURE 23.2

23.3 MULTIPLE-LOAD TRANSFORMER

Program 23-2 determines the general solution
for the input impedance and the voltage levels
of the multiple-load transformer of Figure 23.2.
The general solution for Z_1 is given by

$$Z_1 = \cfrac{1}{\cfrac{1}{(N_1/N_2)^2 \cdot Z_2} + \cfrac{1}{(N_1/N_3)^2 \cdot Z_3}}$$

In Program 23-2, $Z_A = (N_1/N_2)^2 Z_2$, $Z_B = (N_1/N_3)^2 Z_3$ and $Z_1 = 1/(1/Z_A + 1/Z_B)$, as on line 270. The voltages V_2 and V_3 are then determined on lines 290 and 300 by $V_2 = (N_2/N_1)E_1$ and $V_3 = (N_3/N_1)E_1$.

```
10 REM *****  PROGRAM 23-2  *****
20 REM *************************************************
30 REM Program to analyze the operation of the
40 REM multiple-load transformer of Figure 23-2.
50 REM *************************************************
60 REM
100 PRINT "This programs analyzes the operation of"
110 PRINT "the multiple-load transformer of Figure 23-2."
120 PRINT
130 PRINT "Enter the following circuit and transformer data:"
140 PRINT
150 INPUT "Transformer primary turns, N1=";N1
160 INPUT "            secondary turns, N2=";N2
170 INPUT "            secondary turns, N3=";N3
180 PRINT
190 INPUT "Secondary load, Z2=";Z2
200 INPUT "Secondary load, Z3=";Z3
210 PRINT
220 INPUT "And the primary input voltage, E1=";E1
230 PRINT
240 REM Perform calculations
```

```
      ┌ 250 ZA=(N1/N2)^2*Z2
      │ 260 ZB=(N1/N3)^2*Z3
Z₁    │ 270 Z1=1/(1/ZA+1/ZB)
      └ 280 PRINT "The impedance reflected into the primary is";Z1;"ohms"
      ┌ 290 V2=(N2/N1)*E1
      │ 300 V3=(N3/N1)*E1
V     │ 310 PRINT "while the voltage across load Z2, V2=";V2;"volts"
      └ 320 PRINT "and the voltage across load Z3, V3=";V3;"volts"
        330 END
```

READY

RUN

This programs analyzes the operation of
the multiple-load transformer of Figure 23-2.

Enter the following circuit and transformer data:

Transformer primary turns, N1=? <u>200</u>
        secondary turns, N2=? <u>50</u>
        secondary turns, N3=? <u>100</u>

Secondary load, Z2=? <u>10</u>
Secondary load, Z3=? <u>5</u>

And the primary input voltage, E1=? <u>20</u>

The impedance reflected into the primary is 17.7778 ohms
while the voltage across load Z2, V2= 5 volts
and the voltage across load Z3, V3= 10 volts

READY

23.4 IRON-CORE TRANSFORMER

The next two programs deal with the practical iron-core transformer of Figure 23.3. It includes the elements R_p, R_s, X_p, and X_s that do not appear when the transformer is considered to be ideal in nature. Although the assumption that most iron-core transformers are ideal (no internal losses) is an excellent approximation for most applications, Program 23-3 will demonstrate that $\mathbf{V}_g$ must be larger in magnitude to

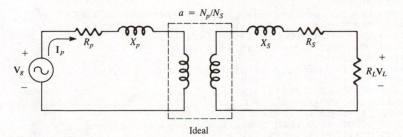

FIGURE 23.3

```
        10 REM *****   PROGRAM 23-3   *****
        20 REM **********************************
        30 REM Program to provide an analysis of
        40 REM the iron-core transformer.
        50 REM **********************************
        60 REM
       100 PRINT "This program determines the generator voltage"
       110 PRINT "required to establish a particular primary current."
       120 PRINT
       130 DIM VG(4),IP(4),ZP(4),ZS(4),ZG(4)
       140 DIM ZL(4),ZE(4),VL(4),V(4),VA(4),VB(4),VC(4)
      ┌150 INPUT "Input current, Ip=";IP(3)
       160 INPUT "at angle";IP(4)
       170 INPUT "Primary impedance, Rp=";ZP(1)
       180 INPUT "                   Xp=";ZP(2)
Input  190 INPUT "turns ratio, a=";A
       200 INPUT "Secondary impedance, Rs=";ZS(1)
       210 INPUT "                     Xs=";ZS(2)
      └220 INPUT "and load resistance, RL=";RL
      ┌230 REM Perform calculations using vector calculations
       240 ZE(1)=ZP(1)+A^2*ZS(1)
       250 ZE(2)=ZP(2)+A^2*ZS(2)
       260 VL(3)=A*IP(3)*RL
       270 VL(4)=IP(4)
       280 V(1)=ZE(1)+A^2*RL
       290 V(2)=ZE(2)
Calc   300 GOSUB 4000 :REM Convert from rectangular to polar form
Vg     310 FOR I=1 TO 4 :REM Calculate IP*ZG
       320 ZG(I)=V(I)
       330 VA(I)=IP(I)  :VB(I)=ZG(I)
       340 NEXT I
       350 GOSUB 4600
       360 FOR I=1 TO 4 :REM Establish VG in polar and rectangular form
       370 VG(I)=VC(I)
      └380 NEXT I
      ┌390 PRINT:PRINT "Required generator voltage is";
Print  400 PRINT VG(3);"at an angle of";VG(4);"degrees."
Results 410 PRINT "Under ideal conditions the magnitude of Vg would be";
      └420 PRINT A*VL(3);"volts, or";VG(3)-A*VL(3);"volts less."
       430 END
      4000 REM Module to convert from rectangular to polar form
      4010 REM Enter with V(1),V(2) - return with V(3),V(4)
      4020 V(3)=SQR(V(1)^2+V(2)^2)
      4030 IF V(1)>0 THEN V(4)=(180/3.14159)*ATN(V(2)/V(1))
      4040 IF V(1)<0 THEN V(4)=180*SGN(V(2))+(180/3.14159)*ATN(V(2)/V(1))
      4050 IF V(1)=0 THEN V(4)=90*SGN(V(2))
      4060 IF V(2)=0 THEN IF V(1)<0 THEN V(4)=180
      4070 RETURN
      4100 REM Conversion from polar to rectangular form
      4110 REM Enter with V(3),V(4) - return with V(1),V(2)
      4120 V(1)=V(3)*COS(V(4)*3.14159/180)
      4130 V(2)=V(3)*SIN(V(4)*3.14159/180)
      4140 RETURN
      4600 REM Module to multiply and divide VA by VB
      4610 REM VC=VA*VB
      4620 VC(3)=VA(3)*VB(3)
      4630 VC(4)=VA(4)+VB(4)
      4640 V(3)=VC(3)  :V(4)=VC(4)
      4650 GOSUB 4100 :REM Convert to rectangular form
      4660 VC(1)=V(1)  :VC(2)=V(2)
      4670 RETURN :REM Return from both multiply and divide
```

READY

```
RUN
.
This program determines the generator voltage
required to establish a particular primary current.

Input current, Ip=? 10
at angle? 0
Primary impedance, Rp=? 1
                    Xp=? 2
turns ratio, a=? 2
Secondary impedance, Rs=? 1
                     Xs=? 2
and load resistance, RL=? 60

Required generator voltage is 2452.04 at an angle of 2.3373 degrees.
Under ideal conditions the magnitude of Vg would be 2400 volts, or
  52.04 volts less.


READY
```

*Example 24.8, ICA

deliver the same voltage (and therefore power) to the resistor R_L if the non-ideal situation is assumed.

The primary equivalent network for the system of Figure 23.3 appears in Figure 23.4, where

$$R_e = R_p + a^2 R_s \text{ (line 240)}$$

and

$$X_e = X_p + a^2 X_s \text{ (line 250)}$$

Note in the program that the input data was requested in a form that isolated the real and imaginary components and permitted a direct calculation of R_e and X_e on lines 240 and 250. The magnitude of the voltage across $a^2 R_L$ is determined on line 260 using

$$aV_L = I_p (a^2 R_L)$$

or

$$V_L = aI_p R_L$$

Since $\mathbf{Z}_L$ is resistive, the angle of $\mathbf{V}_L$ is the same as that of $\mathbf{I}_p$ (line 270). The total real part of the input impedance is defined by line 280 as $R_e + a^2 R_L$, and the total imaginary part as X_e on line 290. Recall that $V(1)$ and $V(2)$ are the

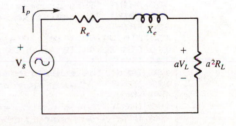

FIGURE 23.4

input operators for the rectangular form of the input for the subroutine of 4000.

Returning to Figure 23.4, we find that

$$\mathbf{V}_g = \mathbf{I}_p (R_e + a^2 R_L + jXe)$$
$$= \mathbf{I}_p (\underbrace{R_e + a^2 R_L}_{V(1)} + \underbrace{jXe}_{V(2)})$$
$$= \mathbf{I}_p (\underbrace{V(3) \: \underline{/V(4)}}_{ZG(I)})$$

Line 330 defines the operator $VA(I)$ as equal to $IP(I)$ and $VB(I) = ZG(I)$. The polar product is then determined by the subroutine at 4600 and stored as $VC(I)$. Lines 360 through 380 establish the polar and rectangular form of

$\mathbf{V}_G$ ($\mathbf{V}_G = VG(1) + jVG(2) = VG(3)\underline{/VG(4)}$. The results are printed out by lines 390 through 420.

Note in the provided run the inductive phase shift of 2.3373° introduced by the reactances X_p and X_s, and the requirement of 52.04 additional volts at the input to provide the same voltage to the load.

Program 23-4 will tabulate the impact of the magnitude of R_L on the load voltage, generator voltage, and the difference between the two. The actual tabulating process is introduced by adding only four steps to Program

23-3. Lines 220 and 230 set up the heading and define the range of values for R_L for the calculations to follow. Line 420 prints out the result in the format of 410, and line 430 sends the program back to line 230 for the next value of R_L. Note in the printout the difference between the generator voltage V_g and the generator voltage V_g' required under ideal conditions (R_e and X_e = 0). The larger the load (a^2R_L factor) compared to R_e and X_e, the less the additional generator voltage required to deliver the same load voltage compared to ideal conditions.

```
          10 REM *****   PROGRAM 23-4   *****
          20 REM *****************************************************
          30 REM Program to provide an analysis of the iron-core
          40 REM transformer for a range of load resistance values.
          50 REM *****************************************************
          60 REM
         100 PRINT "This program tabulates the impact of a change"
         110 PRINT "in resistive load values for the iron-core transformer."
         120 PRINT
         130 DIM VG(4),IP(4),ZP(4),ZS(4),ZG(4)
         140 DIM ZL(4),ZE(4),VL(4),V(4),VA(4),VB(4),VC(4)
       ┌ 150 INPUT "Input current, Ip=";IP(3)
       │ 160 INPUT "at angle";IP(4)
       │ 170 INPUT "Primary impedance, Rp=";ZP(1)
Input  │ 180 INPUT "                  Xp=";ZP(2)
       │ 190 INPUT "turns ratio, a=";A
       │ 200 INPUT "Secondary impedance, Rs=";ZS(1)
       └ 210 INPUT "                     Xs=";ZS(2)
Heading  215 PRINT
         220 PRINT " RL        VL        Vg       Vd=Vg-Vg'   PL"
       ┌ 230 FOR RL=1 TO 12
       │ 240 REM Perform calculations using vector operations
       │ 250 ZE(1)=ZP(1)+A^2*ZS(1)
       │ 260 ZE(2)=ZP(2)+A^2*ZS(2)
       │ 270 VL(3)=IP(3)*A*RL
       │ 280 VL(4)=IP(4)
       │ 290 V(1)=ZE(1)+A^2*RL
       │ 300 V(2)=ZE(2)
R_L    │ 305 PL=VL(3)^2/RL
Loop   │ 310 GOSUB 4000 :REM convert from rectangular to polar
Calc.  │ 320 FOR I=1 TO 4
&      │ 330 ZG(I)=V(I)
Print  │ 340 VA(I)=IP(I) :VB(I)=ZG(I)
Results│ 350 NEXT I
       │ 360 GOSUB 4600 :REM Calculate IP*ZG
       │ 370 FOR I=1 TO 4
       │ 380 VG(I)=VC(I)
       │ 390 NEXT I
       │ 400 VD=VG(3)-A*VL(3)
       │ 410 F$="###    ###.#    ###.##   ##.##    ####.#"
       │ 420 PRINT USING F$,RL,VL(3),VG(3),VD,PL
       └ 430 NEXT RL
         440 END
```

```
4000 REM Module to convert from rectangular to polar form
4010 REM Enter with V(1),V(2) - return with V(3),V(4)
4020 V(3)=SQR(V(1)^2+V(2)^2)
4030 IF V(1)>0 THEN V(4)=(180/3.14159)*ATN(V(2)/V(1))
4040 IF V(1)<0 THEN V(4)=180*SGN(V(2))+(180/3.14159)*ATN(V(2)/V(1))
4050 IF V(1)=0 THEN V(4)=90*SGN(V(2))
4060 IF V(2)=0 THEN IF V(1)<0 THEN V(4)=180
4070 RETURN
4100 REM Conversion from polar to rectangular form
4110 REM Enter with V(3),V(4) - return with V(1),V(2)
4120 V(1)=V(3)*COS(V(4)*3.14159/180)
4130 V(2)=V(3)*SIN(V(4)*3.14159/180)
4140 RETURN
4600 REM Module to multiply and divide VA by VB
4610 REM VC=VA*VB
4620 VC(3)=VA(3)*VB(3)
4630 VC(4)=VA(4)+VB(4)
4640 V(3)=VC(3)  :V(4)=VC(4)
4650 GOSUB 4100 :REM Convert to rectangular form
4660 VC(1)=V(1)  :VC(2)=V(2)
4670 RETURN :REM Return from both multiply and divide
```

READY

RUN

This program tabulates the impact of a change
in resistive load values for the iron-core transformer.

Input current, Ip=? <u>10</u>
at angle? <u>0</u>
Primary impedance, Rp=? <u>1</u>
                   Xp=? <u>2</u>
turns ratio, a=? <u>2</u>
Secondary impedance, Rs=? <u>1</u>
                     Xs=? <u>2</u>

RL	VL	Vg	Vd=Vg-Vg'	PL
1	20.0	134.54	94.54	400.0
2	40.0	164.01	84.01	800.0
3	60.0	197.23	77.23	1200.0
4	80.0	232.59	72.59	1600.0
5	100.0	269.26	69.26	2000.0
6	120.0	306.76	66.76	2400.0
7	140.0	344.82	64.82	2800.0
8	160.0	383.28	63.28	3200.0
9	180.0	422.02	62.02	3600.0
10	200.0	460.98	60.98	4000.0
11	220.0	500.10	60.10	4400.0
12	240.0	539.35	59.35	4800.0

READY

23.5 AIR-CORE TRANSFORMER

The input impedance of an air-core transformer is not related to the secondary by the a^2 factor of the iron-core transformer. Rather, the equation has the following complex format:

$$\mathbf{Z}_i = \mathbf{Z}_p + \frac{(\omega M)^2}{\mathbf{Z}_s + \mathbf{Z}_L}$$

with

$$\mathbf{Z}_p = R_p + jX_{L_p}$$

$$\mathbf{Z}_s = R_s + jX_{L_s}$$

as defined by Figure 23.5.

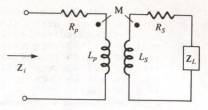

FIGURE 23.5

The first step (lines 230 - 270) is to determine the polar and rectangular form of the sum $\mathbf{Z}_s + \mathbf{Z}_L$. The magnitude of $(\omega M)^2$ is then determined on line 280 and treated as a polar vector with a magnitude of $(\omega M)^2$ and an angle of $0°$,

```
  10 REM *****  PROGRAM 23-5  *****
  20 REM ***********************************************************
  30 REM Program to calculate Zi for an air-core transformer
  40 REM ***********************************************************
  50 REM
 100 DIM V(4),VA(4),VB(4),VC(4)
 110 DIM ZP(4),ZS(4),ZL(4),N(4),D(4),ZI(4)
 120 PRINT "This program calculates Zi for an air-core"
 130 PRINT "transformer."
 140 PRINT
 150 INPUT "Value of XM=wM=";XM
 160 INPUT "Primary  impedance is, Rp=";ZP(1)
 170 INPUT "                       Xp=";ZP(2)
 180 INPUT "Secondary impedance is,Rs=";ZS(1)
 190 INPUT "                       Xs=";ZS(2)
 200 INPUT "and load impedance is, RL=";ZL(1)
 210 INPUT "                       XL=";ZL(2)
 220 PRINT
 230 D(1)=ZS(1)+ZL(1)
 240 D(2)=ZS(2)+ZL(2)
 250 V(1)=D(1) :V(2)=D(2)
 260 GOSUB 4000 :REM Rectangular to polar form conversion
 270 D(3)=V(3) :D(4)=V(4)
 280 N(3)=XM^2 :N(4)=0 :REM Calculate N/D
 290 VA(3)=N(3) :VA(4)=N(4)
 300 VB(3)=D(3) :VB(4)=D(4)
 310 GOSUB 4700
 320 FOR I=1 TO 4 :REM Add ZP+N/D
 330 VA(I)=VC(I) :VB(I)=ZP(I)
 340 NEXT I
 350 GOSUB 4400
 360 FOR I=1 TO 4 :REM Identifying the polar and rectangular forms of ZI
 370 ZI(I)=VC(I)
 380 NEXT I
 390 PRINT
 400 PRINT "The input impedance of the air-core transformer is";ZI(1);
 410 IF ZI(2)>=0 THEN PRINT "+j";
 420 PRINT ZI(2)
 430 PRINT "which is";ZI(3);"at an angle of";ZI(4);"degrees"
 440 END
```

Input (lines 150–210)
Calc. Z_i (lines 230–380)
Print Z_i (lines 400–430)

```
 ┌─4000 REM Module to convert from rectangular to polar form
 │  4010 REM Enter with V(1),V(2) - return with V(3),V(4)
 │  4020 V(3)=SQR(V(1)^2+V(2)^2)
 │  4030 IF V(1)>0 THEN V(4)=(180/3.14159)*ATN(V(2)/V(1))
 │  4040 IF V(1)<0 THEN V(4)=180*SGN(V(2))+(180/3.14159)*ATN(V(2)/V(1))
 │  4050 IF V(1)=0 THEN V(4)=90*SGN(V(2))
 │  4060 IF V(2)=0 THEN IF V(1)<0 THEN V(4)=180
 ▼  4070 RETURN
    4100 REM Conversion from polar to rectangular form
    4110 REM Enter with V(3),V(4) - return with V(1),V(2)
    4120 V(1)=V(3)*COS(V(4)*3.14159/180)
    4130 V(2)=V(3)*SIN(V(4)*3.14159/180)
    4140 RETURN
    4400 REM Module to Add or Subtract VC=VA + or - VB
    4410 REM Enter with VA,VB - return with VC
    4420 VC(1)=VA(1)+VB(1)
    4430 VC(2)=VA(2)+VB(2)
    4440 V(1)=VC(1)  :V(2)=VC(2)
    4450 GOSUB 4000 :REM Convert to polar form
    4460 VC(3)=V(3)  :VC(4)=V(4)
    4470 RETURN
    4500 REM Enter here for subtraction operation
    4510 REM VC=VA-VB
    4520 VC(1)=VA(1)-VB(1)
    4530 VC(2)=VA(2)-VB(2)
    4540 GOTO 4440 :REM Convert to polar form and return
    4600 REM Module to multiply and divide VA by VB
    4610 REM VC=VA*VB
    4620 VC(3)=VA(3)*VB(3)
    4630 VC(4)=VA(4)+VB(4)
    4640 V(3)=VC(3)  :V(4)=VC(4)
    4650 GOSUB 4100 :REM Convert to rectangular form
    4660 VC(1)=V(1)  :VC(2)=V(2)
    4670 RETURN :REM Return from both multiply and divide
    4700 REM Enter here for divide
    4710 REM VC=VA/VB
    4720 VC(4)=VA(4)-VB(4)
    4730 IF VB(3)=0 THEN VC(3)=1E10 :REM VC(3) very large
    4740 IF VB(3)<>0 THEN VC(3)=VA(3)/VB(3)
    4750 GOTO 4640 :REM Convert to rectangular form and return


READY

RUN
.
This program calculates Zi for an air-core
transformer.

Value of XM=wM=? 360
Primary  impedance is, Rp=? 3
                       Xp=? 2.4E3
Secondary impedance is,Rs=? 0.5
                       Xs=? 400
and load impedance is, RL=? 40
                       XL=? 0
```

The input impedance of the air-core transformer is 35.4721 +j 2079.2878
which is 2079.5904 at an angle of 89.0227 degrees

READY

*Example 24.9,ICA

Conversion Modules

to permit the polar division of lines 290 through 310. The next step is to add $\mathbf{Z}_p$ to the results of the division using the rectangular form of each. Lines 320 through 350 perform this operation before identifying the polar and rectangular component of $\mathbf{Z}_i$ on lines 360 through 380. The results are printed out in the polar and rectangular form by lines 400 through 430. Note the highly inductive nature of the input impedance for the provided run.

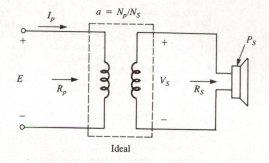

FIGURE 23.6

23.6   TESTING ROUTINE

The last program (23-6) of the chapter will test your ability to analyze the network of Figure

23.6. The values of E, R_s, and a are limited to the ranges indicated by lines 510 through 530. Note the excellent choice of values randomly generated in the program.

```
10 REM *****  PROGRAM 23-6  *****
20 REM *****************************************************
30 REM Program to test the calculations applied to insure
40 REM maximum power transfer using the ideal transformer.
50 REM *****************************************************
60 REM
100 PRINT "This program tests the calculations employed in"
110 PRINT "using a transformer to provide maximum power"
120 PRINT "transfer to a load." :PRINT
130 GOSUB 500 :REM Randomly select network values
140 PRINT "For the transformer system of Figure 23-6:"
150 PRINT "The applied voltage E=";E;"volts"
160 PRINT "The speaker resistance Rs=";RS;"ohms"
170 PRINT "and the turns ratio a=";A
180 PRINT
190 PRINT "Provide the following calculated results:"
200 PRINT
210 INPUT "Primary resistance, Rp=";RA
220 RP=A^2*RS
230 IF ABS(RP-RA)<ABS(0.01*RP) THEN PRINT "Correct" :GOTO 250
240 PRINT "No, it's";RP;"ohms"
250 PRINT
260 INPUT "Voltage across speaker, Vs=";VA
270 VP=E
280 VS=VP/A
290 IF SGN(VS)<>SGN(VA) THEN GOTO 310
300 IF ABS(VS-VA)<ABS(0.01*VS) THEN PRINT "Correct" :GOTO 320
310 PRINT "No, it's";VS;"volts"
320 PRINT
330 INPUT "Current drawn from input supply, Ip=";IA
340 IP=E/RP
350 IF SGN(IP)<>SGN(IA) THEN GOTO 370
360 IF ABS(IP-IA)<ABS(0.01*IP) THEN PRINT "Correct" :GOTO 380
370 PRINT "No, it's";IP;"amps"
380 PRINT
390 INPUT "Power delivered to speaker, Ps=";PA
400 PS=VS^2/RS
410 IF ABS(PS-PA)<ABS(0.01*PS) THEN PRINT "Correct" :GOTO 430
420 PRINT "No, it's";PS;"watts"
430 PRINT
```

```
    440 END
    500 REM Module to select circuit values
    510 E=2*INT(1+10*RND(XV))  :REM E from 2 to 20
    520 RS=INT(1+25*RND(XV))  :REM Rs from 1 to 25
    530 A=INT(2+19*RND(XV))  :REM A from 2 to 20
    540 RETURN
```

READY

RUN

This program tests the calculations employed in
using a transformer to provide maximum power
transfer to a load.

For the transformer system of Figure 23-6:
The applied voltage E= 6 volts
The speaker resistance Rs= 1 ohms
and the turns ratio a= 3

Provide the following calculated results:

Primary resistance, Rp=? 9
Correct

Voltage across speaker, Vs=? 3
No, it's 2 volts

Current drawn from input supply, Ip=? 0.667
Correct

Power delivered to speaker, Ps=? 4
Correct

READY

RUN

This program tests the calculations employed in
using a transformer to provide maximum power
transfer to a load.

For the transformer system of Figure 23-6:
The applied voltage E= 12 volts
The speaker resistance Rs= 12 ohms
and the turns ratio a= 9

Provide the following calculated results:

Primary resistance, Rp=? 972
Correct

Voltage across speaker, Vs=? 1.5
No, it's 1.3333 volts

Current drawn from input supply, Ip=? 0.12
No, it's .012 amps

Power delivered to speaker, Ps=? .15
No, it's .1481 watts

READY

EXERCISES

Write a program to perform each of the following tasks.

1. For the system of Figure 23.1, determine V_P, V_S, I_P, I_S, and the power to the speaker if E, R_S, and a are provided. Assume maximum power transfer conditions. For the first run, use $E = 28V$, $R_S = 1.6k\Omega$, and $a = 4$.

2. Using the results of problem 23-1, tabulate the power to the speaker versus the resistance of the speaker for the range $\frac{R_s}{4a^2}$ to $\frac{4R_s}{a^2}$ in increments of $\frac{R_s}{4a^2}$. For the first run, use the same *values for E, R_S, and a provided in problem* 23-1.

3. Using the results of problem 23-2, plot $P_{speaker}$ versus $R_{speaker}$ for the system of Figure 23.1. For the first run, use the same input data and range indicated in problem 23-2. Label both axes and indicate the range of each variable.

4. For the iron-core transformer of Program 23-3, determine the magnitude of the load voltage for a given value of V_g. For the first run, use $\mathbf{V}_g = 2400\underline{/0°}$, $R_p = 1\Omega$, $X_p = 2\Omega$, $R_S = 1\Omega$, $X_S = 2\Omega$, $R_L = 60$ ohms, and $a = 2$.

5. Tabulate V_L versus R_L for a fixed value of V_g for the iron-core transformer. Use a range of R_L that extends from $10R_S$ to $100R_S$ in increments of $5R_S$. For the first run, use the same input values provided in problem 23-4 (except for R_L, of course).

6. Plot the results of problem 23-5. Be creative in establishing the scale of the horizontal and vertical axes.

7. On the same graph, plot the results of V_L versus R_L from problem 23-6 and V_L versus R_L, if we assume ideal conditions and set R_P, X_P, R_S, and X_S to zero. Compare the resulting curves. Be careful in choosing your full scale deflection.

8. Given the air-core transformer of Program 23-5, determine the currents $\mathbf{I}_p$ and $\mathbf{I}_S$ and the voltage $\mathbf{V}_L$, if $\mathbf{V}_g$ is provided. For the first run, use the same input data as in the provided run of Program 23-5. In addition, use $\mathbf{V}_g = 120\underline{/0°}$.

9. Develop a test routine for the required analysis of problem 23-1.

10. Develop a test routine for the air-core transformer of Program 23-5.

Two-Port Parameters

24

24.1 INTRODUCTION

The complete analysis of the ac response of a transistor network, using the hybrid equivalent network, results in a series of equations that give us the chance to demonstrate the time-saving capabilities of computer analysis. The conversion equations between the various 2-port parameters are a further example of how the computer can be used to perform those mundane, lengthy tasks that must be done accurately.

24.2 HYBRID-PARAMETERS

The hybrid equivalent network for the common-emitter transistor configuration appears in Figure 24.1 with a resistive load.

The general equations for the current gain A_i, voltage gain A_v, input impedance Z_i, and output impedance Z_o are:

$$A_i = \frac{I_2}{I_1} = \frac{h_{fe}}{1 + h_{oe}R_L}$$

$$A_v = \frac{E_2}{E_1} = \frac{-h_{fe}R_L}{h_{ie}(1 + h_{oe}R_L) - h_{re}h_{fe}R_L}$$

$$Z_i = h_{ie} - \frac{h_{re}h_{fe}R_L}{1 + h_{oe}R_L}$$

$$Z_o = \frac{1}{h_{oe} - \frac{h_{re}h_{fe}}{h_{ie} + R_s}}$$

The factor $1 + h_{oe}R_L$ appears in three of the equations and, therefore, is calculated first on line 190 of Program 24-1. The four quantities of interest are then determined by lines 200

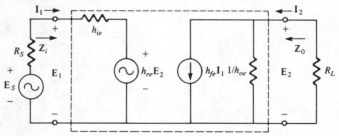

Common-Emitter Transistor Hybrid Equivalent Network

FIGURE 24.1

```
10 REM *****   PROGRAM 24-1   *****
20 REM **********************************************
30 REM Program to perform calculations using hybrid
40 REM h-parameters with a resistive load RL
50 REM **********************************************
60 REM
100 PRINT "Enter the following hybrid parameter data:"
110 INPUT "hi=";HI
120 INPUT "hr=";HR
130 INPUT "hf=";HF
140 INPUT "ho=";HO
150 INPUT "Source resistance is, Rs=";RS
160 INPUT "and load resistance is, RL=";RL
170 PRINT
180 REM Perform calculations
```

```
      ┌ 190  D=1+HO*RL
      │ 200  AI=HF/D
Calc. │ 210  AV=-HF*RL/(HI*D-HR*HF*RL)
      │ 220  ZI=HI-(HR*HF*RL/D)
      └ 230  ZO=1/(HO-HR*HF/(HI+RS))
        240  PRINT
  Aᵢ    250  PRINT "Current gain Ai=";AI
  Aᵥ    260  PRINT "Voltage gain Av=";AV
        270  PRINT
  Zᵢ    280  PRINT "Input impedance Zi=";ZI;"ohms"
  Zₒ    290  PRINT "Output impedance Zo=";ZO;"ohms"
        300  PRINT
        310  END
```

READY

RUN
*

```
Enter the following hybrid parameter data:
hi=? 1E3
hr=? 4E-4
hf=? 50
ho=? 25E-6
Source resistance is, Rs=? 1E3
and load resistance is, RL=? 2E3


Current gain Ai= 47.619
Voltage gain Av=-99.0099
**
Input impedance Zi= 961.9048 ohms
Output impedance Zo= 6.6667E+04 ohms
```

READY

*Example 25.5, ICA
**Example 25.6, ICA

through 230 and printed out by lines 250 through 290.

The computer solution becomes significantly more difficult if you change the load from one that is purely resistive to one that can have both a real and reactive component. The rectangular and polar forms of the load impedance are first determined by lines 220 through 240 of Program 24-2, before the real and imaginary parts of the expression $1 + h_{oe}\mathbf{Z}_L$ are determined by line 250 in the following manner:

$$
\begin{aligned}
1 + h_{oe}\mathbf{Z}_L &= 1 + h_{oe}(R + jX) \\
&= 1 + h_{oe}R + jh_{oe}X \\
&= 1 + h_{oe}V(1) + jh_{oe}V(2)
\end{aligned}
$$

The polar form of $1 + h_{oe}\mathbf{Z}_L$ is also calculated on line 260, and both the polar and rectangular forms are stored as $D(1)$ through $D(4)$ by line 270. On lines 280 and 290, the division operators of the subroutine at 4700 are defined as

$$
\frac{VA(3)\,\underline{/VA(4)}}{VB(3)\,\underline{/VB(4)}} = \frac{HF\,\underline{/0°}}{D(3)\,\underline{/D(4)}} = \frac{h_{fe}\,\underline{/0°}}{V(3)\,\underline{/V(4)}}
$$

and the division to determine A_i is carried out by the subroutine. Lines 320 through 350 will then print out the current gains in both forms.

Although the expression for the denominator of A_v is more complex, it also has a

```
       10 REM *****   PROGRAM 24-2   *****
       20 REM *********************************************
       30 REM Program to perform calculations using hybrid
       40 REM h-parameters with a load ZL.
       50 REM *********************************************
       60 REM
      100 DIM V(4),VA(4),VB(4),VC(4),N(4),D(4),ZL(4)
      110 DIM A(4)
      120 PRINT "Enter the following hybrid parameter data:"
      130 INPUT "hi=";HI
      140 INPUT "hr=";HR
Input 150 INPUT "hf=";HF
      160 INPUT "ho=";HO
      170 INPUT "Load resistance R=";ZL(1)
      180 INPUT "                  X=";ZL(2)
      190 PRINT
      200 REM Perform calculations
      210 REM Determine hf/(1+ho*ZL)
      220 V(1)=ZL(1) :V(2)=ZL(2)
      230 GOSUB 4000 :REM Convert from rectangular to polar form
      240 ZL(3)=V(3) :ZL(4)=V(4)
Calc. 250 V(1)=1+HO*ZL(1) :V(2)=HO*ZL(2)
  A_i 260 GOSUB 4000 :REM Convert from rectangular to polar form
      270 FOR I=1 TO 4 :D(I)=V(I) :NEXT I
      280 VA(3)=HF :VA(4)=0
      290 VB(3)=D(3) :VB(4)=D(4)
      300 GOSUB 4700
      310 REM Prepare to output the current gain
Print 320 FOR I=1 TO 4
  A_i 330 A(I)=VC(I)
      340 NEXT I
      350 PRINT "Current gain, Ai:" :GOSUB 5960
      360 PRINT :REM Determine denominator of Av
      370 D(1)=HI+(HI*HO-HR*HF)*ZL(1)
      380 D(2)=(HI*HO-HR*HF)*ZL(2)
      390 V(1)=D(1) :V(2)=D(2)
      400 GOSUB 4000 :REM Convert to polar form
Calc. 410 D(3)=V(3) :D(4)=V(4)
  A_v 420 REM Determine -hf*ZL/(Denominator)
      430 VA(3)=HF*ZL(3)
      440 IF ZL(4)>0 THEN VA(4)=ZL(4)-180
      450 IF ZL(4)<=0 THEN VA(4)=ZL(4)+180
      460 VB(3)=D(3) :VB(4)=D(4)
      470 GOSUB 4700
      480 FOR I=1 TO 4 :REM Prepare to output the voltage gain
      490 A(I)=VC(I)
  A_v 500 NEXT I
      510 PRINT "Voltage gain, Av:"
      520 GOSUB 5960
      530 END

      4000-5990
      as in earlier programs


READY
```

```
RUN

Enter the following hybrid parameter data:
hi=? 1E3
hr=? 4E-4
hf=? 50
ho=? 25E-6
Load resistance R=? 3E3
              X=? 4E3

Current gain, Ai:
 46.3117 at an angle of-5.3146 degrees
which is 46.1126 -j 4.2895

Voltage gain, Av:
 246.2576 at an angle of-127.9987 degrees
which is-151.6065 -j 194.0574


READY
```

definite real and imaginary part, as shown below.

$$h_{ie} (1 + h_{oe}(R + jX)) - h_{re}h_{fe}(R + jX)$$

$$= h_{ie} + (h_{ie}h_{oe} - h_{re}h_{fe})R + j(h_{ie}h_{oe} - h_{re}h_{fe})X$$

$$= h_{ie} + (h_{ie}h_{oe} - h_{re}h_{fe})ZL(1) + j(h_{ie}h_{oe} - h_{re}h_{fe})ZL(2)$$

$$= D(1) + jD(2) \qquad \text{(lines 370 and 380)}$$

The rectangular-to-polar conversion operators are then defined on line 390 for the subroutine at 4000, followed by the storage of the polar form as $D(3)\underline{/D(4)}$.

On lines 430 through 460, the operators for the division operations of line 4700 are defined, and the result for A_v printed out by lines 480 through 520. Lines 440 and 450 simply include the impact of the negative sign in the A_v equation, because for inductive loads:

$$- VA(3)\underline{/VA(4)} = + VA(3)\underline{/VA(4) - 180°}$$

$$= HF \cdot ZL(3)\underline{/ZL(4) - 180°}$$

Even though a phase shift of $\pm 180°$ will include the effect of a minus sign, the decision operation of $+ 180°$ or $- 180°$ was added to keep the angle for A_v less than $180°$.

Note that the provided inductive load resulted in a negative angle for both the current gain and the voltage gain. Consider also the magnitude of each gain for fairly typical values of h_{ie}, h_{re}, h_{fe}, and h_{oe}.

24.3 PARAMETER CONVERSION

The conversion of the **Z** parameters to **Y** parameters lets us use the subroutines between 4000 and 5990. The common denominator in the conversion equations appearing below clearly defines the denominator as the first factor to be determined in Program 24-3.

$$\mathbf{Y}_{11} = \frac{\mathbf{Z}_{22}}{\Delta_Z}, \ \mathbf{Y}_{12} = \frac{-\mathbf{Z}_{12}}{\Delta_Z}, \ \mathbf{Y}_{21} = \frac{-\mathbf{Z}_{21}}{\Delta_Z},$$

$$\mathbf{Y}_{22} = \frac{\mathbf{Z}_{11}}{\Delta_Z}$$

with $\qquad \Delta_Z = \mathbf{Z}_{11}\mathbf{Z}_{22} - \mathbf{Z}_{12}\mathbf{Z}_{21}$

On lines 190 through 380, the operator-subscripted variables for the rectangular and polar forms of each impedance are determined. The subroutine at 1500 will perform the conversion, and the subroutine at 1900 will print out

```
                 10 REM *****  PROGRAM 24-3  *****
                 20 REM ******************************************
                 30 REM Program to convert from Z to Y parameters
                 40 REM ******************************************
                 50 REM
                100 DIM V(4),VA(4),VB(4),VC(4),N(4),D(4)
                110 DIM ZA(4),ZB(4),ZC(4),ZD(4)
                120 DIM YA(4),YB(4),YC(4),YD(4)
                130 PRINT "This program provides the conversion from the"
                140 PRINT "Z-parameters to the Y-parameters."
                150 PRINT
                160 PRINT "Enter the following Z-parameter data:"
                170 PRINT
                180 REM Module to accept input of Z-parameters
                190 PRINT:PRINT "Z11=:"
                200 GOSUB 5500
Z11             210 FOR I=1 TO 4
                220 ZA(I)=V(I)
                230 NEXT I
                240 PRINT:PRINT "Z12=:"
                250 GOSUB 5500
Z12             260 FOR I=1 TO 4
                270 ZB(I)=V(I)
                280 NEXT I
                290 PRINT:PRINT "Z21=:"
                300 GOSUB 5500
Z21             310 FOR I=1 TO 4
                320 ZC(I)=V(I)
                330 NEXT I
                340 PRINT:PRINT "Z22=:"
                350 GOSUB 5500
Z22             360 FOR I=1 TO 4
                370 ZD(I)=V(I)
                380 NEXT I
                390 PRINT
                400 GOSUB 1500 :REM Convert from Z to Y parameters
                410 GOSUB 1900 :REM Print resulting Y parameters
                420 END
               1500 REM Module to convert Z-parameters to Y-parameters
               1510 FOR I=1 TO 4 :REM Z11*Z22
Z11 · Z22      1520 VA(I)=ZA(I) :VB(I)=ZD(I)
               1530 NEXT I
               1540 GOSUB 4600
               1550 FOR I=1 TO 4 :REM Z12*Z21
               1560 D(I)=VC(I) :VA(I)=ZB(I) :VB(I)=ZC(I)
Z12 · Z21      1570 NEXT I
               1580 GOSUB 4600
               1590 FOR I=1 TO 4 :REM D=Z11*Z22-Z12*Z21
Z11 · Z22      1600 VA(I)=D(I) :VB(I)=VC(I)
 - Z12 · Z21   1610 NEXT I
               1620 GOSUB 4500
               1630 FOR I=1 TO 4 :REM Y11=Z22/D
               1640 D(I)=VC(I)
Y11            1650 VA(I)=ZD(I) :VB(I)=D(I)
               1660 NEXT I
               1670 GOSUB 4700
               1680 FOR I=1 TO 4 :REM Y12=-Z12/D
               1690 YA(I)=VC(I)
Y12            1700 VA(I)=ZB(I)
               1710 NEXT I
               1720 VA(3)=-VA(3) :GOSUB 4700
```

```
        ┌1730 FOR I=1 TO 4 :REM Y21=-Z21/D
        │1740 YB(I)=VC(I)
Y21     │1750 VA(I)=ZC(I)
        │1760 NEXT I
        └1770 VA(3)=-VA(3) :GOSUB 4700
        ┌1780 FOR I=1 TO 4 :REM Y22=Z11/D
        │1790 YC(I)=VC(I) :VA(I)=ZA(I)
        │1800 NEXT I
Y22     │1810 GOSUB 4700
        │1820 FOR I=1 TO 4
        │1830 YD(I)=VC(I)
        │1840 NEXT I
        └1850 RETURN
        ┌1900 REM Module to print values of Y-parameters
        │1910 PRINT:PRINT "Y11=";
        │1920 FOR I=1 TO 4 :V(I)=YA(I) :NEXT I :GOSUB 5960
        │1930 PRINT "Y12=";
Print   │1940 FOR I=1 TO 4 :V(I)=YB(I) :NEXT I :GOSUB 5960
Y       │1950 PRINT "Y21=";
        │1960 FOR I=1 TO 4 :V(I)=YC(I) :NEXT I :GOSUB 5960
        │1970 PRINT "Y22=";
        │1980 FOR I=1 TO 4 :V(I)=YD(I) :NEXT I :GOSUB 5960
        └1990 RETURN
          ┌ 2000-5990
          │ as in earlier programs
          │
          │
          ▼
```

RUN

This program provides the conversion from the
Z-parameters to the Y-parameters.

Enter the following Z-parameter data:

Z11=:
(1) Enter rectangular form, or (2) Enter polar form:? <u>1</u>
X=? <u>3</u>
Y=? <u>4</u>

Z12=:
(1) Enter rectangular form, or (2) Enter polar form:? <u>2</u>
Magnitude=? <u>5</u>
at angle? <u>90</u>

Z21=:
(1) Enter rectangular form, or (2) Enter polar form:? <u>2</u>
Magnitude=? <u>5</u>
at angle? <u>90</u>

Z22=:
(1) Enter rectangular form, or (2) Enter polar form:? <u>1</u>
X=? <u>3</u>
Y=? <u>-4</u>

```
Y11= .1 at an angle of-53.1301 degrees
which is .06 -j .08
Y12=-.1 at an angle of 90.0001 degrees
which is-2.7756E-18 -j .1
Y21=-.1 at an angle of 90.0001 degrees
which is-2.7756E-18 -j .1
Y22= .1 at an angle of 53.1302 degrees
which is .06 +j .08
```

READY

the results. The denominator is determined by lines 1510 through 1620, and the **Y** parameters from lines 1630 through 1850. Note, in particular, lines 1720 and 1770, which add the negative sign to the results for $\mathbf{Y}_{12}$ and $\mathbf{Y}_{21}$, which are determined by the subroutines at lines 1680 and 1730 respectively.

The provided run reveals that if $\mathbf{Z}_{12} = \mathbf{Z}_{21}$, then $\mathbf{Y}_{12} = \mathbf{Y}_{21}$, and the choice of input impedances has resulted in the same magnitude (although the sign changes) for each of the admittance parameters.

24.4 TEST ROUTINE

The last program (24-4) of the chapter will test your ability to calculate the voltage gain and input impedance for the network of Figure 24.1 for randomly selected values of hybrid parameters and load resistance. The presence of $1 + h_{oe}R_L$ in both equations dictates that it be calculated first and the result applied on lines 250 and 310.

```
10 REM ***** PROGRAM 24-4 *****
20 REM ********************************************************
30 REM Program to test the calculations using hybrid
40 REM parameters with a resistive load, RL.
50 REM ********************************************************
60 REM
100 PRINT "This program tests the calculations involving"
110 PRINT "the use of hybrid parameters to determine Av and Zi"
120 PRINT "with a resistive load."
130 PRINT
140 GOSUB 350 :REM Randomly select hybrid and resistor values
150 PRINT "For hybrid parameters of:"
160 PRINT "hi=";HI
170 PRINT "hr=";HR
180 PRINT "hf=";HF;" and,"
190 PRINT "ho=";HO
200 PRINT
210 PRINT "and load resistance, RL=";RL;"ohms"
220 PRINT
230 INPUT "What is the value of voltage gain, Av=";AN
240 D=1+HO*RL
250 AV=-HF*RL/(HI*D-HR*HF*RL)
260 IF SGN(AN)<>SGN(AV) THEN GOTO 280
270 IF ABS(AN-AV)<ABS(0.01*AV) THEN PRINT "Correct" :GOTO 290
280 PRINT "No, it's";AV
290 PRINT
300 INPUT "What is the value of Zi";AN
310 ZI=HI-(HR*HF*RL/D)
320 IF ABS(AN-ZI)<ABS(0.01*ZI) THEN PRINT "Correct" :GOTO 340
330 PRINT "No, it's";ZI;"ohms"
340 END
```

```
350 REM Subroutine to select values
360 HI=1000*INT(1+5*RND(XV)) :REM hi from 1E3 to 5E3
370 HR=4E-4*INT(1+2.5*RND(XV)) :REM hr from 4E-4 to 10E-4
380 HF=50*INT(1+4*RND(XV)) :REM hf from 50 to 200
390 HO=25E-6*INT(1+2*RND(XV)) :REM ho from 25E-6 to 50E-6
400 RL=1000*INT(1+5*RND(XV)) :REM RL from 1000 to 5000 ohms
410 RETURN
```

READY

RUN

This program tests the calculations involving
the use of hybrid parameters to determine Av and Zi
with a resistive load.

For hybrid parameters of:
hi= 3000
hr= 4E-04
hf= 100 and,
ho= 5E-05

and load resistance, RL= 5000 ohms

What is the value of voltage gain, Av=? -160
No, it's-140.8451

What is the value of Zi? 2600
No, it's 2840 ohms

READY

RUN

This program tests the calculations involving
the use of hybrid parameters to determine Av and Zi
with a resistive load.

For hybrid parameters of:
hi= 3000
hr= 1.2E-03
hf= 50 and,
ho= 2.5E-05

and load resistance, RL= 1000 ohms

What is the value of voltage gain, Av=? -16.5
Correct

What is the value of Zi? 2925
Correct

READY

EXERCISES

Write a program to perform each of the following tasks.

1. Determine the impedance (Z) parameters
for the π network of Figure 24.2. For the first
run, use $Z_1 = 5\underline{/0°}$, $Z_2 = 10\underline{/90°}$, and Z_3
$= 20\underline{/-90°}$.

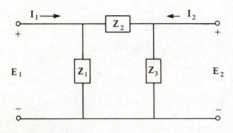

FIGURE 24.2

2. Determine the admittance (Y) parameters
for the π network of Figure 24.2. Use the same
values as in problem 24-1 for the first run.

3. Determine the hybrid (h) parameters for
the π network of Figure 24.2. Use the same
values as in problem 24-1 for the first run.

4. Convert the resulting h parameters of problem 24-3 to Y parameters, and compare results
with problem 24-2.

5. Determine Z_i and Z_o for Figure 24.1 with an
impedance load. For the first run, use the same
input values as in the provided run of Program
24-2.

6. Develop a test routine for the calculations
performed in problem 24-5.

Appendix A

Summary of BASIC mathematical operations

Arithmetic operations:
- + add
- − subtract
- * multiply
- / divide
- ^ exponentiate (raise to a power)

Relational operators:
- < less than
- > greater than
- = equal to
- <= or =< less than or equal to
- >= or => greater than or equal to
- <> not equal to

Hierarchy:
- ( ) parentheses
- ^ exponentiation
- *,/ multiply or divide
- +,− add or subtract
- precedence order is from left to right for operations on the same level

Appendix B

Glossary of BASIC commands

Examples

DATA Item list
    Hold data for access by
    READ command

`DATA 1,-7,"HELLO"`

DIM Array name (N1,N2, .. Nk)
    Allocate storage for an array
    of dimension or size N1 or
    N1 × N2, . . .

`DIM A(50),D(100),T$(20,5)`

END End execution of the program with return to command mode

```
FOR   var = exp TO exp STEP exp          FOR K = 1 TO N+2
      Open a FOR/NEXT loop
      (STEP is optional, being 1 if
      not otherwise specified)

GOSUB   line number                      GOSUB 4000
      Directs program to continue
      program at specified line
      number, with continuation
      after present line on
      RETURN command

GOTO   line number                       GOTO 120
      Directs program continue
      from indicated line number

IF   exp THEN statement                  IF A$ = "YES" THEN PRINT
      Test expression (exp)                 "GOOD"
      If exp is TRUE then perform
         statement
      If exp is FALSE then con-
         tinue program at next
         line

IF   exp THEN statement-1 ELSE           IF A$ = "YES" THEN PRINT
      statement-2                           "GOOD" ELSE PRINT
      If exp is TRUE then perform           "SORRY"
         statement-1

      If exp is FALSE then per-
         form statement-2
      Continue at next program
         line, unless otherwise
         directed

INPUT   "message string";var             INPUT "CHOICE=";C
      Print message string, print ?,
      and then accept input of
      specified variable(s)

NEXT   variable                          NEXT K
      Close FOR/NEXT loop for
      specified variable
```

ON exp GOSUB line-1, line-2,
   . . . line-k
     Evaluate expression;
     If INT (exp equals one or
        numbers 1 thru k con-
     tinue program execution at ap-
     propriate line number; other-
     wise go to next line

              ON C GOSUB 100,250,430

ON exp GOTO line-1, line-2,
   . . . line-k
     Same as ON . . GOSUB ex-
     cept program will return
     after present line upon sub-
     routine RETURN command

              ON C GOTO 50,190,400

PRINT exp
     Output the value of the
        expression
     Output any message strings
        specified in the specified
        position
    , used as separator causes
        advance to next print zone
        - typically about 8 spaces
    ; used as separator causes
        advance of typically one
        space after numeric output
        and one space (or none)
        after string output

              PRINT A,B;C^2

PRINT TAB(N); exp
     First tab over to position N
        then perform rest of
        print command

              PRINT TAB(5);D

PRINT USING string, exp
     Print the exp using a format
        specified by the string
    ### specifies a 3-place
        integer
    ##.# specifies a real num-
        ber of 2 integer places
    and 1 fractional place

              PRINT USING F$,a,t

```
READ   variable list              READ A,T,L$
         Align values to the specified
         variable(s), starting with the
         current DATA element

RETURN
         Return program execution
         after GOSUB command call-
         ing this subroutine

STOP
         Stop executing program,
         with print of current line
         number and return to com-
         mand mode
```

Arithmetic Functions:

```
ABS (exp)    Returns absolute value

ATN (exp)    Return arctangent value in radians

COS (exp)    Returns cosine of exp (assumes exp in radians)

EXP (exp)    Returns the natural exponential value

INT (exp)    Returns the largest integer not greater than the expression
             value

LOG (exp)    Returns the natural logarithm (base e) of exp (exp must be
             positive)

RND          various forms RND(0)
                           RND(1)
                           RND(X)
             Returns a pseudo-random number between 0 and 1 (On
             some RND(N) returns an integer random number between 1
             and N)

SIN (exp)    Returns −1 for negative exp
                     0 for exp = 0
                   + 1 for positive exp

SIN (exp)    Returns sine of the expression (assumes exp is in radians)

SQR (exp)    Returns square root of exp (exp must be non-negative)

TAN (exp)    Returns the tangent of exp (assumes exp is in radians)
```